Studies in Computational Intelligence

Volume 563

Series editor

Janusz Kacprzyk, Polish Academy of Sciences, Warsaw, Poland
e-mail: kacprzyk@ibspan.waw.pl

About this Series

The series "Studies in Computational Intelligence" (SCI) publishes new developments and advances in the various areas of computational intelligence—quickly and with a high quality. The intent is to cover the theory, applications, and design methods of computational intelligence, as embedded in the fields of engineering, computer science, physics and life sciences, as well as the methodologies behind them. The series contains monographs, lecture notes and edited volumes in computational intelligence spanning the areas of neural networks, connectionist systems, genetic algorithms, evolutionary computation, artificial intelligence, cellular automata, self-organizing systems, soft computing, fuzzy systems, and hybrid intelligent systems. Of particular value to both the contributors and the readership are the short publication timeframe and the world-wide distribution, which enable both wide and rapid dissemination of research output.

More information about this series at http://www.springer.com/series/7092

Ronald R. Yager · Marek Z. Reformat
Naif Alajlan
Editors

Intelligent Methods for Cyber Warfare

Editors
Ronald R. Yager
Iona College
Machine Intelligence Institute
New Rochelle, NY
USA

Marek Z. Reformat
Department of Electrical and Computer Engineering
University of Alberta
Edmonton, AB
Canada

Naif Alajlan
Department of Computer Engineering
College of Computer and Information Sciences
King Saud University
Riyadh
Saudi Arabia

ISSN 1860-949X ISSN 1860-9503 (electronic)
ISBN 978-3-319-34482-9 ISBN 978-3-319-08624-8 (eBook)
DOI 10.1007/978-3-319-08624-8

Springer Cham Heidelberg New York Dordrecht London

Printed on acid-free paper

Springer is part of Springer Science+Business Media (www.springer.com)

Preface

The cyberspace has an overwhelming influence on everyday activities of individuals and organizations. It has become a crucial part of society, industry, and government. It is used as a repository of different types of information, including critical ones, and as a means to oversee many activities and undertakings of social as well as industrial nature. Any type of malevolence directed on cyberspace creates disruptions affecting large numbers of people, organizations, and companies. Any minor attack on the Internet, as the most prominent part of cyberspace, can cause uncountable damage and loss of information, and has the ability to distress people's activities, and paralyze corporations and governments.

This vulnerability of the cyberspace can be used to intentionally confuse, disrupt, and even stop normal functions of societies and organizations. In the light of our increased dependence on the proper and sound operation of the cyberspace, mechanisms and systems, preventing any disruption and malicious actions on the Internet is of critical importance. In other words, safety and security of cyberspace has become essential.

By Cyberwarfare, we mean situations in which one entity, individual or organization, attacks another entity through cyberspace for the purpose, among other things, of stealing information, effecting the performance of its adversaries computer environment or sabotaging physical or information centric systems. With the pervasive use of computers and the Internet, organizations have become more and more vulnerable to cyber attacks.

The techniques emerging from areas of Computational Intelligence and Machine Learning are destined to find their way to cyberwarfare-related applications. The currently developed and mature techniques of fuzzy logic, artificial neural networks, evolutionary computing, prediction and classification, decision-making techniques, game theory, and also information fusion can be used to address a broad range of issues and challenges related to prevention, detection and impact analysis of different intrusions and attacks on cyber infrastructure.

This volume gives readers a glimpse, by all means far from being comprehensive, on new and emerging ways that Computational Intelligence and Machine Learning methods can be applied to issues related to cyberwarfare. The book includes a number of chapters that can be conceptually divided into three topics: chapters describing different data analysis methodologies with their applications to cyberwarfare issues, chapters presenting a number of instruction detection approaches, and chapters dedicated to analysis of possible cyber attacks and their impact. The book starts with a number of chapters related the topic of data and information processing methods. Machine Learning, fuzziness, decision-making, and information fusion are examples of topics and methods targeted by the following chapters.

The first chapter entitled "Malware and Machine Learning" by Charles LeDoux and Arun Lakhotia provides an overview of Machine Learning techniques and their application to detect different forms of malware. The similarities between malware that contains inherent patterns and similarities due to code and code pattern reuse, and Machine Learning that operates by discovering inherent patterns and similarities create an opportunity to explore a synergetic effect created via combining these two fields. The authors provide an overview of machine learning methods and how they are being applied in malware analysis. They describe the major issues together with an elucidation of the malware problems that machine learning is best equipped to solve.

Recognizing fuzzy logic-based techniques are some of the most promising approaches for crisis management is stimulated by Dan E. Tamir, Naphtali D. Rishe, Mark Last, and Abraham Kandel in their chapter "Soft Computing Based Epidemical Crisis Prediction". They focus on epidemical crisis prediction as one of the most challenging examples of decision making under uncertain information. According to the authors, the key for improving epidemical crises prediction capabilities is the ability to use sound techniques for data collection, information processing, and decision making under uncertainty. They point out that complex fuzzy graphs can be used to formalize the techniques and methods used for the data mining. Additionally, they assert that the fuzzy-based approach enables handling events of low occurrence via low fuzzy membership/truth-values, and updating these values as information is accumulated or changed.

An approach called ACP—Artificial societies, Computational experiments, and Parallel execution—is described and used for security-related purposes in the chapter "An ACP-Based Approach to Intelligence and Security Informatics" authored by Fei-Yue Wang, Xiaochen Li, and Wenji Mao. The authors focus on behavioral modeling, analysis and prediction in the domain of security informatics. Especially, they look at group behavior prediction and present two methods of doing it. The first approach uses plan-based inference that takes into consideration agents' preferences. The second approach uses graph theory and incorporates a graph search algorithm to forecast complex group behavior. The results of experimental studies to demonstrate the effectiveness of the proposed methods are presented.

Attacks on open information sources are addressed in the chapter "Microfiles as a Potential Source of Confidential Information Leakage" by Oleg Chertov and Dan Tavrov. In particular, they look at microfiles as an important source of information in cyberwarfare. They illustrate, using real data, that ignoring issues that ensure group anonymity can lead to leakage of confidential information. They show that it is possible to define fuzzy groups of respondents and obtain their distribution using appropriate fuzzy inference system. They discuss methods for protecting distributions of crisp as well as fuzzy groups of respondents.

The issues related to open source intelligence are addressed by Daniel Ortiz-Arroyo who authored the chapter "Decision Support in Open Source Intelligence". The decision support system presented here has been developed within the framework of the FP7 VIRTUOSO project. At the beginning, the author describes the overall scope and architecture of the VIRTUOSO platform. Further, he provides details of the main components of the constructed decision support system. These components employ computational intelligence techniques—soft-fusion and fuzzy logic—to integrate and visualize, together with other VIRTUOSO tools, diverse sources of information, and to provide access to the knowledge extracted from these sources. Some applications of this system in cyber-warfare are described.

The process of integration of data and information is addressed in "Information Fusion Process Design Issues for Hard and Soft Information: Developing an Initial Prototype" by James Llinas. The author provides a thorough description of challenges and requirements imposed on data and information fusion systems providing support for decision-making processes in the military/defense domains. In these domains, the nature of decision-making ranges from conventional military-like to socio-political—also characterized as "hard" and "soft" decisions. Because of this, the nature of information required for analysis is highly diversified. The heterogeneity of available information is indicated as an important factor driving the data and information fusion process design. Overall, the author offers perspectives on how those new requirements affect the design and development of data and information fusion systems.

An important aspect of instruction detection is addressed in the next three chapters. The first of them—"Intrusion Detection with Type-2 Fuzzy Ontologies and Similarity Measures" by Robin Wikstrom and Jozsef Mezei—targets an issue of embedding experts' knowledge in detecting constantly changing intrusion types. The authors propose a framework based on fuzzy ontology and similarity measures to incorporate experts' knowledge to the process of identification of these anomalies and handling imprecise information. Such a framework allows for identification of attacks that have never been experienced before. The authors present a fuzzy ontology developed based on the intrusion detection needs of a financial institution.

Another approach for instruction detection is proposed by Gulshan Kumar and Krishan Kumar. In the chapter "A Multi-objective Genetic Algorithm Based Approach for Effective Intrusion Detection Using Neural Networks", they propose a novel multiobjective genetic algorithm (MOGA) based approach for effective intrusion detection based on benchmark datasets. The approach generates a pool of non-inferior solutions—detection systems—that optimize trade-offs of multiple conflicting objectives, and creates an ensemble of these solutions to detect intrusions. The approach consists of three phases: (1) a MOGA based generation of solutions leading to creation of a Pareto front of non-inferior individual solutions; (2) a further refinement of the obtained solutions via identification of a Pareto front of ensemble solutions; and (3) an aggregation method used for fusing individual predictions to determine outcome of the ensemble-based detection system. The authors used two benchmark datasets: KDD cup 1999, and ISCX 2012 to demonstrate and validate the performance of the proposed approach for intrusion detection.

The next chapter starts with a brief review of exiting works in the Machine Learning community that offers treatments to cyber insider detection. Following this, the authors of "Cyber Insider Mission Detection for Situation Awareness"—Haitao Du, Changzhou Wang, Tao Zhang, Shanchieh Jay Yang, Jai Choi, and Peng Liu—introduce their own method for early detection of a mission of system's insider. The method uses Hidden Markov Models to estimate insider's levels of activities. Fuzzy rules and Ordered Weighted Average are used to fuse multiple facets of information about an intruder. Experimental results based on simulated data show that the integrated approach detects the insider mission with high accuracy and in a timely manner, even in the presence of obfuscation techniques.

Research activities that address an important objective aiming at better understanding of cyberwarfare scenarios, as well as impacts that different attacks have on multiple aspects of systems are presented next. Here, we have three versatile and important contributions.

The first of them "A Game Theoretic Engine for Cyber warfare" is dedicated to the application of game-theoretic principles to the cyberwarfare domain. Allen Ott, Alex Moir, and John T. Rickard—the authors of the chapter—look at application of a game theory to investigate behavior of an attacker and defender. They use a well-known Themistocles engine that has been developed and used over the past decade in cyberwarfare analysis, for the modeling and analysis of cyberwarfare offensive and defensive tactics. It is shown that generated courses of actions (COAs) for both offensive and defensive cyberwarfare scenarios are consistent with the move choices made by independent experts who monitor the game. The authors indicate future extensions such as fuzzification of move scores using both type-1 and interval type-2 membership functions, as well as utilization of hierarchical linguistic weighted power means for the aggregation of COA scores. All this will enable to handle inherent imprecision associated with the costs/benefits of individual moves.

The impact of cyber threats on a military network subjected to attacks is addressed in the next chapter "Mission Impact Assessment for Cyber warfare". The authors—Jared Holsopple, Shanchieh Jay Yang, and Moises Sudit—estimate impact of such threats on operations of a military system during its mission. They propose application of a tree-based structure, called a Mission Tree, to model relationships between missions, tasks, and assets. These relationships are modeled using Order Weighted Aggregators (OWAs), which address a diverse set of relationship types. The Mission Tree is capable of providing a quantitative estimate of and impact by propagating the impact "up," from the leaves to the root, through the tree. An important aspect of the impact assessment process proposed in the chapter is related to constant changes of missions or tasks performed by the system. The authors explore how scheduled or non-scheduled changes should affect the mission tree and influence the impact assessment process.

Economical consequences of attacks are subject of the next chapter by Suchitra Abel entitled "Uncertainty Modeling: The Computational Economists' View on Cyberwarfare". The author analyzes factors affecting the security of internet-based business. A casual model-based security system that focuses on core characteristics of contemporary internet-based businesses is presented. It extends traditional utility-based models with Bayesian causal networks. These networks represent relationships between variables, internal, and external, influencing business activities in normal and under attack conditions.

We hope that the readers will enjoy this book and that they will benefit from the useful and interesting methods and techniques conveyed by the authors in the broad domain of cyberwarfare.

New Rochelle, USA — Ronald R. Yager
Riyadh, Saudi Arabia — Naif Alajlan
Edmonton, Canada — Marek Z. Reformat

Contents

Malware and Machine Learning

Charles LeDoux and Arun Lakhotia

Abstract Malware analysts use Machine Learning to aid in the fight against the unstemmed tide of new malware encountered on a daily, even hourly, basis. The marriage of these two fields (malware and machine learning) is a match made in heaven: malware contains inherent patterns and similarities due to code and code pattern reuse by malware authors; machine learning operates by discovering inherent patterns and similarities. In this chapter, we seek to provide an overhead, guiding view of machine learning and how it is being applied in malware analysis. We do not attempt to provide a tutorial or comprehensive introduction to either malware or machine learning, but rather the major issues and intuitions of both fields along with an elucidation of the malware analysis problems machine learning is best equipped to solve.

1 Introduction

Malware, short for malicious software, is the weapon of cyber warfare. It enables online sabotage, cyber espionage, identity theft, credit card theft, and many more criminal, online acts. A major challenge in dealing with the menace, however, is its sheer volume and rate of growth. Tens of thousands of new and unique malware are discovered *daily*. The total number of new malware has been growing exponentially, doubling every year over the last three decades.

Analyzing and understanding this vast sea of malware manually is simply impossible. Fortunately for the malware analyst, very few of these unique malware are truly novel. Writing software is a hard problem, and this remains the case whether said software is benign or malicious. Thus, malware authors often reuse code and code

C. LeDoux (✉) · A. Lakhotia
Center for Advanced Computer Studies, University of Louisiana at Lafayette,
PO Box 44330, Lafayette, LA 70504, USA
e-mail: charles.a.ledoux@gmail.com

A. Lakhotia
e-mail: arun@louisiana.edu

R.R. Yager et al. (eds.), *Intelligent Methods for Cyber Warfare*,
Studies in Computational Intelligence 563, DOI 10.1007/978-3-319-08624-8_1

patterns in creating new malware. The result is the existence of inherent patterns and similarities between related malware, a weakness that can be exploited by malware analysts.

In order to capitalize on this inherent similarity and shared patterns between malware, the anti-malware industry has turned to the field of Machine Learning, a field of research concerned with "teaching" computers to recognize concepts. This "learning" occurs through the discovery of indicative patterns in a group of objects representing the concept being taught or by looking for similarities between objects. Though humans too use patterns in learning, such as using color, shape, sound, and smell to recognize objects, machines can find patterns in large swaths of data that may be gibberish to a humans, such as the patterns in sequences of bits of a collection of malware. Thus, Machine Learning has a natural fit with Malware Analysis since it can more rapidly learn and find patterns in the ever growing corpus of malware than humans.

Both Machine Learning and Malware Analysis are very diverse and varied fields with equally diverse and varied ways in which they overlap. In this chapter, we seek to provide a guiding, overhead cartography of these varied landscapes, focusing on the areas and ways in which they overlap. We do not seek to provide a comprehensive tutorial or introduction to either Malware or Machine Learning research. Instead, we strive to elucidate the major ideas, issues, and intuitions for each field; pointing to further resources when necessary. It is our intention that a researcher in either Malware Analysis or Machine Learning can read this chapter and gain a high-level understanding of the other field and the problems in Malware that Machine Learning has, is, and can be used to solve.

2 A Short History of Malware

The theory of malware is almost as old as the computer itself, tracing back to lectures by von Neumann in late 1940s on self-reproducing automata [1]. These early malware, if they can be called as such, did nothing significantly more than demonstrate self-reproduction and propagation. For example, one of the earliest malware to escape "into the wild" was called Elk Cloner and would simply display a small poem every 50th time an infected computer was booted:

```
Elk Cloner: The program with a personality

It will get on all your disks
It will infiltrate your chips
Yes it's Cloner!

It will stick to you like glue
It will modify ram too
Send in the Cloner!
```

The term computer *virus* was coined in early 1980s to describe such self-replicating programs [2]. The use of the term was influenced by the analogy of computer malware to biological viruses. A biological virus comes alive after it infects a living organism. Similarly, the early computer viruses required a host—typically another program—to be activated. This was necessitated by the limitations of the then computing infrastructure which consisted of isolated, stand-alone, machines. In order to propagate, that is infect currently uninfected machines, a computer virus necessarily had to copy itself in various drives, tapes, and folders that would be accessed by different machines. In order to ensure that the viral code was executed when it reached the new machine, the virus code would attach itself to, i.e. infect, another piece of code (a program or boot sector) that would be executed when the drive or tape reached another machine. When the now infected code would later execute, so would the viral code, furthering the propagation.

The early viruses remained mostly pranks. Any damage they caused, such as crashing a computer or exhausting disk space, was largely unintentional and a side effect of uncontrolled propagation. However, the number and spread of viruses quickly grew to enough of a nuisance that it led to the development of first anti-virus companies in the late 1980s. Those early viruses were simple enough that they could be detected by specific sequences of bytes, a la signatures.

The advent of networking, leading to the Internet, changed everything. Since data could now be transferred between computers without using an external storage device, so could the viruses. This freedom to propagate also meant that a virus no longer needed to infect a host program. A new class of malware called worm emerged. A worm was a stand alone program that could propagate from machine to machine without necessarily attaching to any other program.

Malware writing too quickly morphed from simple pranks into malicious vandalism, such as that done by the ILOVEYOU worm. This worm came as an attachment to an email with the (unsurprising) subject line "ILOVEYOU". When a user would open the attachment, the worm would first email itself to the user's contacts and then begin destroying data on the current computer. There were a number of similar malware created, designed only to wreak havoc and gain underground notoriety for their authors. These "graffiti" malware, however, soon gave way to the true threat: malware designed to make money and steal secrets.

Malware today has little if any resemblance to the malware of past. For one, gone are the simple days of pranks and vandalism conducted by bored teenagers and budding hackers. Modern malware is an well-organized activity forming a complete underground economy with its own supply chain. Malware is now a tool used by large underground organizations for making money and a weapon used by governments for espionage and attacks. Malware targeted towards normal, everyday computers can be designed to steal bank and credit card information (for direct theft of money), harvest email addresses (for selling to spammers), or gain remote control of the computer. The major threat from malware, however, comes from malware targeted not towards the average computer, but towards a particular corporation or government. These malware are designed to facilitate theft of trade or national secrets, steal crucial information (such as sensitive emails), or attack infrastructure. For example,

Stuxnet was malware designed to attack and damage various nuclear facilities in Iran. These malware often have large organizations (such as rival corporations) or even governments behind them.

3 Types of Malware

Whenever there is a large amount of information or data, it helps to categorize and organize it so that it can be managed. Classification also aids in communication between people, giving them a common nomenclature. The same is true of malware. The industry uses a variety of methods to classify and organize malware. The classification is often based on the method of propagation, the method of infection, and the objective of the malware. There is, however, no known standard nomenclature that is used across the industry. Classifications sometimes also come with legal implications. For instance, can a program that inserts advertisements as you browse the web be termed as malicious. What if the program was downloaded and installed by the user, say after being enticed by some free offering? To thwart legal notices the industry invented the term *potentially unwanted program* or PUP to refer to such programs.

Though there is no accepted standard for classification of malware in the industry, there is a reasonable agreement on classifying malware on their method of propagation into three *types*: virus, worm, and trojan (short for Trojan horse).

Virus, despite being often used as a synonym for malware, technically refers to a malware that attaches a copy of itself to a host, as described earlier. Propagation by infecting removable media was the only method for transmission available prior to the Internet, and this method is still in use today. For instance, modern viruses travel by infecting USB drives. This method is still necessary to reach computer systems that are not connected to the Internet, and is hypothesized as the way Stuxnet was transmitted.

A **trojan** propagates the same way its name sake entered the city of Troy, by hiding inside something that seems perfectly innocent. The earliest trojan was a game called ANIMAL. This simple game would ask the user a serious of questions and attempt to guess what animal the user was thinking of. When the game was executed, a hidden program, named PERVADE, would install a copy of itself and ANIMAL to every location the user had access to. A common modern example of a trojan is a fake antivirus, a program that purports to be an anti-virus system but in fact is a malware itself.

A **worm**, as mentioned earlier, is essentially a self-propagating malware. Whereas a virus, after attaching itself to a program or document, relies on an action from a user to be activated and spread, a worm is capable of spreading between network connected computers all by itself. This is typically accomplished one of two ways: exploiting vulnerabilities on a networked service or through email. The worm CODE RED was an example of the first type of worm. CODE RED exploited a bug in a specific type of server that would allow a remote computer to execute code on the server. The worm would simply scan the network looking for a vulnerable server. Once found, it would attempt to connect to the server and exploit the known bug. If successful, it would create another instance of the worm that repeated the whole process. The ILOVEYOU worm, discussed earlier, is an example of an email worm and spread as an email attachment. When a user opened the attachment, the worm would email a copy of itself to everyone in the user's contact list and damage the current machine.

While the above methods of propagation are the mostly commonly known, they by no means represent all possible ways in which malware can propagate. In general, one of two methods are employed to get a malware onto a system: exploit a bug in software installed on the computer or exploit the trust (or ignorance) of the user of the computer through social engineering. There are many different types of software bugs that allow for arbitrary code to be executed and almost as many ways to trick a user into installing a malware. Complicating matters further, There is no technical reason for a malware to limit its use to only one method of propagation. It is entirely conceivable, as was demonstrated by Stuxnet, for a malware to enter a network through email or USB, and then spread laterally to other machines by exploiting bugs.

4 Malware Analysis Pipeline

The typical end goal of malware analysis is simple: automatically detect malware as soon as possible, remove it, and repair any damage it has done. To accomplish this goal, software running on the system being protected (desktop, laptop, server, mobile device, embedded device, etc.) uses some type of "signatures" to look for malware. When a match is made on a "signature", a removal and repair script is triggered. The various portions of the analysis "pipeline" all in one way or another support this end goal [3, 4].

The general phases of creating and using these signatures are illustrated by Fig. 1. Creating a signature and removal instructions for a new malware occurs in the "Lab." The input into this malware analysis pipeline is a feed of suspicious programs to be analyzed. This feed can come from many sources such as honeypots or other companies. This feed first goes through a triage stage to quickly filter out known programs and assign an analysis priority to the sample. The remaining programs are then analyzed to discover what it looks like and what it does. The results of the analysis phase are used to create a signature and removal/repair instructions which are then verified for correctness and performance concerns. Once verified, these signatures are propagated to the end system and used by a scanner to detect, remove, and repair malware.

Each of the various phases of the anti-malware analysis process is attempting to accomplish a related, but independent task and thus has its own unique goals and performance constraints. As a result, each phase can independently be automated and optimized in order to improve the performance of the entire analysis pipeline.

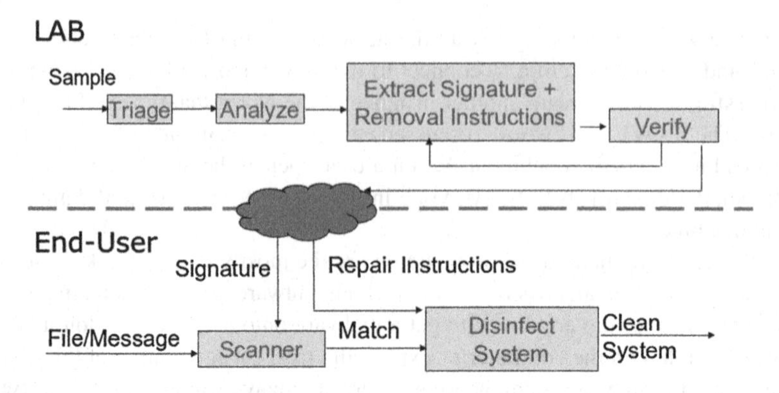

Fig. 1 Phases of the malware analysis pipeline

In fact, it is almost a requirement that automation techniques be tailored for the specific phase they are applied in, even if the technique could be applied to multiple phases. For example, a machine learning algorithm designed to filter out already analyzed malware in the triage stage will most likely perform poorly as a scanner. While both the triage stage and the scanner are accomplishing the same basic task, detect known malware, the standard by which they are evaluated is different.

4.1 Triage

The first phase of analysis, triage, is responsible for filtering out already analyzed malware and assigning analysis priority to the incoming programs. Malware analysts receive a *very* large number of new programs for analysis every day. Many of these programs, however, are essentially the same as programs that have already been analyzed and for which signatures exist. A time stamp or other trivial detail may have been changed causing a hash of the binary to be unique. Thus, while the program is technically unique, it does not need to reanalyzed as the differences are inconsequential. One of the purposes of triage is to filter these binaries out.

In addition to filtering out "exact" matches (programs that are essentially the same as already analyzed programs), triage is typically also tasked with assigning the incoming programs into malware families when possible. A malware family is a group of highly related malware, typically originating from common source code. If an incoming program can be assigned to a known malware family, any further analysis does not need to start with zero a priori knowledge, but can leverage general knowledge about the malware family, such as known intent or purpose.

A final purpose of the triage stage is to assign analysis priority to incoming programs. Humans still are and most likely will remain an integral part of the analysis pipeline. Like any other resource, what the available human labor is expended upon must be carefully chosen. Not all malware are created equal; it is more important that some malware have signatures created before others. For example, malware that only affects, say, Microsoft Windows 95 will not have the same priority as malware that affects the latest version of Windows.

The performance concerns for the triage phase are (1) ensuring that programs being filtered out truly should be removed and (2) efficient computation in order to achieve very high throughput. Programs filtered out by triage are not subjected to further analysis and thus it is very important that they do not actually need further analysis. Especially dangerous is the case of malware being filtered out as a benign program. In this case, that particular malware will remain undetectable. Marking a known malware or a benign program as malware for further processing, while undesirable, is not disastrous as it can still be filtered out in the later processing stages. Along the same lines, it is sufficient that malware be assigned to a particular family with only a reasonably high probability rather than near certainty. Finally, speed is of the utmost importance in this stage. This stage of the analysis pipeline examines the largest number of programs and thus requires the most efficient algorithms. Computationally expensive algorithms at this stage would cause a backlog so great that analysts would never be able to keep up with malware authors.

4.2 Analysis

In the analysis phase, information about what the program being analyzed does, i.e. its behavior, is gathered. This can be done in two ways: statically or dynamically.

Static analysis is performed without executing the program. Information about the behavior of the program is extracted by *disassembling* the binary and converting it back into human readable machine code. This is not high level source code, such as C++, but the low level assembly language. An assembly language is the human readable form of the instructions being given directly to the processor. ARM, PowerPC, and ×86 are the better known examples of assembly languages. After disassembly, the assembly code (often just called the malware "code" for short) can be analyzed to determine the behavior of the program. The methods for doing this analysis constitute an entire research field called program analysis and as such are outside the scope of this chapter. Nielson et al. [5] have a comprehensive tutorial to this field.

Static analysis can *theoretically* provide perfect information about the behavior of a program, but in practice provides an over approximation of the behaviors present. Only what is in the code is what can be executed, thus the code contains everything the program can do. However, extracting this information from a binary can be difficult, if not impossible. Perfectly solving many of the problems of static analysis is undecidable.

As an example of the problems faced by static analysis, binary disassembly is itself an undecidable problem. Binaries contain both data and code and separating the two from each other is undecidable. As a result some disassemblers treat the entire binary, including data, as if it were code. This results in a proper extraction of most of the original assembly code, along with much code that never originally existed.

There are many other methods of disassembly, such as the recent work by Schwarz et al. [6]. While these methods significantly improve on the resulting disassembly, none can guarantee correct disassembly. For instance, it is possible that there exists "dead code" in the original binary, i.e. code that can never be reached at runtime. In an ideal disassembly, such code ought to be excluded. Thus all of static analysis operates on approximations. Most disassemblers used in practice do not guarantee either over approximation or under approximation.

Dynamic analysis, in contrast with static analysis, is conducted by actually executing the program and observing what it does. The program can be observed from either within or without the executing environment. From within uses the same tools and techniques software developers use to debug their own programs. Tools that observe the operating system state can be utilized and the analyzed program run in a debugger. Observation from without the execution environment occurs by using a specially modified virtual machine or emulator. The analyzed program is executed within the virtual environment and the tools providing the virtualization observe and report the behavior of the program.

Dynamic analysis, as opposed to static analysis, generally provides an under approximation of the behaviors contained in the analyzed program, but guarantees that returned behaviors can be exhibited. Behaviors discovered by dynamic analysis are obviously guaranteed to be possible as the program was observed performing these behaviors. *Only* the observed behaviors can be returned, however. A single execution of a program is not likely to exhibit all the behaviors of the program as only a single path of execution through the binary is followed per run. A differing execution environment or differing input may reveal previously unseen behaviors.

4.3 Signatures and Verification

While the most common image conjured by the phrase "malware signatures" is specific patterns of bytes (often called strings) used by an Anti-Virus system to detect a malware, we do not use the term in that restricted sense. What we mean by signature is any method utilized for determining if a program is malware. This can include the machine learning system built to recognize malware, a set of behaviors marked as malicious, a white list (anything not on the white list is marked as malicious), and more. The important thing about a signature is that it can be used to determine if a program is malware or not.

Along with the signatures, instructions for how to remove malware that has infected the system and repair any damage it has done must also be created. This is usually done manually, utilizing the results of the analysis stage. Observe what the malware did, and then reverse it. One major concern here is ensuring that the repair instructions do not cause even more damage. If the malware changed a registry key, for example, and the original key is unknown, it may be safest to just leave the key alone. Changing it to a different value or removing it all together may result in corrupting the system being "protected." Thus repair instructions are often very conservative, many times only removing the malware itself.

Once created, the signatures need to be verified for correctness and, more importantly, for accuracy. Even more important than creating a signature that matches the malware is creating a signature that *only* matches the malware. Signatures that also match benign programs are worse than useless; they are acting like malware themselves! Saying that benign programs are actually malware, called a false positive, is an error that cannot be tolerated once the signatures have been deployed to the scanner.

4.4 Application

Once created, the signatures are deployed to the end user. At the end system, new files are scanned using the created signatures. When a file matches a signature, the associated repair instructions followed.

The functionality of the scanner will depend on the type of signature created. String based signatures will use a scanner that checks for existence of the string in the file. A scanner based on Machine Learning signatures will apply what has been learned through ML to detect malware. A rule based scanner will check if the file matches its rules, and so on and so forth.

5 Challenges in Malware Analysis

One of the fundamental problems associated with every step of the malware analysis pipeline is the reliance on incomplete approximations. In every stage of the pipeline, the exact solution is generally impossible. Triage cannot perfectly identify every part of every program that has already been identified. Analysis will generate either potentially inaccurate or incomplete information. All types of signatures are limited. Even verification is limited by what can be practically tested.

Naturally, malware authors have developed techniques that directly attack each stage of the analysis pipeline and shift the error in the inherent approximations to their favor. Packing and code morphing are used against triage to increase the number of "unique" malware that must be analyzed. Packing, tool detection, and obfuscation are used against the analysis stage to increase the difficultly of extracting any meaningful information.

While the ultimate goal of the malware authors is obviously to completely avoid detection, simply increasing the difficulty of achieving detection can be considered a "win" for the malware authors. The more resources consumed in analyzing a single malware, the less total malware that can be analyzed and detected. If this singular cost is driven high enough, then detection of any but the most critical malware simply becomes too expensive.

5.1 Code Morphing

The most common and possibly the most effective attack against the malware analysis pipeline targets the first stage: triage. The attack is to simply inundate the pipeline with as many unique malware as possible. Unique is not used here to mean novel, i.e. does something unique; here it simply means that the triage stage considers it something that has not been analyzed before. Analysis stages further down the pipe from Triage are allowed to be more expensive because it is assumed Triage has filtered out already analyzed malware, severely reducing the number of malware the expensive processes are run on. By slipping more malware past Triage and forcing the more expensive processes to run, the cost of analysis can be driven up, possibly prohibitively high.

One of the ways this attack is accomplished is through automated morphing of the malware's code into a different but semantical equivalent form. Such malware is often called metamorphic or polymorphic. Before infecting a new computer, a rewriting engine changes what the code looks like through such means as altering control flow, utilizing different instructions, and adding instructions that have no semantic effect. The changes performed by the rewriting engine only change the look or syntax of the code and leave its function or semantics intact. The result is that each "generation" of metamorphic malware is functionally equivalent, but the code can be radically different.

While several subtle variations in definitions exist, we view the difference between metamorphic and polymorphic malware as where the rewriting engine lies. Metamorphic malware contains its own, internal rewriting engine, that is, the malware binary rewrites itself. Polymorphic malware, on the other hand, have a separate mutating engine; a separate binary rewrites the malware binary. This mutating engine can either be distributed with the malware (client side) or kept on a distributing server and simply distribute a different version of malware every time (server side).

Metamorphic malware is more limited than polymorphic malware in the transformations it can safely perform. Any rewriting engine is going to contain limitations as to what it can safely take as input. If the engine is designed to modify the control flow of the program, for example, it will only be able to rewrite programs for which it can identify the existing control flow. Since metamorphic malware contains its own rewriting engine, the output of the rewriting engine must be constrained to acceptable input. Without this constraint, further mutations would not be possible. Polymorphic malware, however, does not contain this constraint. Since the rewriting engine is separate and can thus always operate over the exact some input, the output does not need to be constrained to only acceptable input.

5.2 Packing

Packing is a process whereby an arbitrary executable is taken and encrypted and compressed into a "packed" form that must be uncompressed and decrypted, i.e. "unpacked", before execution. This packed version of the executable is then packaged as data inside another executable that will decompress, decrypt, and run the original code. Thus, the end result is a new binary that looks very different from the original, but when executed performs the exact same task, albeit with some additional unpacking work. A program that does packing is referred to a packer and the newly created executable is called the packed executable.

Packing directly attacks Triage and static analysis. While packing a binary does not modify any of the malware's code, it drastically modifies the binary itself, potentially even changing a number of statistical properties. If there is some randomization within the packing routine, a binary that appears truly unique will result every time the exact same malware is packed. Unless the Triage stage can first unpack the binary, it will not be able to match it to any known malware.

Packing does more than simply complicate the triage stage, it also directly attacks any use of static analysis. As discussed in Sect. 4.2, the first step in static analysis is usually to disassemble the binary. Packing, however, often encrypts the original binary, preventing direct disassembly. A disassembler will not be able to meaningfully interpret the stored bits unless it is first unpacked and the original binary recovered.

The need to unpack a program (recover the original binary) is usually not a straight forward task—hence the existence of a challenge. As one might expect, there exists very complex packers intentionally designed to foil unpacking. Some packers, for example, only decrypt a single instruction at a time while others never fully unpack the binary and instead run the packed program in a virtual machine with a randomly created instruction set.

It might seem that simply detecting that an executable was packed would be sufficient to determine that it was malware. There are, however, legitimate uses for packing. First, packing is capable of reducing the overall size of the binary. The compression rate of the original binary is often large enough that even with the additional unpacking routine (which can be made fairly small), the packed binary is smaller in size than the original binary. Of course, when size is the only concern, the encryption part of packing is unnecessary. So, perhaps detecting encryption is sufficient? Unfortunately, no. Encryption has a legitimate application in protecting intellectual property. A software developer may compress and encrypt the executables they sell and ship to prevent a competitor from reversing the program and discovering trade secrets.

5.3 *Obfuscation*

While packing attempts to create code that cannot be interpreted at all, obfuscation attempts to make extracting meaning from the code, statically or dynamically, as difficult as possible. In general, obfuscation refers to writing or transforming a program into a form that hides its true functionality. The simplest example of a source code obfuscation is to give all variables meaningless names. Without descriptive names, the analyst must determine the purpose of each variable. At the binary level, examples of obfuscation include adding dead code (valid code that is never executed), interleaving several procedures within each other, and running all control flow through a single switch statement (called control flow flattening). An in depth treatment of code obfuscation, including methods for deobfuscating the code, is given by Collberg and Nagra [7].

5.4 *Tool Detection*

A major problem in dynamic analysis is malware detecting that it is being analyzed and modifying its behavior. Static analysis has a slight advantage in that the analyzed malware has no control over the analysis process. In dynamic analysis, however, the malware is actually being executed and so can be made capable of altering its behavior. Thus, malware authors will often check to see if any of the observation tools often used by malware analysis are present, and if so, perform only benign activities. For example, a malware may check to see if it is being run by a debugger and if so, exit. This effectively makes the malware invisible to dynamic analysis.

There are two types of checks that can be done by malware: check for a class of tool and check for a specific tool. There are specific types of tools normally used to observe malware in dynamic malware analysis such as debuggers and virtualization. When these tools are used, they usually leave some detectable artifact in the system. For example, in both cases of using a debugger or a virtualized environment, it will be necessity be the case that executing at least some instructions will take longer that if running unobserved. If a malware can detect this discrepancy, through a timer, perhaps, it can detect it is being observed.

Easier than checking for a class of tools, however, is to just check for the specific set of the most widely used tools. Finding a single check that can detect all tools of a particular type is difficult, and the test can be unreliable. A (usually) simpler test is to check for the existence of a specific tool. For example, an executing program could check if it is being run under one of the most common debuggers, Olly Debug, by looking for a process named *ollydbg.exe*. As a natural limitation of software, the number of mature, commonly used commercial or open source analysis tools available is relatively limited. Thus, a malware author can implement a number of simple checks and prevent a large portion of analysis. Naturally, the tool authors can remove the detected artifact, but completely eliminating every trace of a executing program is near impossible. As tool authors remove one artifact, malware authors can use another, resulting in a never ending game of cat and mouse.

5.5 *Difficulty Obtaining Verification Data*

An important part of the verification stage is obtaining "ground truth" information. This ground truth can simply be thought of as the correct answer. If evaluating a new technique's ability to detect malware, then the ground truth would be the labeling of each executable as malware or benign. If evaluating a classifier's ability to separate malware into families, the ground truth would consist of the labeling of each executable with the family it belongs to. The ground truth is needed to determine the correctness of the labels assigned by a machine learning system and to measure its performance.

Obtaining this ground truth is typically an expensive and error prone process. This is because the ground truth usually must be determined manually. Creating the labeling for executables often requires a human analyst to examine the program and give an expert opinion. This takes time and, as with any human judgment, is subject to potential errors. While this may not seem like an issue when the labeling is simply "malware" and "not malware," the challenge increases significantly when labeling a malware with its family. This task may involve manually, albeit with support of tools, viewing and comparing large amounts of information—such as disassembled code, strings, and API calls—for correct labeling. This is complicated by the fact that different malware families share similar characteristics, such as using the same method to trap keystrokes. In such cases, human judgment, inaccuracies, and fatigue may lead to errors in labeling.

6 Machine Learning Concepts

In general, the purpose of machine learning algorithms is to "teach" a program to recognize some type of concept [8]. The concept learned and the way it is taught are of course specific to both the exact machine learning algorithm and the application of the program. In the malware domain, these concepts can be as broad as "malware" or as focused as "implements x." They can be as abstract as "worm" or concrete as "written by Bob Doe." The concepts that can be recognized, and the applications of this recognition, are practically limitless.

The set of all concepts that can be learned by a particular machine learning algorithm is referred to as the *concept space*. This is not a set picked out by a researcher, but the theoretical set of *all* possible concepts for the particular machine learning algorithm selected. A single algorithm is not capable of learning just any concept from the universe of all concepts, but only a comparatively small subset of this infinite universe.

6.1 Features

Machine learning algorithms do not directly digest raw malware, but rather first extract *features* that provide an abstract view of the malware. These features can be thought of as the "language" of the classifier; a way for describing malware to a given machine learning algorithm. For example, features for representing fruit may include things such as shape, color, and whether or not it is firm. An apple would then have the features "shape = round, color = red, firm = yes" while a banana would have the features "shape = long, color = yellow, firm = no". An example feature type used in malware is the set of system calls made. The types of features used to represent malware are discussed in more detail in Sect. 7.

Defining the type of feature used by a machine learning algorithm is a crucial design decision as the feature space (the set of all possible features that can be taken as input) defines the concept space (the set of all possible concepts that can be learned). Only concepts that are capable of being represented or described by the type of features selected by the designer can be learned by the algorithm. For example, features that describe a binary will not, barring black magic, represent the concept "apple." Thus, it is important that the defined features create a concept space containing the concept to learn. Even more important, however, is that the concept space contain as little else as possible. If the concept space is too large, learning the desired concept becomes a "needle in the haystack" problem.

6.2 Classification and Clustering

In malware, the two basic tasks machine learning is used for are classification and clustering. Classification attaches one of a predetermined set of labels to an unknown program. Each of these labels represents a class or category of objects. Thus, assigning a label to a program is akin to marking that program as belonging to a specific class; hence the term classification. The simplest example of classification is labeling a program as malicious or benign. Clustering, on the other hand, partitions the given group of programs into clusters of related programs. The criteria for "related" is usually a similarity function that measures how much two programs resemble each other.

The general pipeline for performing classification in malware analysis is given in Fig. 2. This process has two stages: learning and classification. Both stages begin by extracting the feature representation of the malware. In the learning phase, the malware additionally contain attached labels providing the "correct answer" for classification. In other words, malware, and consequently features, are labeled with the concept they exemplify. A learning algorithm takes these features and labels and creates a model to represent the learned concepts. This model can be thought of as a function that takes a feature representation of a malware as input and outputs the concept that the malware best matches, i.e. a label. Classification, then, is simply

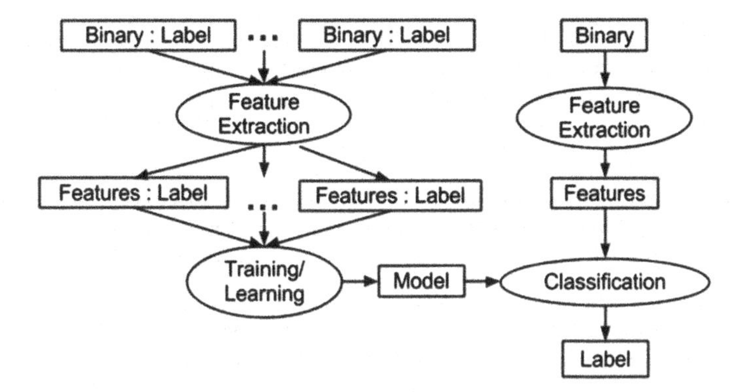

Fig. 2 General classification process

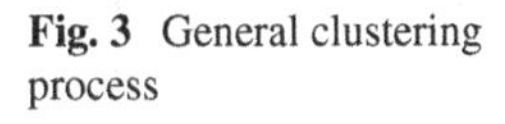

Fig. 3 General clustering process

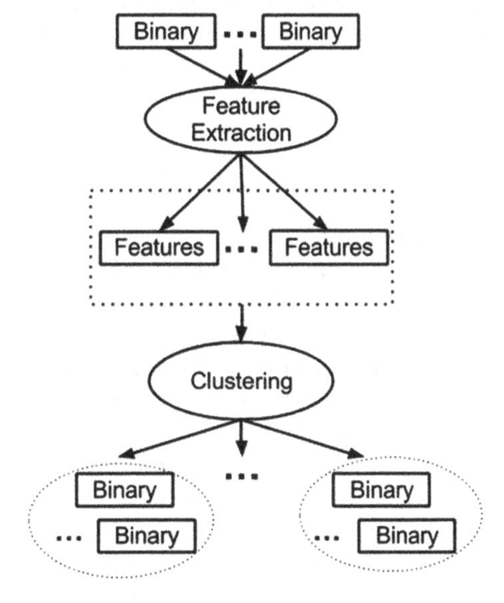

applying this function to an unknown malware. The details of what a model is and how it operates are specific to the exact type of learning algorithm used and thus out of scope for this chapter.

The basic idea of clustering is to put malware into clusters such that malware within a given cluster are more closely related to each other than with malware outside the cluster. The basic process used to accomplish this is given in Fig. 3. Clustering, like classification, first begins by extracting feature representations of the malware. Clustering, however, does not contain a learning phase or utilize labels. The set of features is given directly to a clustering algorithm that partitions the input set into clusters of related malware. Like classification, the exact methods of defining "related" and partitioning are varied and out of scope for this discussion.

While both classification and clustering effectively partition a given group of malware, there are a number of key differences between the two tasks. Classification uses a predefined (usually human defined) set of labels; clustering uses the number of groupings that best fit the given notion of "related." Classification focuses on attaching labels to a *single* malware at a time; clustering operates on whole *groups* of malware. Labels in classification directly correspond to concepts (the concept to learn is "named" by the label); concepts represented by clusters are not named and are thus not always apparent (besides the overly-general "related to each other" concept).

As an illustrative example, let us consider the difference between classifying and clustering a group of binaries into malware families. The first difference is in the method of processing the binaries. After the learning stage, classification will examine each binary one at a time and attach a family name to it. Clustering, on the other hand, will immediately begin operating on the entire collection at once, returning a partitioning of the binaries into groups without labels. The second difference is the number of families that the binaries can be grouped into. In classification, this number is pre-specified and the learning algorithm can learn no more and *no less* than this number. Clustering, however, can theoretically create as many groups as there are binaries (one binary per group). It may, for example, split what an analyst considered one family into two sub-families. The final major difference is in interpreting the learned concepts. It is obvious how the labels of classification correspond to malware families; the labels were created by the analyst after all. The same is not necessarily true of clustering. Not only can the number of families differ from what is expected by the analyst, but none of the groups are labeled and so discovering which groups map to which family is not always straightforward.

A special case of clustering often used in malware analysis, called near neighbor search, is to find programs that are similar to a given "query" program. Knowing what an unknown malware is similar to has a number of uses, the most immediate being a leveraging of existing knowledge. If the unknown malware is 90 % similar to an already existing malware, only the 10 % dissimilar portion must be analyzed. Thus, for this and other reasons it is common to want to find the group of known malware that are very (very here being a relative term) similar to an unknown malware of interest. This is conceptually the same as creating a clustering of similar malware and discarding all clusters except the one containing the unknown malware. (Efficient near neighbor search algorithms, of course, do not actually create the clusters that will be discarded.)

6.3 Types of Learning

While the utilized feature type defines the concept space (Sect. 6.1), it is the type of learning algorithm that defines how this concept space is searched. There are a large number of machine learning algorithms, but they can generally be broken up into several categories based on how they learn concepts: Supervised Learning, Unsupervised Learning, Semi-Supervised Learning, and Ensemble Learning.

Supervised Learning (Sect. 8) can be described as "learning by example." Supervised Learning is often considered synonymous with classification (as supervised learning is often the type of learning algorithm used for classification) and follows the classification pipeline laid out in Fig. 2. The labeled features are fed to the learning algorithm as examples of what each concept looks like. The model is built based upon these examples and the concepts learned correspond the to labels provided.

Unsupervised Learning (Sect. 9) is learning without the "correct answer" given by labels. This form of learning is often considered synonymous with clustering and follows the clustering pipeline put forth in Fig. 3. Instead of forming a model of predefined concepts, unsupervised learning groups objects together based on a given concept of "relatedness" or "similarity." The idea is that objects within a cluster should be more closely related (more similar) to each other than objects outside the cluster. After clustering, every group of objects will represent some concept, though there are no labels or names attached to the represented concept.

Semi-supervised Learning (Sect. 11) combines both supervised and unsupervised learning. Semi-Supervised learning operates over a group of object where some, but not all, of the objects contain "correct answer" labels. The typical application of this sort of learning algorithm is to perform clustering and use the labels that do exist to improve the final clustering result.

Ensemble Learning (Sect. 12) is learning from a collection of classifiers or clusters. For a classifier ensemble, a number of classifiers are trained with their associated models. To classify a new malware, the results of applying all the created models are combined into a single classification. Cluster ensembles work similarly. A number of clusterings of the data are independently created and then merged to create a final clustering.

While Classification and Clustering are often used as synonyms for Supervised and Unsupervised Learning, a distinction can be made. Classification and clustering are tasks, while supervised and unsupervised learning are types of learning algorithms. It is possible to perform classification using an unsupervised learning algorithm. Given a set of malware, clusters can be formed in an unsupervised method (without using any labels), labels attached to the clusters after they have been formed, and classification done by assigning a new malware the label of the most similar cluster. This classification would be unsupervised learning because the concepts (the clusters) were learned without any labels being utilized. The labels were simply names attached to already learned concepts.

7 Malware Features

As initially discussed in Sect. 6.1, the type of feature defined and utilized in any machine learning application is of utmost importance. Features are the input to the machine learning algorithm and define the concept space, i.e. the space of all possible concepts that can be modeled by this algorithm. If the desired concept cannot be modeled, the features are useless. Similarly, if the concept space is too large, then learning how to model the desired concepts is analogous to finding a needle in a haystack. Thus the great importance of defining high quality features.

Just as the types of malware analysis can be divided into static and dynamic, so too can malware features. Static features [9] are features extracted from the binary of the malware without executing it, i.e. through static analysis. While this can refer to the actual bits of the binary or structural information contained in the header, it more commonly refers to features extracted from the disassembled binary. A disassembled binary is created by converting the bits of the binary back into human readable machine code (assembly language, not high level source such as C). Various transformations on this disassembly are then performed to create many different types of features.

Dynamic features [10], on the other hand, are features extracted by executing the malware and observing what it does, i.e. through dynamic analysis. There are several levels of abstraction that dynamic features can exist at ranging from a trace of the instructions executed in the processor to a predefined set of behaviors watched for.

Static and dynamic features have the issues discussed in Sect. 4.2. It is usually impossible to perfectly extract the precise of feature representation for a given malware and thus approximations are used. Static features often result in an over approximation and dynamic features often result in an under approximation. Take for example, the defined feature type "the set of system calls that can be made by the program." Static analysis will usually extract almost all of the actual system calls that belong in the true set of features, but will also include potentially many system calls that the program will never actually execute. Features extracted by dynamic analysis are guaranteed to be in the true feature set, but not all system calls will be observed and recorded.

7.1 Binary Based Features

The simplest static features are structural features based directly on the raw binary. That is, features that treat the binary as nothing more than an executable file and do not attempt to extract any information more abstract than what the structure of the binary is. These types of features have been based on sequences of bytes in the binary, information contained in the binary's header, and the strings visible in the binary.

7.1.1 N-Gram and N-Perm

One of the most common types of binary based static features is the *byte n-gram*. N-grams are a feature commonly used in text classification and they are created by sliding a window n characters long across a document and recording the unique strings

Table 1 Byte code 2-grams and 2-perms

Byte code	2-grams	2-perms
0f0e	'0f0e' '0e0f'	'0e0f' '0fb4' '0eb4'
0f0e	'0eb4' 'b40f'	
b40f		

found (a technique often referred to as shingling or windowing in text classification literature). For example, the 2-grams for the string 'ababc' are 'ab', 'ba', and 'bc'. In the malware context, the n-grams are created over the byte code representation of the binary. N-grams were introduced to malware classification by members of the IBM TJ Watson Research Center [11–14].

A modification on n-grams proposed by Karim et al. [15] was *n-perms*. N-grams are extremely brittle, especially when n in large. Simply removing, adding, or swapping a few instructions can have a major impact on the set of n-grams retrieved. Karim et al. [15] proposed to treat all permutations of a single n-gram as equal. Thus, in the example given earlier, the string 'ababc' with 2-grams 'ab', 'ba', and 'bc' would only have 2-perms of 'ab' and 'bc'. The 2-grams 'ab' and 'ba' would be considered equal and only 'ab' stored.

Examples of byte code and the corresponding n-grams and n-perms are given in Table 1. The byte code is the hexadecimal representation of a binary. The 2-grams are created by sliding a window of size two across the individual bytes and recording unique sequences. The 2-grams are shown in the order encountered. The 2-perms are created by internally sorting each 2-gram. So, for example, the 2-gram '0f0e' becomes the n-perm '0e0f'. This has the desired effect of normalizing all permutations. In this example, there are four 2-grams, but only three 2-perms.

7.1.2 PE Header Based

A common source of structural information for creating features is the *PE header*. The PE Header is the data structure in a Windows executable that holds the information needed by the loader to place the program into memory and begin execution. Thus most of the information in the PE header relates to the locations and sizes of important pieces of the binary. For example, the loader needs to know the number, location, and size of the sections of the binary.[1] There are also a number of important data structures that the loader needs to be able to find, such as the import table. The import table is a list of the various external libraries that will need to be loaded into this program's address space.

Using the structural information provided by the PE Header in the triage analysis stage turns out to be quite effective. The information in the PE Header is very quick to extract (the OS does this every time the program is executed) and robust against

[1] Binaries are not one big blob, but separated into sections of logically related code and data. At a minimum, there will be two sections: one designated for data and the other for code.

minor changes in the binary. Compiling a program twice may result in binaries unique by hash because of trivial changes, but both binaries would contain almost identical PE Header information. This approach was proposed by Wang et al. [16].

Walenstein et al. [17] examined this effectiveness of using header information as features. They were especially interested in learning how much the information in the header contributed to the accuracy obtained when using byte based n-grams. When n-grams are computed over the entire binary, the header information is implicitly captured. What they found was that not only was header information quite useful in discovering malware variants, but that it also contributed a good deal to the usefulness of byte based features. Walenstein et al. [17], however, cautioned that it would be trivial for malware authors to remove or modify most, if not all, of the identifying information in the PE header and so this information should not be solely relied upon.

A non-structural feature that has been used from the PE header is the set of libraries and functions listed in the import table [18]. The import table is a data structure pointed to by the PE header and contains the list of functions from external libraries that are required by the program. Since external libraries are obviously external to the binary, hard coded addresses of the library functions cannot be used in the program. This is where the import table comes in. For each required function, the import table contains an entry with the name of the function and an address for the function. At load time, the loader pulls the libraries into the program's address space and writes the correct address to each entry in the import table.

The list of imported library functions contains a wealth of information. When dealing with well known libraries, such as libc or the set of Windows system calls, the intended task of each function is, well, known. For example, if functions from crypt32.dll (dll is short for dynamic link library) are imported, it is likely that this program performs cryptography tasks. Thus, if the import table hasn't been tampered with (usually a bad assumption in malware, but still true often enough), it can provide a nice high level idea of the types of behaviors the program is intended to perform.

7.1.3 Strings

A final type of feature based on the binary is *strings*. This was first explored by Schultz et al. [18] and is quite simply any sequence of bytes that can be validly interpreted as a printable sequence of characters. This idea was later refined by Ye et al. [19] to only include "interpretable" strings, that is only strings that make semantic sense. Ye et al. [19] posited that strings such as "8ibvxzciojrwqei" provided little useful information.

Table 2 illustrates the difference between strings and interpretable strings. The column to the left is the byte sequences found in the binary. The middle column is the result of treating the byte sequences as ASCII characters. Finally, the last column indicates whether the string would be considered interpretable.

Table 2 Strings and interpretable strings

Bytes	String	Interpretable?
44 65 74 65 63 74 65 64 20 6d 65 6d 6F 72 79 20 6C 65 61 6b 79 81 0A 00	"Detected Memory leaks! n"	✓
47 65 74 4C 61 73 74 41 63 74 69 76 65 50 6F 70 75 70 00	GetLastActivePopup	✓
31 39 32 2e 31 36 38 2e 38 33 2e 31 35 33	192.168.83.153	✓
3b 33 2b 23 3e 36 26 1e	;3+# 6.&	χ
77 73 72 65 77 6e 61 66 34 79 6F 77 33 69 37 35	wsrewnaf4yow3i75	χ

7.2 Disassembly Based Features

While structural information is useful for finding malware that *looks* the same, it isn't always useful for discovering malware that *behaves* the same. Using the list of imported library functions is a good, but limited first step. Ideally, we could convert the binary back into its original source code, complete with comments, and extract features from there. This, however, is a hopeless endeavor as much of the information is lost when the source code is compiled. Comments and variable names, for example, are usually completely tossed away and thus unrecoverable. As a next-best option, a disassembly of the binary (described below) is used.

Many static features begin with a disassembly stage, but perform further abstractions on the disassembly before extracting features. These types of features will be discussed in later sections. Here, we describe what disassembly is and the features that are extracted directly from it.

7.2.1 Disassembly

Disassembling a binary refers to extracting its assembly code (often simply referred to as "code" or "disassembly" for short). Assembly is a low level programming language that describes in human readable terms the actions of the processor. Recovering the assembly code is a matter of parsing the binary and mapping the bytes back into the human readable terms. The mapping between bytes and assembly instructions is one to one, but due to the complexity of the most prevalent architecture (the ×86 family), statically accomplishing this task is not as straightforward as it may seem. It is further complicated by malware authors taking deliberate steps to break disassembly [20, 21]. Despite these complications, robust tools have been developed that can provide surprisingly accurate results.

It may seem that an easy way to extract the disassembly of a binary is to do it dynamically instead of statically. In other words, execute the binary and record the instructions as interpreted by the processor. This has been done [22–24], but what results isn't a full disassembly of the binary, but rather an *instruction trace*. Only the instructions that were actually executed by the processor will be disassembled and recorded. A single execution path through dynamic analysis is likely to only encounter a small percentage of the full binary. When it is desired to extract the full disassembly of the binary, it is still best done statically.

An example of disassembly is given below in Table 3. At the far left is the C program from which the binary was compiled. It is a simple program that does nothing useful; it just increments a variable ten times. Next are the instruction addresses and the bytes that comprise the individual instructions. The disassembled instructions are presented next to and on the same line as the byte code representation of the instructions.

The first three instructions set up the stack (the place in memory the variables are stored). The registers ebp and esp are the base pointer and stack pointer. The base pointer points to the base of the stack and the stack pointer points to the top. The next three instructions (addresses 80483ba–80483c8) initialize the variables i and j. The variable j in placed on the stack first (memory address [ebp-8]), i second (memory address [ebp-4]). Variable i is set to zero twice because this is the case in the source code (when it is initialized and when the loop starts). The top of the loop is at address 80483d1 where j is incremented by 1, followed by an increment of i and a check if i is equal to 9 (less than 10 condition in the source). If it is not, control flow is returned back to the top of the loop. If i is equal to 9, then the final instructions are executed, exiting the program.

Table 3 Example disassembly

Source code	Address	Byte code	Disassembly
int main() {	80483b4	55	push ebp
int j = 0;	80483b5	89 e5	mov ebp, esp
int i = 0;	80483b7	83 ec 10	sub esp, 0x08
for (i=0; i< 10; i++){	80483ba	c7 45 f8 00 00 00 00	mov [ebp-8], 0
j++;	80483c1	c7 45 fc 00 00 00 00	mov [ebp-4], 0
}	80483c8	c7 45 fc 00 00 00 00	mov [ebp-4], 0
}	80483cf	eb 08	jmp 80483d9
	80483d1	83 45 f8 01	add [ebp-8], 1
	80483d5	83 45 fc 01	add [ebp-4], 1
	80483d9	83 7d fc 09	cmp [ebp-4], 9
	80483dd	7e f2	jle 80483d1
	80483df	c9	leave
	80483e0	c3	retn

7.2.2 Opcodes and Mnemonics

One of the most common types of features directly based on the disassembly is opcodes [25–29] or mnemonics [30, 31] of the disassembled instructions, usually in ngram or nperm form. The opcode of an instruction is the first byte or two that tells the CPU what type the instruction is (ex. add, move, jump) and the types of operands to expect (register, memory address). A mnemonic is simply the human readable symbol that represents the opcode. Several opcodes will map to the same mnemonic because a mnemonic does not contain the operand information of the opcode. For example, the mnemonic for adding two values together is `add`, but there are around eight different opcodes for this one `add` mnemonic.

Examples of opcodes and mnemonics are given in Table 4. The assembly and corresponding byte codes are given first, followed by the instruction's opcode and mnemonic. All the instructions listed have unique opcodes, but there are only two unique mnemonics: push and mov. For push instructions, the opcode tells the processor which register to push onto the stack. The opcode will be 50+ a number representing which register was pushed. In a mov, the opcode indicates the kind of move that will occur. For example, opcode 8b indicates that the move will be from register to register, while opcode 89 tells the processor that the move will be from a register into a memory location. The next byte in the instruction tells the processor which register combinations will be used. The byte 'ec' explicitly indicates that the contents of esp will be moved into ebp. Byte 45 means the contents of eax will be moved into the address specified by the value currently in ebp plus a one byte offset; the next byte is this offset.

Using opcodes or mnemonics abstracts away the exact operands used for the instructions. This enables code that differs in only the relative jump addresses, for example, to produce the same set of features. The noise of structural based features is thus reduced without much loss of information. Obviously, mnemonics abstracts the features further than opcodes because opcodes still contain information of the types of operands. An instruction adding two registers will have a different opcode but the same mnemonic as an instruction adding a register address and an immediate (constant).

Table 4 Example opcodes and mnemonics

Assembly	Bytes	Opcode	Mnemonic
push eax	50	50	push
push ebp	55	55	push
mov ebp, esp	8b ec	8b	mov
mov [ebp, oxf8], eax	89 45 f8	89	mov

7.3 Control Flow Based

The control flow of a program is the way in which execution, or control, is capable of passing through the program. Control refers to the part of a program that is currently executing, i.e. is in control. Control flow, then, is the various execution paths that can be taken through the program.

Control flow information can either be captured globally or locally. In general, the complete information regarding all possible paths that can be taken through every single part of the program is too great and too complicated to be of use. Thus, global control flow is typically captured by recording how different pieces of the program interact with one another. Within each piece, local control flow will then be separately recorded. In malware, this is done using a call graph and control flow graphs. A call graph records which functions call each other and a control flow graph records for a single function the possible execution paths through that function.

7.3.1 Callgraph

A *callgraph* is a directed graph depicting the calling relationships between the procedures of the program, i.e. which procedures contain calls to which other procedures. A callgraph does not give information on how control flows through the procedure itself, just how it is transferred between procedures. This provides a coarse, high level overview of the flow of control and data through the entire binary. The use of call graphs for malware analysis was first proposed by Carrera and Erdelyi [32] and further refined by Briones and Gomez [33] and Kinable and Kostakis [34].

An example call graph with the C source it was derived from is given in Fig. 4. The call graph was derived from source code because this makes the concept easier to comprehend, but the exact same concepts apply at the assembly level as well. This source has five functions: main, doAwesome, doMoreAwesome, doWork, and doAwesomeWork. Each of these five functions is represented by a single node on the call graph. An edge between two node represent a "calls" relationship. For example, the edge between main and doWork represents the relationship "main calls doWork." Loops indicate recursion, such as the loop from doMoreAwesome onto itself. The function doMoreAwesome recursively calls itself until sufficient awesome has been achieved.

While a callgraph can be constructed statically or dynamically, it is usually done statically. This is because the purpose of a call graph is generally to show which functions *can* call which other functions, not to only show that a function *did* call another function in a *single* execution path. Static analysis is more capable than dynamic analysis in extracting this full call graph. Dynamic analysis will only be able to determine calling relationships that it actually witnesses.

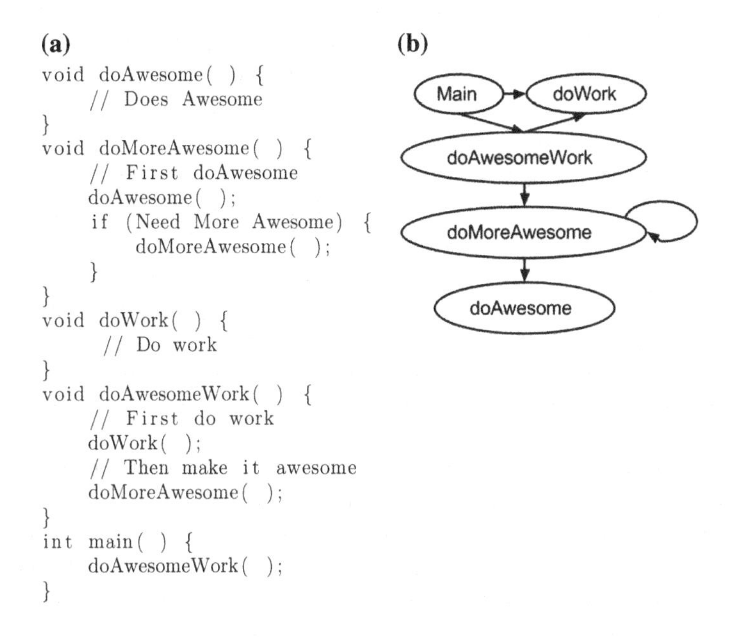

Fig. 4 Example callgraph

7.3.2 Control Flow Graph

At a finer grain than the call graph is the *control flow graph* (CFG) [35]. A CFG is a graph of the control flow *within a single procedure*. Just as a call graph provides a view of control flow at the binary level, a CFG provides this view at a procedure level. The CFG is created by first breaking the sequential code of the procedure into discrete blocks called *basic blocks*. Basic blocks are constructed such that if control reaches the block, every instruction within the block is guaranteed to execute; basic blocks have a single entry and a single exit point. Edges in the CFG represent decisions made about the sequence of instructions to execute. For example, an if statement will create two edges, one representing the "TRUE" path, the other the "FALSE" path.

An example of creating a CFG is given in Fig. 5. The Assembly from which the CFG is created is given on the left and the corresponding CFG on the right. The assembly code initializes a variable at memory location ebp-8 and adds the value of eax to it 10 times. The is equivalent to multiplying eax by 10 and storing the value in the variable. The first basic block, consisting of two moves and a jump, initializes the variables and jumps to the loop condition at the end. The beginning of the new basic block will not be immediately after the jump instruction, but rather where the jump instruction goes to. In this case the basic block will start at the condition check. This condition check will exit if the counter stored at memory location ebp-4 is equal to 9. Otherwise it will go to code that adds eax and increments the counter before returning back to the check.

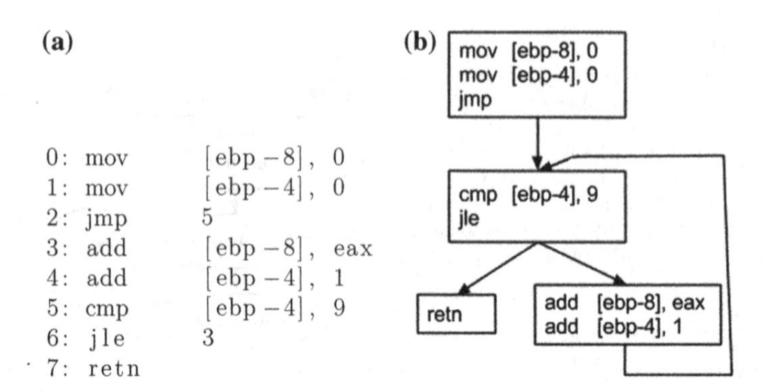

Fig. 5 Example control flow graph (CFG)

This example illustrates how the single entry, single exit property breaks apart the code into basic blocks. Just because instructions are sequential without a branching instruction between them does not mean that they will all be in a basic block together. The add instructions and condition check in addresses 3–6 are sequential, but they are broken into two basic blocks because the jump at address 2 branches into the middle of this sequence.

7.4 Semantics Based

In any programming language there are infinitely many ways to accomplish any given task, even a task as simple as assigning zero to a variable. As discussed in Sect. 5, this fact is often used by malware authors to create differing versions of malware that accomplish the same goal. In response, malware analysts have created features that do not rely on what the code *is* but rather what the code *does*, i.e. its *semantics*.

7.4.1 State Change

In static analysis, semantics is often defined as the effect executing the code will have on the hardware state [36–38], i.e. the change in values stored in registers and memory. Not every change need be recorded but rather the end result of a given portion of code. The size of these portions differ, but generally it is either a basic block [37, 38] or a procedure [36]. Computing the effect of executing a basic block is straightforward as, barring exception handling, execution flows through successive instructions in the block. The semantics of a block follows from the functional composition of semantics of individual instructions. Doing the same for a procedure, however, is not straightforward, especially, as is usually the case, when its CFG has branches and loops.

There are two methods of representing semantics: (a) enumerated concrete semantics and (b) symbolic expressions. In the first method the semantics is represented as a set of pairs of specific, concrete input and output states. For instance, the pair (`eax = 5, ebx = 20`) may represent that when executed with the value 5 in the register `eax`, the program upon termination will have the value 20 in `ebx`. A set of such pairs, of course with significantly more complex input and output states, would represent the semantics of the code. Using symbolic expressions instead, the semantics may be represented succinctly as a single expression. For instance, the expression `ebx = pre(eax)* 4` may represent that the value of `ebx` upon termination is the 4 times the value of `eax` at the start.

The benefit of using enumerated concrete semantics is that two different ways of effecting the same change of state will have exactly same input and output pairs. Alternatively, even a single difference in the input and output pair would imply a different semantics. The challenge, of course, is that a program may have exponentially large space of input, making it infeasible to compute and represent such semantics. On the other hand, symbolic expressions offer the advantage that a single expression may represent the entire semantics. However, since one cannot guarantee that two equivalent program always produce the same expression—this would solve the Halting Problem—differences in the semantic expression need not imply that the underlying code has different semantics.

The two different methods of representing, as well as computing, semantics have been used respectively by Jin et al. [37] and Lakhotia et al. [38]. Jin et al. [37] address the issue of exponentially large input space by randomly sampling the space. They pre-generate a large number of input states and use them to compute the semantics of all basic blocks. Each basic block is executed using the same set of input states and the corresponding output state recorded. A basic block can then be represented by a hash of all of its output states.

Lakhotia et al. [38] use symbolic interpretation to compute the semantics of blocks, without assigning any concrete values. Rather than execute a basic block with specific inputs, [38] execute it with symbolic inputs. Furthermore, they use algebraic properties, such as the distributive law, and simplifications to map expressions into a canonical form when possible. For example, they simplify the expression `(eax + 4) * 5` to `eax * 5 + 20`. This ensures that, for a large set of operations, all expressions that are functionally equivalent resolve to the same symbolic representation. After computing the symbolic output values, Lakhotia et al. [38] further abstract the semantics by converting the concrete register names and numbers into logical variables. This type of feature they call the "juice" of the basic block and term it "GenSemantics."

Tables 5 and 6 respectively illustrate how the features of [37, 38] work. In both tables, the left column contains two functionally identical basic blocks. Both tables contain the same two blocks. The two presented blocks both assign 3 to eax, set the value of edx to ecx, and multiply by 3 the value of ebx + 4. The first block uses a multiply instruction; the second block uses a series of adds to have the same effect.

Table 5 ([37]) shows the result of executing the two basic blocks on two sets of input values. The columns labeled *Input 1* and *Input 2* show the input states, as values

Table 5 Features by Jin et al. [37]

Basic block	Input 1	Output 1	Input 2	Output 2
add ebx,4	eax = 33	eax = 3	eax = 62	eax = 3
mov eax,3	ebx = 4	ebx = 21	ebx = 7	ebx = 33
mul ebx,eax	ecx = 25	ecx = 58	ecx = 72	ecx = 54
mov ecx,edx	edx = 58	edx = 58	edx = 54	edx = 54
add ebx,4				
mov eax,3	eax = 33	eax = 3	eax = 62	eax = 3
mov ecx,ebx	ebx = 4	ebx = 21	ebx = 7	ebx = 33
add ebx,ecx	ecx = 25	ecx = 58	ecx = 72	ecx = 54
add ebx,ecx	edx = 58	edx = 58	edx = 54	edx = 54
mov ecx,edx				

Table 6 Features by Lakhotia et al. [38]

Basic block	Semantics	GenSmantics
add ebx,4		
mov eax,3	eax = 3	A = N1
mul ebx,eax	ebx = pre(ebx) * 3 + 12	B = pre(B) × N1 + N2
mov ecx,edx	ecx = pre(edx)	C = pre(D)
add ebx,4		
mov eax,3		
mov ecx,ebx	eax = 3	A = N1
add ebx,ecx	ebx = pre(ebx) * 3 + 12	B = pre(B) × N1 + N2
add ebx,ecx	ecx = pre(edx)	C = pre(D)
mov ecx,edx		

in the registers eax, ebx, ecx, and edx. The columns labeled *Output 1* and *Output 2* show the corresponding states after executing the blocks.

As can be seen in the example, the outputs of the two blocks are exactly the same when using the same set of input values. It is important that the same list of input values be used across all blocks in all binaries that are going to be compared. As should be obvious, only features created with the same list of inputs can be meaningfully compared. Thus, as in this example, there will often be more generated values than needed because we have to generate enough values so that *all possible* basic blocks will have enough input values.

Table 6 gives [38] features extracted from the same two blocks. The middle column provides the initial semantics extracted through symbolic execution and simplification. The notation `pre(x)` represents the value of x when the basic block was reached; it is a symbolic notation for the input value. The right column contains the generalized semantics.

Chaki et al. [36] tackles the problem of computing the semantics at the level of the procedure instead of at the basic block level. Computing a single expression that represents the semantics of an entire expression becomes significantly more challenging, more so if the semantics is to be in a canonical form to ease comparison. For instance, computing the semantics of a loop would involve computing an expression that represents its fixed point. However, if a single path through the procedure is selected, the semantics can be computed as straightforwardly as in basic blocks. This is what [36] do. Given a single path through the procedure, they compute features very similar to the GenSemantics of [38]. This process is repeated for all bounded paths through the procedure. For unbounded paths, due to cycles, a bounded depth first traversal is used to limit the length of the path traversed. The features from all the traversed paths are unioned together to represent the whole procedure.

7.4.2 API Calls

Another way of defining the semantics of a program is through the set of API calls the program makes. An API call refers to when a program calls a function provided by some library, often the system library. The system library is a library of functions provided by the operating system to perform tasks that only the operating system has permissions to perform directly, such as writing a file to disk. Other well known libraries include the standard C library, encryption libraries, and more. If a program makes an API call to a well known and widely distributed library, then the purpose of that call can easily be determined.

When the order in which API calls are made is preserved in an *API trace*, the semantic information becomes even stronger. Sequences of API calls can be used to potentially determine the presence of high level behaviors [22, 39] such as "walks through a directory" or "copies itself to disk."

The set or trace of API calls provides a "behavioral profile" of the binary being examined. This idea was first proposed by bailey et al. [40] and later formalized by Trinius et al. [41]. Trinius et al. [41] designed what they called a *Malware Instruction Set* (MIST) for representing various layers of semantic information present in API calls. The layers are ordered from most to least abstract. Each "instruction" in MIST represents a single API call.

The first layer of a MIST instruction contains a category for the API call and the operation performed by the call. The category represents the type of object the API call operates on. For example, an API call that writes a file to disk would be in a "File System" category, one that opens a socket in a "Network" category, and an API that edits a registry value would be in a "Registry" category. The operation is, as the name implies, what was done to the object. The "File System" category, for example, has operations such as "open_file", "read_file", "write_file". Every category has a pre-specified set of possible operations. Both the category and operation are derived from the name of the API call.

Subsequent layers of a MIST instruction are derived from the parameters to the API call and ordered by the expected variability of their values. For example, an API call that manipulates a file will contain as one of its parameters the name of the file to manipulate. From this file path, we can separate out two parts, the path and the name.

The path will have a lower variability than the file name. A binary that writes to the windows directory is likely to write to the windows directory on every execution. However, the names of the files may well be different for every execution, since its common for malware to randomly generate filenames. Thus, the file path has a lower expected variability than the file name and is put at a higher level.

7.5 Hybrid Features

In an effort to increase the accuracy of classification, various authors have utilized feature fusion: combining multiple types of features to created hybrid features.

The first to use feature fusion was Masud et al. [42]. They combined features from three different levels of abstraction: binary n-grams, derived assembly features (DAF), and system calls. Binary n-grams are the same as described in Sect. 7.1. The feature on the next level of abstraction is the DAF. A DAF is basically just the disassembled binary n-grams. A DAF is created by extending the n-gram on the front and the back as needed to fit instruction boundaries and then disassembling these instructions. At the highest level of abstraction are system calls. The names of the system calls present in the header of the executable were extracted and used.

Lu et al. [43] combine static and dynamic features. The static features chosen were the names of system calls present in the binary of the executable and the dynamic features were expert defined behaviors commonly seen in malicious programs. Several examples of the behaviors searched for are packing the executable, DLL injection, and hiding files. There were twelve such features defined. These static and dynamic feature sets were fused by taking the union of the feature sets.

Islam et al. [44] simply combined two features they considered useful in an attempt to make an even more useful feature set. They take the union of function length frequency and printable string feature sets. Function length frequency is the number of functions a program has of a given length. It was found by Islam that malware within the same family tend to have functions of around the same length, thus motivating the use of function length frequency as a feature. Printable strings are simply sequences of bytes which can be interpreted as valid ASCII strings.

LeDoux et al. [45] takes a different approach. Whereas the above cited works all combined mutually exclusive feature sets, LeDoux et al. fuses features that are very similar, but collected in different ways. The hypothesis being that the differences present between two feature sets that should be identical is itself a feature and can help identify malware that is intentionally obfuscated against one of the methods of collecting features.

8 Supervised Learning

Supervised Learning describes a class of learning algorithms that build a model used in classification (see Sect. 6.2). These algorithms learn the concepts represented by the model by examining labeled examples of the concepts. The label provides the "correct answer," that is the concept that is being exemplified. This training data (the labeled examples) is often referred to as the ground truth. After the model has been learned, a never before seen example of a learned concept can be classified by matching it to the learned model.

The method of learning and classifying used by Supervised Learning can conceptually be thought of as teaching a child different types of animals by showing him flash cards and telling him the name of the animal. "This is a horse. This is a dog. This is another horse. This is a cat." By looking at many examples of horses, dogs, and cats, the child builds up an internal representation of each animal type. Then when shown a picture of a never before seen dog, and asked to "classify it", the child will be able to respond "It is a dog."

One of the limitation of classifiers built using supervised learning is that they will *only* be able to attach a label it has already learned and *must* attach one of these labels. Supervised learning can only explicitly learn concepts it is told about through the labels in the training data. Neither is it able to provide an "I don't know" answer. It must attach one of its learned labels. To extend our earlier example further, if our child is only shown pictures of horses, cats, and dogs, and then asked to "classify" a picture of a lion, he will respond "cat," not lion. If shown a picture of a tractor, however, the child will be able to respond "I don't know." A classifier built using supervised learning would respond "horse."

Another potential issue with supervised learning specific to the malware domain is the reliance on the ground truth labels. In malware analysis, ground truth can often only be reliably determined through manual reverse engineering and analysis of the binary by a human expert. The expert makes a judgment call based upon his experience and assigns an appropriate label to the binary. This process is expensive in terms of time and resources required, thus restricting the number of labels that can be reasonably generated. The required level of expertise for performing this task limits the reasonable size of ground truth even further (crowd sourcing can't solve this problem). As an added detriment, any process that relies on human decisions is prone to error, regardless of the level of expertise of the human. Training a Supervised Classifier on improperly labeled data can lead to very poor performance.

The main application of supervised learning to malware analysis has been in attempting to built automated malware detectors [46], in triage [47], and in evaluating the quality of malware features. The original intention of supervised learning in malware was to build a classifier that could act as the scanner on the end system. However, the false positive rate of these classifiers tend to be too high. A false positive is when the classifier labels a benign program as malware. On the end system, the false positive rate must be extraordinarily low as constantly labeling valid programs as malware will at best annoy the user into not using the scanner and at worst break the system by removing critical system files.

While not useful as a scanner, supervised learning has been of use in triage [47]. Supervised Learning can be used to build a classifier that identifies if a new, incoming program belongs to any of the already known and analyzed malware families. Knowing which family malware belongs to makes any manual analysis needed an easier prospect because the analyst does not have to start from nothing, but can leverage knowledge of what that family of malware does and how it usually operates. A higher false positive rate can be tolerated in the triage stage than in a scanner because the results are still verified later on.

A final use of supervised learning in malware is for controlled test and comparison of the quality of malware features. Due to the fact that Supervised Learning required that data with ground truth exists, determining performance metrics for classification is straightforward. A portion of the labeled data is set aside for testing purposes and the rest of the data used for creating the model. The testing data is classified using the created model and labels assigned by the model are compared against the ground truth labels to determine accuracy of classification. To compare the quality of features, the accuracy of classification can be compared. Features that are of a higher quality will result in a higher accuracy.

9 Unsupervised Learning

Unsupervised Learning [34, 48–51] algorithms learn concepts without the use of any labeled data. Labeled data may be used to perform a post hoc evaluation of the learned concepts, but they are in no way used for learning. Concepts are learned in unsupervised algorithms by clustering together related objects in such a way that objects within a cluster are more similar to each other than objects outside the cluster. Each cluster then represents a single concept. What this concept is, however, is not always easy to discover.

In order to determine which binaries are related and so belong in the same cluster, unsupervised learning algorithms rely on measuring the *similarity* of malware. A comprehensive overview of the various similarity functions that can be used to compare binaries is given by Cesare and Xiang [52]. It is important to note that similarity metrics do not directly measure the similarity of two binaries, but rather the similarity of the binary's *features*. The same pair of binaries may have very different similarities depending on the choice of feature type. Take, for example, two malware with dissimilar implementation but an almost identical set of performed system calls. Features that rely mostly on the implementation, such as opcode n-grams, will result in measuring a low similarity. Features that only look at system calls, however, will result in the exact opposite: high similarity.

One of the major issues with clustering is that many of the most common clustering algorithms require a priori knowledge of the number of clusters to create; the number of required clusters is a parameter to the clustering algorithm. When it is known how many malware families are present in the data set, it is common practice to use this number as the number of clusters. However, while the number of families present in a laboratory sample of malware may be determined, the same cannot be said of the sets of malware encountered "in the wild." If the incorrect number of clusters is selected, performance will quickly degrade.

Fortunately, not all unsupervised algorithms require prespecifying the number of clusters to create. Representative of these algorithms, and the most commonly used, is *hierarchical clustering*. Hierarchical clustering can be either agglomerative or divisive. Agglomerative Hierarchical clustering works by iteratively combining the two most similar items into a cluster. Items include both the individual objects and already created clusters. So a single iteration could consist of putting two objects into a cluster of size two, adding an object to an already existing cluster, or combining two clusters together. This process is repeated until everything has been put into a single cluster. Each step is tracked, resulting in a hierarchical tree of clusters called a dendrogram. Divisive hierarchical clustering works similarly, but starts with everything in a single cluster and iteratively splits clusters until everything is in a cluster of size one.

Hierarchical clustering changes the problem from needing to know a priori the optimal number of clusters to knowing when to stop the agglomerative (or divisive) process, i.e. where to cut the tree. This is determined using a type of metric known as a clustering validity index [53]. While a number of such indices exist, they are all a measure of some statistic regarding the similarity of objects within the clusters, the distance between objects in different clusters, or a combination of the two. A number of these metrics were evaluated by [50] for performance in the malware domain using a particular feature type.

10 Hashing: Improving Clustering Efficiency

In applications of clustering in malware, there are two inherent performance bottle necks: large feature space and the requirement to compute all pairwise similarities. The number of features generated for each individual malware is usually quite large. A single malware can have thousands, even hundreds of thousands of features representing itself. This has a two fold effect. First, it drives up the time cost of doing a single similarity computation. Second, storing this massive amount of features requires an equally massive amount of memory. Either a specialized computer with the required amounts of memory must be used, or the majority of the time spent computing similarities will be in swapping the features to and from disk.

Compounding the time and space requirements of clustering is the complexity of a typical clustering algorithm. Even efficient algorithms run at $O(n^2)$ due to the need to compute the similarity between every pair of objects.[2] This is tolerable for small to

[2] Objects don't need to be compared with themselves and similarity functions are (typically) symmetric, so the actual number of comparisons required is $(n^2 - n)/2)$.

medium sized datasets, but severely limits the scalability of clustering. Parallelism can help up to a point, but handling millions of malware requires more efficient approaches.

The solutions being explored for both of the above problems are the same: hashing. Feature hashing based on Bloom filters [54] is being used to both reduce the memory overhead and the time a single similarity comparison takes. The number of required pairwise comparisons is being reduced or even eliminated by performing clustering based on a shared hash value. A few explorations have been made in using cryptographic hashes [51, 55, 56]. More commonly, however, Locality Sensitive Hashing is being used as a "fuzzy hash" to reduce the complexity of clustering [37, 48]. In some instances, constant time is even achieved [37]!

10.1 Feature Hashing

To reduce the memory requirement and time for a single similarity comparison, [54] present a method for hashing features into a very small representation using Bloom filters [57]. A Bloom filter is a probabilistic data structure used for fast set membership testing. It consists of a bit vector of size m with all bits initially set to zero and k hash functions, $h_1, h_2, \ldots .h_k$, that hash objects into integers uniformly between 1 and m, inclusive. In practice, these k hash functions are approximated by simply splitting the MD5 into k even chunks, taking the modulo of each chunk with m, and then treating each resulting value as the result of one of k hash functions. To insert object $\times$ into the bloom filter, for each $h_i(x)$, set the bit at position $h_i(x)$ to one. Object y can then be tested for membership by checking that all bits at positions $h_i(x)$ for each i from 1 to k is set to one.

Jang et al. [54] perform feature hashing by inserting all features for a single binary into a bloom filter with only one hash function. The decision to use only one hash function was made in order to facilitate fast and intuitive similarity comparisons. Similarity between malware can be quickly approximated by measuring the number of shared and unique bits in each binary's corresponding bloom filter. In addition, only the bloom filters need to be stored, not every single feature, thus reducing the total storage requirement. Jang et al. [54] experienced compression rates of up to 82 times!

10.2 Concrete Hashes

Cryptographic hashes such as MD5 and SHA-1 are widely used to filter out exact binary duplicates. However, hashes at this level are not robust; changing a single bit can result in a dramatically different hash. Several approaches have been taken to improve this robustness by taking hashes of different types of features instead of the raw bits of the binary.

Wicherski [51] attempted to use the information stored in the PE header of a binary to define a hash for the purpose of filtering out duplicate binaries. The information in the PE header was utilized to make the hash robust against minor changes in the binary. For example, if the only difference between two binaries is a time stamp in the header, the binaries should be considered equal. To create this hash, [51] takes a subset of bits from several PE header fields, concatenates them together (in a fixed, reproducible ordering) and takes the SHA-1 of this bit sequence.

Wicherski [51] does not use all bits from all PE header fields, but rather judicially selects both the fields and bits used in the construction of the hash. The fields used are image flags such as whether the binary is a DLL, the Windows subsystem the binary is to be run in, the initial size of the stack and the heap, the initial address each section of the binary is loaded into memory at, the size of each section, and the flags for each section that indicate permissions and alignment. For each PE field included in the hash, only a subset of the full bits in the field are utilized. The exact bit range used for each field is chosen such that a change in one of bits indicates a major structural change in the binary. For example, the first 8 bits of the 32 bit initial stack size are almost always 0, and there is often minor changes in the lower bits in polymorphic malware.

In a different approach, [55, 56] use cryptographic hashes of functions in order to discover when malware is sharing code. The only modification made by Cohen and Havrilla [56] to the code before hashing is that all constants, such as jump and memory access addresses, are zeroed out. A cryptographic hash of the string representation of the code of the procedure is then taken. LeDoux et al. [55] use the abstractions defined by Lakhotia et al. [38] and described in Sect. 7.4. For each basic block in a procedure, the feature of Lakhotia et al. [38] is computed. A hash of the procedure is created by sorting and concatenating all the basic block features together and taking a cryptographic hash.

10.3 Locality Sensitive Hashing

To improve upon the robustness of concrete cryptographic hashes, MinHash, a type of Locality Sensitive Hash (LSH) [58] is being utilized in several different ways. A MinHash is a hash functions with the property that hashes of two arbitrary objects will be equivalent with a probability equal to the Jaccard Similarity between the objects. In other words, if two binaries are 90 % similar, there is a 90 % chance that they will produce the same LSH. Jaccard Similarity is a method for measuring the similarity of two sets and is defined as $|A \cap B|/|A \cup B|$, the size of the intersection divided by the size of the union. Two arbitrary MinHashes, then, will collide with a probability equal to the Jaccard Index of the two hashed objects, $P(MinHash(A) = MinHash(B)) = |A \cap B|/|A \cup B|$.

To create a MinHash of a set, an ordering over the universe of objects that can be placed in the set must be defined. This ordering can be arbitrary, but it must be fixed. A very simple example ordering is just "the order in which I first encountered the object" (this ordering does, however, need to be remembered across all sets that are to be compared). After this ordering is defined, several random permutations of the ordering are selected. A MinHash is then the minimum elements of the set as defined by the selected random order permutations.

For an illustrative example of MinHash, take the sets $A = 1, 2, 3$ and $B = 4, 5, 6$ with the ordering defined numerically. To create the MinHash, we first randomly select a number of permutations of the ordering, let's say these are 5, 4, 2, 1, 3 and 3, 1, 4, 5, 2. Then the MinHash of set A is 2,1 and that of B is 5, 4.

There are two ways in which MinHashes are being used in malware analysis: for direct clustering and as a filter to reduce the complexity of clustering. The computed MinHashes can be used to define a clustering that has constant time complexity. Two binaries are considered to be in the same cluster if and only if they have the same MinHash. This was an approach taken by Jin et al. [37].

Rather than directly clustering, MinHashes can also be used as a filter to drastically reduce the size of n so that the clustering complexity of $O(n^2)$ becomes bearable. Bayer et al. [48] first utilized this approach in the malware space. They defined many MinHashes by selecting a different set of random permutations for each MinHash. Malware were first put into an initial cluster if *any* of the MinHashes matched. Bayer et al. [48] would then computed the Jaccard Similarity for only pairs of binaries *in the same cluster*. The resulting similarities were then used to create a final clustering. Since n now refers to the number of object in a single cluster, $O(n^2)$ becomes a tolerable performance.

11 Semi-supervised Learning

Halfway between Supervised and Unsupervised Learning is the wonderful world of Semi-supervised Learning. As the name implies, Semi-supervised Learning utilizes both labeled and unlabeled data. It is usually used for clustering with the available labels helping decide both the number and shape of clusters to create.

While there are a number of existing Semi-supervised learning algorithms [59], there currently exists only one instance of such an application in malware. Santos et al. [60] use a Semi-Supervised Learning algorithm known as "Learning with Local and Global Consistency" (LLGC) [61] to classify binaries as malicious or benign.

LLGC first starts with a directed, weighted graph representation of the data. The nodes of the graph are binaries being clustered and the weighted edges are the similarity values between the binaries. Every node keeps track of how much it "believes" it belongs to a particular label by attaching a weight to the different labels. When the graph is first constructed, unlabeled data has zero belief in any of the labels and labeled data has perfect belief in its label. So a binary labeled "malware" will start with a weight of 1 for the label "malware" and a weight of 0 for the label "benign." An unlabeled binary will have a weight of 0 for both labels.

LLGC relies on two assumptions to create clusters. The first is that similar objects are likely to have the same label ("Local" consistency). The second is that objects in the same cluster are likely to have the same label ("Global" consistency). LLGC uses these two assumptions to "spread" beliefs through the initial graph. Nodes with higher similarity with each other will be incrementally updated so that their weights assigned to labels more closely mirror each other. The incremental updates continue until weights attached to labels converge. Each node is then assigned the label it has the highest weight attached to, i.e. the label it has the highest belief in.

12 Ensembles

Similar to the way in which features can be combined to create a more accurate classification, different learning algorithms can be combined to produce a single, more accurate learner. Such a learner is referred to as an *ensemble*. There are several ways in which classifiers can be combined [62]: voting, stacking, bagging, and boosting.

One of the simplest methods of creating an ensemble is *voting* [63–65]. This simply consists of creating a number of independent classifiers, running them separately, obtaining the different outputs, and then using some type of voting mechanism to determine which of the outputs to accept as the final answer. This voting mechanism most often takes the form of majority vote (simply select the answer that the most classifiers returned). Other forms of voting include weighted majority vote, a veto vote [64] where a "special" classifier can veto the decision of the majority, and a trust-based vote [65] where voting takes into account how much each individual classifier is "trusted" to provide a correct answer.

The above voting strategies really only apply to learning algorithms that perform classification and do not apply to clustering. A majority vote, for example, doesn't make sense for clustering as the number of possible ways to cluster is practically unbounded. It is unlikely that any two partitioning created by two different clustering algorithms will be the same. Instead, a *consensus partition* [66, 67] is used. Consensus partitioning can conceptually be thought of as a voting scheme for clustering. Like in voting ensembles, the first step is to cluster the data several times using independent methods each time. In the "combining phase" a new partitioning of the data is created that maximizes the "consensus" between each independent clustering. There are a number of ways that have been proposed to create this consensus clustering using graph based, combinatorial, or statistical methods. Various ways of creating this consensus clustering are covered by Strehl and Ghosh [68] and Topchy et al. [69].

Stacking [43, 70–72] is a type of ensemble in which a set of classifiers are connected in series, with each classifier taking as input the output of the classifier before it in the series. A simple example of such a classifier is [43] and their system called SVM-AR, depicted in Fig. 6. In this system, [43] first use a supervised classifier to decide if an executable is malicious or benign. After a decision is made, a trained rule-based classifier is used to "check" the results. If the supervised classifier said the executable was malware, the executable is checked against the rules for determining if an executable is benign. If any of these rules match, then final decision is "benign." A similar process is applied if the supervised classifier says the executable is benign.

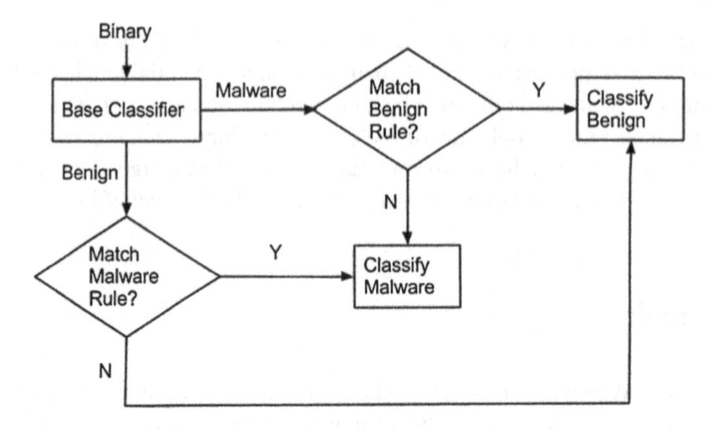

Fig. 6 SVM-AR by Lu et al. [43], an example stacking ensemble

Bagging [19] is an ensemble technique that instead of combining several different kinds of learners, uses the same learning algorithm, but trains the algorithm on random bootstrapped samples of the training data. There is a class of learning algorithms known as unstable algorithms. In unstable learning algorithms, slight changes in the training data can lead to very different models. Bagging seeks to solve this by combining many models created on slightly different data. The differing data is created through bootstrapped sampling of the original training data. A bootstrap sample is a random sample taken *with replacement*. Thus each bootstrapped sample will contain duplicates and be missing some of the original data. The learned models are usually combined through voting.

Boosting [46, 72] is meant to be an improvement over bagging. Like bagging, boosting combines many differently trained instances of the same learning algorithm. Where boosting differs is in the way the various training data sets are selected. Bagging selects training sets at random. Boosting selects training sets such that training focuses on data the already trained classifiers are getting incorrect. This results in building classifiers that specialize on specific portions of the data. Each individually trained classifier may have weak performance overall, but be extremely accurate on a specific subset of the data. Combining a number of such specialized classifiers (by voting, usually) results in more accurate ensemble overall.

References

1. Neumann, J.V. : Theory of Self-reproducing Automata. IEEE Trans. Neural Networks. **5**(1), 3–14 (1994)
2. Cohen, F.: Computer viruses. PhD thesis, University of Southern California (1985)
3. Measuring and optimizing malware analysis: An open model. L.L.C,Technical report, Securosis (2012)
4. Schon, B., Dmitry, G., Joel, S.: Automated sample processing, Technical Report, Mcafee AVERT, Auckland, New Zealand (2006)

5. Nielson, F., Nielson, H.R., Hankin, C.: Principles of Program Analysis. Springer, Berlin (1999). ISBN 9783540654100
6. Schwarz, B., Debray, S., Andrews, G.: Disassembly of executable code revisited. In: Proceedings of Ninth Working Conference on Reverse Engineering, IEEE, 2002, pp. 45–54
7. Collberg, C., Nagra, J.: Surreptitious Software: Obfuscation, Watermarking, and Tamperproofing for Software Protection. Pearson Education (2010). ISBN 9780321549259
8. Mitchell, T.M.: Machine Learning. McGraw-Hill, New York (1997). ISBN 0070428077 9780070428072 0071154671 9780071154673
9. Shabtai, A., Moskovitch, R., Elovici, Y., Glezer, C.: Detection of malicious code by applying machine learning classifiers on static features: a state-of-the-art survey. Inf. Sec. Tech. Rep. **14**(1), 1629 (2009)
10. Egele, M., Scholte, T., Kirda, E., Kruegel, C.: A survey on automated dynamic malware analysis techniques and tools. ACM Comput. Surv. **44**(2), 6:1–6:42 (2008). ISSN 0360–0300. doi:10.1145/2089125.2089126
11. Arnold, W. Tesauro, G.: Automatically generated WIN32 heuristic virus detection. In: 2000 Virus Bulletin International Conference, pp. 51–60. The Pentagon, Abingdon, Oxfordshire, OX14 3YP, England, Virus Bulletin Ltd (2000)
12. Kephart, J.O., Arnold, B.: Automatic extraction of computer virus signatures. In: Ford, R. (ed.) 4th Virus Bulletin International Conference, pp. 178–184, Abingdon, England, Virus Bulletin Ltd (1994)
13. Kephart, J.O., Arnold, B.: A biologically inspired immune system for computers. In: Fourth International Workshop on the Synthesis and Simulation of Living Systems, pp.130–139 (1994)
14. Kephart, J.O., Sorkin, G.B., Arnold, W.C., Chess, D.M., Tesauro, G.J., White, S.R.: Biologically inspired defenses against computer viruses. In: IJCAI 95, pp. 985–996 (1995)
15. Karim, M.E., Walenstein, A., Lakhotia, A., Parida, L.: Malware phylogeny generation using permutations of code. J. Comput. Virol. **1**(1), 13–23 (2005)
16. Wang, T.-Y., Wu, C.-H., Hsieh, C.-C.: Detecting unknown malicious executables using portable executable headers. In: Fifth International Joint Conference on INC, IMS and IDC, NCM 09, pp. 278–284 (2009). doi:10.1109/ncm.2009.385
17. Walenstein, A., Hefner, D.J., Wichers, J.: Header information in malware families and impact on automated classifiers. In: 2010 5th International Conference on Malicious and Unwanted Software (MALWARE), p. 1522 (2010). doi:10.1109/malware.2010.5665799
18. Schultz, M.G., Eskin, E., Zadok, F., Stolfo, S.J.: Data mining methods for detection of new malicious executables. In: Proceedings of 2001 IEEE Symposium on Security and Privacy, S P 2001, pp. 38–49 (2001). doi:10.1109/secpri.2001.924286
19. Ye, Y., Chen, L., Wang, D., Li, T., Jiang, Q., Zhao, M.: SBMDS: an interpretable string based malware detection system using SVM ensemble with bagging. J. Comput. Virol. **5**(4), 283–293 (2008). ISSN 1772–9890, 1772–9904. doi:10.1007/s11416-008-0108-y
20. Kruegel, C., Robertson, W., Valeur, F., Vigna, G.: Static disassembly of obfuscated binaries. In: Proceedings of the 13th USENIX Security Symposium, pp. 255–270. Usenix (2004)
21. Linn, C., Debray, S.: Obfuscation of executable code to improve resistance to static disassembly. In: Proceedings of the 10th ACM Conference on Computer and Communications Security, pp. 290–299, ACM Press, New York, NY, USA (2003)
22. Christodorescu, M., Jha, S., Kruegel, C.: Mining specifications of malicious behavior. In: Proceedings of the 1st India Software Engineering Conference, ISEC '08, p. 514, New York, NY, USA (2008). ACM. ISBN 978-1-59593-917-3. doi:10.1145/1342211.1342215
23. Debray, S. Patel, J.: Reverse engineering self-modifying code: Unpacker extraction. In: 2010 17th Working Conference on Reverse Engineering (WCRE), pp. 131–140 (2010). doi:10.1109/WCRE.2010.22
24. Sharif, M., Lanzi, A., Giffin, J., Lee, W.: Automatic reverse engineering of malware emulators. In: 2009 30th IEEE Symposium on Security and Privacy, pp. 94–109 (2009). doi:10.1109/SP.2009.27
25. Alazab, M., Kadiri, M.A., Venkatraman, S., Al-Nemrat, A.: Malicious code detection using penalized splines on OPcode frequency. In: Cybercrime and Trustworthy Computing Workshop (CTC), 2012 Third, pp. 38–47 (2012). doi:10.1109/CTC.2012.15

26. Bilar, D.: Opcode as predictors for malware. Int. J. Electron. Sec. Digit. Forensics **1**(2), 156–168 (2007)
27. Hu, X., Bhatkar, S., Griffin, K., Shin, K.G.: MutantX-S: scalable malware clustering based on static features. In: USENIX Annual Technical Conference (USENIX ATC 13), pp. 187–198 (2013)
28. Moskovitch, R., Feher, C., Tzachar, N., Berger, E., Gitelman, M., Dolev, S., Elovici, Y.: Unknown malcode detection using opcode representation. Intell. Secur. Inform. **48**, 204–215 (2008)
29. Runwal, N., Low, R.M., Stamp, M.: Opcode graph similarity and metamorphic detection. J. Comput. Virol. **8**(1–2), 37–52 (2012). ISSN 1772–9890, 1772–9904, doi:10.1007/s11416-012-0160-5
30. Chouchane, M.R., Lakhotia, A.: Using engine signature to detect metamorphic malware. In: Proceedings of the 4th ACM Workshop on Recurring Malcode, WORM '06, pp. 73–78, New York, NY, USA (2006). ACM. ISBN 1-59593-551-7. doi:10.1145/1179542.1179558
31. Hu, X., Chiueh, T.-C., Shin, K.G.: Large-scale malware indexing using function-call graphs. In: Proceedings of the 16th ACM Conference on Computer and Communications security, pp. 611–620 (2009)
32. Carrera, E., Erdelyi, G.: Digital genome mapping: advanced binary malware analysis. In: Proceedings of the 2004 Virus Bulletin Conference, pp. 187–197 (2004)
33. Briones, I., Gomez, A.: Graphs, entropy and grid computing: automatic comparison of malware. Virus Bulletin, 1–12 (2008). http://pandalabs.pandasecurity.com/blogs/images/PandaLabs/2008/10/07/IsmaelBriones-VB2008.p
34. Kinable, J., Kostakis, O.: Malware classification based on call graph clustering. J. Comput. Virol. **7**(4), 233–245 (2011). ISSN 1772–9890, 1772–9904, doi:10.1007/s11416-011-0151-y
35. Kruegel, C., Kirda, E., Mutz, D., Robertson, W., Vigna, G.: Polymorphic worm detection using structural information of executables. In: Valdes, A., Zamboni, D. (eds.) Recent Advances in Intrusion Detection, no. 3858. Lecture Notes in Computer Science, pp. 207–226. Springer, Berlin (2006). ISBN 978-3-540-31778-4, 978-3-540-31779-1
36. Chaki, S., Cohen, C., Gurfinkel, A.: Supervised learning for provenance-similarity of binaries. In: Proceedings of the 17th ACM SIGKDD International Conference on Knowledge Discovery and Data Mining, KDD '11, p. 1523, New York, NY, USA, 2011. ACM. ISBN 978-1-4503-0813-7. doi:10.1145/2020408.2020419
37. Jin, W., Chaki, S., Cohen, C., Gurfinkel, A., Havrilla, J., Hines, C., Narasimhan, P.: Binary function clustering using semantic hashes. In: Proceedings of the 11th International Conference on Machine Learning and Applications (ICMLA), vol. 1, pp. 386–391 (2012). doi:10.1109/ICMLA.2012.70
38. Lakhotia, A., Preda, M.D., Giacobazzi, R.: Fast location of similar code fragments using semantic 'juice'. In: Proceedings of the 2nd ACM SIGPLAN Program Protection and Reverse Engineering Workshop, PPREW '13, p. 5:15:6, New York, NY, USA (2013). ACM. ISBN 978-1-4503-1857-0. doi:10.1145/2430553.2430558
39. Pfeffer, A., Call, C., Chamberlain, J., Kellogg, L., Ouellette, J., Patten, T., Zacharias, G., Lakhotia, A., Golconda, S., Bay, J., Hall, R., Scofield, D.: Malware analysis and attribution using genetic information. In: Proceedings of the 7th IEEE International Conference on Malicious and Unwanted Software (MALWARE 2012), pp. 39–45, IEEE Computer Society Press, Fajardo, Puerto Rico, Oct. (2012)
40. Bailey, M., Oberheide, J., Andersen, J., Mao, Z.M., Jahanian, F., Nazario, J.: Automated classification and analysis of internet malware. In: RAID07: Proceedings of the 10th International Conference on Recent Advances in Intrusion Detection, pp. 178–197, Berlin, Heidelberg, Springer-Verlag (2007)
41. Trinius, P., Willems, C., Holz, T., Rieck, K.: A malware instruction set for behavior-based analysis, Technical Report, University of Mannheim (2009). http://citeseerx.ist.psu.edu/viewdoc/download
42. Masud, M.M., Khan, L., Thuraisingham, B.: A hybrid model to detect malicious executables. In: IEEE International Conference on Communications, ICC 07, pp. 1443–1448 (2007). doi:10.1109/icc.2007.242

43. Lu, Y.B., Din, S.C., Zheng, C.F., Gao, B.J.: Using multi-feature and classifier ensembles to improve malware detection. J. CCIT **39**(2), 57–72 (2010)
44. Islam, R., Tian, R., Batten, L., Versteeg, S.: Classification of malware based on string and function feature selection. In: Cybercrime and Trustworthy Computing, Workshop, p. 917 (2010)
45. LeDoux, C., Walenstein, A., Lakhotia, A.: Improved malware classification through sensor fusion using disjoint union. In: Information Systems, Technology and Management, pp. 360–371, Grenoble, France. Springer, Berlin Heidelberg (2012). ISBN 978-3-642-29166-1. doi:10.1007/978-3-642-29166-1_32
46. Kolter, J.Z., Maloof, M.A.: Learning to detect and classify malicious executables in the wild. J. Mach. Learn. Res. **7**, 2721–2744 (2006)
47. Walenstein, A., Venable, M., Hayes, M., Thompson, C., Lakhotia, A.: Exploiting similarity between variants to defeat malware. In: Proceedings of BlackHat Briefings DC 2007 (2007)
48. Bayer, U., Comparetti, P.M., Hlauschek, C., Kruegel, C., Kirda, E.: Scalable, behavior-based malware clustering (2009). http://citeseerx.ist.psu.edu/viewdoc/download?doi=10.1.1.148.7690&rep=rep1&type=pdf
49. Gurrutxaga, I., Arbelaitz, O., Ma Perez, J., Muguerza, J., Martin, J.I., Perona, I.: Evaluation of malware clustering based on its dynamic behaviour. In: Roddick, J.F., Li, J., Christen, P., Kennedy, P.J. (eds.) Seventh Australasian Data Mining Conference (AusDM 2008), Crpit, vol. 87, pp. 163–170, Glenelg, South Australia, Acs (2008)
50. Wang, Y., Ye, Y., Chen, H., Jiang, Q.: An improved clustering validity index for determining the number of malware clusters. In: 3rd International Conference on Anti-counterfeiting, Security, and Identification in Communication, 2009, ASID 2009, pp. 544–547. doi:10.1109/ICASID.2009.5277000
51. Wicherski, G.: peHash: a novel approach to fast malware clustering. In: Proceedings of LEET09: 2nd USENIX Workshop on Large-Scale Exploits and Emergent Threats (2009)
52. Cesare, S., Xiang, Y.: Software Similarity and Classification. Springer, Heidelberg (2012)
53. Legany, C., Juhsz, S., Babos, A.: Cluster validity measurement techniques. In: Proceedings of the 5th WSEAS International Conference on Artificial Intelligence, Knowledge Engineering and Data Bases, AIKED'06, pp. 388–393, Stevens Point, Wisconsin, USA (2006). World Scientific and Engineering Academy and Society (WSEAS). ISBN 111-2222-33-9
54. Jang, J., Brumley, D., Venkataraman, S.: BitShred: feature hashing malware for scalable triage and semantic analysis. In: Proceedings of the 18th ACM Conference on Computer and Communications Security, CCS '11, pp. 309–320, ACM, New York, NY, USA (2011). ISBN 978-1-4503-0948-6. doi:10.1145/2046707.2046742
55. LeDoux, C., Lakhotia, A., Miles, C., Notani, V., Pfeffer, A.: FuncTracker: discovering shared code to aid malware forensics extended abstract (2013)
56. Cohen, C., Havrilla, J.S.: Function hashing for malicious code analysis. In: CERT Research Annual Report 2009, pp. 26–29. Software Engineering Institute, Carnegie Mellon University (2009)
57. Bloom, B.H.: Space/time trade-offs in hash coding with allowable errors. Commun. ACM **13**(7), 422–426 (1970)
58. Rajaraman, A., Ullman, J.D.: Mining of Massive Datasets. Cambridge University Press, Cambridge (2012). ISBN 9781139505345
59. Zhu, X.: Semi-supervised learning literature survey, Technical Report, Computer Sciences, University of Wisconsin-Madison (2005). http://citeseerx.ist.psu.edu/viewdoc/download?doi=10.1.1.99.9681&rep=rep1&type=pdf. Accessed 14 Mar 2013
60. Santos, I., Nieves, J., Bringas, P.: Semi-supervised learning for unknown malware detection. In: International Symposium on Distributed Computing and Artificial Intelligence, pp. 415–422 (2011)
61. Zhou, D., Bousquet, O., Lal, T.N., Weston, J., Schlkopf, B.: Learning with local and global consistency. Adv. Neural Inf. Process. Syst. **16**, 321–328 (2004)
62. Kuncheva, L.I.: Combining Pattern Classifiers: Methods and Algorithms. Wiley-Interscience, Hoboken (2004). ISBN 0471210781

63. Dahl, G., Stokes, J.W., Deng, L., Yu, D.: Large-scale malware classification using random projections and neural networks. In: Proceedings IEEE Conference on Acoustics, Speech, and Signal Processing, pp. 3422–3426 (2013)
64. Shahzad, R., Lavesson, N.: Veto-based malware detection. In: 2012 Seventh International Conference on Availability, Reliability and Security (ARES), pp. 47–54 (2012). doi:10.1109/ARES.2012.85
65. Shahzad, R.K., Lavesson, N.: Comparative analysis of voting schemes for ensemble-based malware detection. Wireless Mob. Netw. Ubiquitous Comput. Dependable Appl. **4**, 76–97 (2013)
66. Ye, Y., Li, T., Chen, Y., Jiang, Q.: Automatic malware categorization using cluster ensemble. In: Proceedings of the 16th ACM SIGKDD International Conference on Knowledge Discovery and Data Mining, pp. 95–104 (2010)
67. Zhuang, W., Ye, Y., Chen, Y., Li, T.: Ensemble clustering for internet security applications. IEEE Trans. Syst. Man Cybern. Part C Appl. Rev. **42**(6), 1784–1796 (2012). ISSN 1094-6977. doi:10.1109/TSMCC.2012.2222025
68. Strehl, A., Ghosh, J.: Cluster ensembles a knowledge reuse framework for combining multiple partitions. J. Mach. Learn. Res. **3**, 583–617 (2003). ISSN 1532–4435. doi:10.1162/153244303321897735
69. Topchy, A., Jain, A.K., Punch, W.: Clustering ensembles: models of consensus and weak partitions. IEEE Trans. Pattern Anal. Mach. Intell. **27**(12), 1866–1881 (2005). ISSN 0162–8828. doi:10.1109/TPAMI.2005.237
70. Barr, S.J., Cardman, S.J., Martin, D.M.Jr.: A boosting ensemble for the recognition of code sharing in malware. J. Comput. Virol. **4**(4), 335–345 (2008). ISSN 1772–9890, 1772–9904, doi:10.1007/s11416-008-0087-z
71. Menahem, E., Shabtai, A., Rokach, L., Elovici, Y.: Improving malware detection by applying multi-inducer ensemble. Comput. Stat. Data Anal. **53**(4), 1483–1494 (2009). ISSN 0167–9473. doi:10.1016/j.csda.2008.10.015
72. Zabidi, M., Maarof, M., Zainal, A.: Ensemble based categorization and adaptive model for malware detection. In: 2011 7th International Conference on Information Assurance and Security (IAS), pp. 80–85 (2011). doi:10.1109/ISIAS.2011.6122799

Soft Computing Based Epidemical Crisis Prediction

Dan E. Tamir, Naphtali D. Rishe, Mark Last and Abraham Kandel

Abstract Epidemical crisis prediction is one of the most challenging examples of decision making with uncertain information. As in many other types of crises, epidemic outbreaks may pose various degrees of surprise as well as various degrees of "derivatives" of the surprise (i.e., the speed and acceleration of the surprise). Often, crises such as epidemic outbreaks are accompanied by a secondary set of crises, which might pose a more challenging prediction problem. One of the unique features of epidemic crises is the amount of fuzzy data related to the outbreak that spreads through numerous communication channels, including media and social networks. Hence, the key for improving epidemic crises prediction capabilities is in employing sound techniques for data collection, information processing, and decision making under uncertainty and exploiting the modalities and media of the spread of the fuzzy information related to the outbreak. Fuzzy logic-based techniques are some of the most promising approaches for crisis management. Furthermore, complex fuzzy graphs can be used to formalize the techniques and methods used for the data mining. Another advantage of the fuzzy-based approach is that it enables keeping account of events with perceived low possibility of occurrence via low fuzzy membership/truth-

This material is based in part upon work supported by the National Science Foundation under GrantsI/UCRC IIP-1338922, AIR IIP-1237818, SBIR IIP-1330943, III-Large IIS-1213026, MRI CNS-0821345, MRI CNS-1126619, CREST HRD-0833093, I/UCRC IIP-0829576, MRI CNS-0959985, FRP IIP-1230661.

D.E. Tamir (✉)
Department of Computer Science, Texas State University, San Marcos, TX, USA
e-mail: dan.tamir@txstate.edu

N.D. Rishe · A. Kandel
School of Computing and Information Sciences, Florida International University, Miami, FL, USA
e-mail: rishen@cs.fiu.edu

A. Kandel
e-mail: abraham.kandel.fiu@gmail.com

M. Last
Department of Information Systems Engineering, Ben-Gurion University of the Negev, Beer-Sheva, Israel
e-mail: mlast@bgu.ac.il

R.R. Yager et al. (eds.), *Intelligent Methods for Cyber Warfare*,
Studies in Computational Intelligence 563, DOI 10.1007/978-3-319-08624-8_2

values and updating these values as information is accumulated or changed. In this chapter we introduce several soft computing based methods and tools for epidemic crises prediction. In addition to classical fuzzy techniques, the use of complex fuzzy graphs as well as incremental fuzzy clustering in the context of complex and high order fuzzy logic system is presented.

1 Introduction

In the context of this Chapter, the term *epidemical* (or *epidemic*) referrers to a disease which spreads widely and attacks many persons at the same time. An epidemical crisis is an epidemic that spreads very fast and affects numerous people in different countries and continents. In this sense, the term *epidemical crisis* and *pandemic* are almost synonyms.

Epidemical crisis prediction (ECP) is a special case of disaster prediction and management (DPM). DPM is one of the most challenging examples of decision making under uncertain information. One of the main issues related to ECP/DPM is the amount of a priori information available, i.e., the amount of surprise affiliated with the epidemical crisis as well as the risk and devastation that follow the crisis outbreak. A closer look at these parameters shows that the velocity and acceleration (and higher derivatives) of these parameters are highly important. Another important aspect includes secondary adverse effects (i.e., secondary epidemical crises) that are triggered by the initial disaster.

One of the main concerns about epidemical crises is the amount of surprise that accompanies the outbreak. Emergencies may produce a wide range of surprise levels. The terror attack of 9/11 is an example of a disaster with very high level of surprise. On the other hand the landfall of a hurricane in Florida in the middle of a hurricane season is not as surprising.

Even in the extreme cases where the nature of the disaster is known, preparedness plans are in place, and analysis, evaluation, and simulations of the disaster management procedures have been performed, the amount and magnitude of "surprises" that accompany the real disaster pose an enormous demand. In the more severe cases, where the entire disaster is an unpredicted event, the disaster management and response system might fast run into a chaotic state. Hence, the key for improving disaster preparedness and mitigation capabilities is in employing sound techniques for data collection, information processing, and decision making under uncertainty.

Analysis of epidemical crises presents three types of challenges: the first is the ability to predict the occurrence of epidemical crises, the second is the need to produce a preparedness plan, and the third is the actual real time response activities related to providing remedies for a currently occurring disaster.

As a special case of DPM, ECP is a highly challenging example of decision making under uncertain information. Epidemical outbreaks might pose various degrees of surprise as well as various degrees of the "derivatives" of the surprise. One of the unique features of epidemical crises is the amount of fuzzy data related to an outbreak that spreads through numerous communication channels and social networks.

The key for improving epidemical crises prediction and management capabilities is in employing sound techniques for data collection, information processing, and decision making under uncertainty and in exploiting the modalities and media of the spread of the fuzzy information related to the outbreak. Hence, fuzzy logic-based techniques are some of the most promising approaches for crisis management. Furthermore, complex fuzzy graphs can be used to formalize the techniques and methods used for the data mining. Another advantage of the fuzzy-based approach is that it enables keeping account of events with perceived low possibility of occurrence via low fuzzy membership/truth-values and updating these values as information is accumulated or changed.

In this chapter we introduce several soft computing based methods and tools for epidemical crises prediction. In addition to classical fuzzy techniques, the use of complex fuzzy graphs as well as incremental fuzzy clustering in the context of complex and high order fuzzy logic systems is presented.

The rest of this chapter elaborates on some of the important aspects of ECP and DPM, concentrating on the related uncertainty which can be addressed via fuzzy logic-based tools as well as the geospatial-temporal analytics/Big Data aspects of the problem. Section 2 provides the background, Sect. 3 provides an overview of several fuzzy logic-based tools for ECP, and Sect. 4 concludes and proposes future research.

2 Background

The topic of predicting epidemical crises falls under the more general subject of disaster prediction management and mitigation (DPM). In this section, we discuss DPM (Sect. 2.1) and elaborate on some of the distinguishing factors of ECP (Sect. 2.2). In addition, we describe the geospatial-temporal analytics of the correlation of environmental factors and incidence of disease and report on a set of tools and concept demonstrations that show the solvability of Big Data problems involving geospatial data correlated with publically available medical data.

Epidemical crises occur with different degrees of unpredictability and severity, which are manifested in two main facets. First, the actual occurrence of the pandemics might be difficult (potentially impossible) to predict. Second, regardless of the level predictability of the crisis, it is very likely that it will be accompanied by secondary effects. Hence, epidemical crises are a major source of "surprise" and uncertainty and their mitigation and management require sound automatic and intelligent handling of uncertainty.

Often, the stakeholders of ECP programs are classifying the unpredictability of epidemical crises as two types of unknowns: unknown unknowns and known unknowns. The first type of unknowns (unknown unknowns) is often referred to by the metaphor of a *black swan*, coined by Taleb [30, 31], while the second type of unknowns (known unknowns) is referred to as a *gray swan*. Arguably, however, there is no such animal as a black swan and every swan is a gray swan depending on the amount of surprise it carries and the potential devastation associated with it. In almost all the cases of epidemical crises recorded so far there was some *a priori* information concerning the disaster yet this information was filtered, ignored, or just did not pass a threshold of being classified as significant.

2.1 Severity and Predictability of Epidemical Crises

In his excellent books [30, 31], Nassim Nicholas Taleb describes the following features of blackswan events:

1. The blackswan is an outlier; lying outside of the space of regular expectations.
2. It has very low predictability and it carries an extremely adverse impact.
3. The unknown component of the event is far more relevant than the known component.
4. Finally, we can explain the event *post factum* and through those explanations make it predictable in retrospect.

In this sense, the black swan represents a class of problems that can be referred to as the "unknown unknowns." However, a thorough investigation of many of the events that are widely considered as black swans, e.g., the September 11, 2001 attack in NYC, shows that there had been available information concerning the evolving event; yet, this information did not affect the decision making and response prior to the attack. Hence, the term "not connecting the dots" is often used to describe these phenomena. This brings to the forefront the problem of predicting the occurrence of epidemical crises. More important is the issue of identifying (and not ignoring) anomalies.

This suggests that the term *black swan* is a bit too extreme and one should consider using the term *gray swan* where the gray level relates to the level of surprise. A gray swan represents an unlikely event that can be anticipated and carries an extremely adverse impact. In this respect, gray swans represent a two dimensional spectrum of information. The first dimension represents the predictability of the event where black swans are highly unpredictable and white swans are the norm. The second dimension represents the amount of adverse outcome embedded in the event; with black swans representing the most adverse outcomes. Consequently, the black swan is a special case of a gray swan.

To further elaborate, one type of unpredictability relates to a set of events that can be considered as known unknowns. For example, a hurricane occurring in Florida during the hurricane season should not surprise the responsible authorities. Moreover, often, there is a span of a few days between the identification of the hurricane and the actual landfall. Regardless, even a predictable hurricane landfall carries numerous secondary disastrous events that are hard to predict.

We refer to these secondary events as second generation gray swans. Second generation swans are generated and/or detected while the disaster is occuring. The collapse of the Twin Towers in 9/11 is an example of a second-generation gray swan. A gray swan might (and according to the Murphy laws is likely to) spawn additional gray swans. Generally, second generation swans evolve late and fast. Hence, they introduce a more challenging detection problem and require specialized identification tools, such as dynamic clustering. Detecting relatively slow evolving [first generation] gray swans before the disaster occurs and relatively fast evolving second generation gray swans requires an adequate set of uncertainty management tools.

The following is a partial list of well-known gray swans, each of which has a different degree of surprise as well as different degree of severity.

1. Philippines typhoon disaster
2. Bangladeshi factory collapse
3. Iceland volcano eruption
4. Fukushima—tsunami followed by nuclear radiation and risk of meltdown of nuclear facilities in the area
5. HIV, HSV2, Swine, SARS, and West Nile Virus infection outbreaks
6. 9/11 NYC attack followed by the collapse of the Twin Towers
7. Financial Markets' falls (1987, 2008), Madoff's fraud
8. The 1998 fall and bailout of Long-Term Capital Management L.P. hedge fund
9. Yom Kippur War
10. December 7, 1941—Pearl Harbor
11. Assassinations of Lincoln, Kennedy, Sadat, and Rabin

Fuzzy logic is one of the suggested tools that can help create a better understanding of ECP tools, including, but not limited to, intelligent robotics, learning and reasoning, language analysis and understanding, and data mining. Hence, research in fuzzy logic and uncertainty management is critical for producing a successful ECP programs.

Recently, the academic community and government agencies have effected spurring growth in the field of data mining in Big Data systems.These advances are beginning to find their way into ECP programs and are redefining the way we address potential disaster and mitigate the effects of epidemical crises. Nevertheless, academia, industry, and governments need to engage as a unified entity to advance new technologies as well as apply established technologies in preparation and response to the specific emerging problems of epidemical crises.

2.2 Epidemical Crises Information Spread

Predicting the epidemical outbreak is an important component of the management and mitigation of a pandemic. It can enable early setup of a mitigation plan. Nevertheless, the fact that an epidemical crisis is somewhat predictable does not completely reduce the amount of surprise that accompany the actual occurrence of the crisis.

Hence, any mitigation plan, that is, a set of procedures compiled in order to address the adverse effects of epidemical crises, should be flexible enough to handle additional surprises related to secondary adverse effects of the epidemical crisis. Finally, the real time ramification program is the actual set of procedures enacted and executed as the epidemical crisis occurs. There are two preconditions for successful remedy of epidemical crisis affects. First, the authorities/leadership have to attempt following the original mitigation plan as close as possible while instilling a sense of trust and calmness in the people that experience the crisis and the mitigation provider teams. Second, and as a part of their leadership traits, the authorities must possess the ability to adapt their remedy procedures to the dynamics of the epidemical crisis, potentially providing effective improvisations aimed at handling secondary disastrous events that evolve from the main disaster. A simple example for such events is looting and violence that might accompany a major disaster. For this end, fast automatic assessment of the dynamics is paramount.

One of the distinguishing features of pandemics' outbreaks is the modalities of sharing information in communication and social networks. Often, due to the panic that accompanies outbreaks there is an explosion of data which is characterized by large amount of information and high "velocity" and higher derivatives of velocity of information spread. Using the current terminology to describe the phenomenon: pandemic news are likely to become viral. Notably, many countries would try to completely ban, control, or limit, the news spread. But experience show that these attempts are not likely to be fruitful: citizens of these countries can still find numerous ways to spread the information through social networks. Hence, data mining in electronic versions of newspapers and related media as well as in social networks is an important cyber warfare ammunition. In addition, it is well known (but not published) that many national security agencies are digitizing "hard forms" of publicly available information of other countries such as newspapers. These might be related to official, semi-official, or private media-outlets. This media, however, is easier to control by a country that is trying to conceal an epidemical outbreak. One interesting pattern in this situation is the appearance of a few news reports at the beginning of the outbreak, followed by acceleration in reports, followed by a complete seizure of these reports due to a discovery of the news by the country's leaders, followed by a complete ban on this news.

Interestingly, there is another cyber source of information which relates to restrictions that a country might put on travelers entering the country. For example, during the SARS pandemic several countries has required that people entering the country would go through a fast automatic screening of body temperatures.

2.3 Geospatial-Temporal Analytics of Correlation of Environmental Factors and Incidence of Disease

We have developed tools and concept demonstrations that show the solvability of Big Data problems involving geospatial data correlated with publically available medical data. We bring the Big Data approach to geospatial epidemiology, a field of study focused on describing and analyzing geographic variations of disease spread, considering demographic, environmental, behavioral, socioeconomic, genetic, and infectious risk factors [8]. Our work in this area assists the development of the related field of personalized medicine by correlating clinical, genetic, environmental, demographic, and other background geospatial data.

Our TerraFly GeoCloud [18] system combines several diverse technologies and components in order to analyze and visualize geospatial data. In this system, a user can upload a spatial dataset and display it using the TerraFly Map API [26]. Datasets can be subsequently analyzed using various functions, such as Kriging, a geo-statistical estimator for unobserved locations, and Spatial Clustering, which involves the grouping of closely related spatial objects. Various analysis functions related to spatial epidemiology have been integrated into TerraFly GeoCloud. Analysis functions can be used by selecting the appropriate dataset and function in the interface menu, along with the variables to be analyzed. TerraFly GeoCloud then processes the data and returns a result that can be visualized on the TerraFly Map or on a chart. Results displayed on the map include a legend, which identifies certain range values by color. Certain visualizations are interactive, allowing additional information to be displayed.

Our Spatial Epidemiology System provides four kinds of API algorithms for data analysis and results visualization, based on the TerraFly GeoCloud System: (1) disease mapping (mortality/morbidity map, SMR map); (2) disease cluster determination (spatial cluster, HotSpot analysis tool, cluster and outlier analysis); (3) geographic distribution measurement (mean central, median central, standard distance, distributional trends); and (4) regression (linear regression, spatial autoregression). The system is interfaced with our Health Informatics projects [4, 13, 25, 27–29, 40, 43].

We work on tools and methodologies that will assist in operational and analytical Health Informatics. The TerraFly Geospatial Analytics System (http://terrafly.com) demonstrates correlation of location to environment-related disorders, enabling clinicians to more readily identify macro-environmental exposure events that may alter an individual's health. It also enables applications in targeted vaccine and disease management, including disease surveillance, vaccine evaluation and follow-up, intelligent management of emerging diseases, cross-analysis of locations of patients and health providers with demographic and economic factors, personalized medicine, and other geospatial and data-intensive applications.

3 Tools for Predictions and Evaluations of Fuzzy Events

Fuzzy logic-based techniques are some of the most promising approaches for ECP. The advantage of the fuzzy-based approach is that it enables keeping account on events with perceived low possibility of occurrence via low fuzzy membership/truth-values and updating these values as information is accumulated or changed. Numerous fuzzy logic-based algorithms can be deployed in the data collection, accumulation, and retention stage, in the information processing phase, and in the decision making process. In this section we describe several possible fuzzy tools to try and predict epidemical crises and cope with evolving epidemical crises via sound ECP programs. We consider the following fuzzy logic-based tools:

1. Fuzzy switching mechanisms
2. Fuzzy expectation and variance
3. Fuzzy relational databases (FRDB), fuzzy data-mining and fuzzy social network architectures (FSNA)
4. Neuro Fuzzy-based Logic, and Systems
5. Complex and multidimensional fuzzy Sets, Logic, and Systems
6. Complex fuzzy graphs
7. Dynamic and incremental fuzzy clustering

3.1 Making Decisions with no Data

As an example for this idea we use the fuzzy treatment of the transient behavior of a switching system and its static hazards [12]. Perhaps the major reason for the ineffectiveness of classical techniques in dealing with a static hazard and obtaining a logical explanation of the existence of a static hazard lies in their failure to come to grips with the issue of fuzziness. This is due to the fact that the hazardous variable implies imprecision in the binary system, which does not stem from randomness but from the lack of a sharp transition between members in the class of input states. Intuitively, fuzziness is a type of imprecision that stems from a grouping of elements into classes that do not have sharply defined boundaries—that is, where there is no sharp transition from membership to non-membership. Thus, the transition of a state has a fuzzy behavior during the transition time.

Any fuzzy-valued switching function can be expressed in disjunctive and conjunctive normal forms, in a similar way to two-valued switching functions. A fuzzy-valued switching function over n variables can be represented by a mapping $f\colon [0, 1]^n \rightarrow [0, 1]$. We define a V-fuzzy function as a fuzzy function $f(x)$ such that $f(\xi)$ is a binary function for every binary n-dimensional vector ξ. It is clear that a V-fuzzy function f induces a binary function F such that $F\colon [0, 1]^n \rightarrow [0, 1]$ determined by $F(\xi) = f(\xi)$ for every binary n-dimensional vector ξ.

If a V-fuzzy function f describes the complete behavior of a binary combinational system, its steady-state behavior is represented by F, the binary function induced by f. Let $f(x)$ be an n-dimensional V-fuzzy function, and let ξ and ρ be adjacent binary n-dimensional vectors. The vector $T_{\xi_j}^{\rho}$ is a static hazard of f iff $f(\xi) = f(\rho) \neq f(T_{\xi_j}^{\rho})$.

If $f(\xi) = f(\rho) = 1$ then $T_{\xi_j}^{\rho}$ is a 1-hazard. If $f(\xi) = f(\rho) = 0$ then $T_{\xi_j}^{\rho}$ is a 0-hazard. If f is V-fuzzy and $T_{\xi_j}^{\rho}$ is a static hazard, then $f(T_{\xi_j}^{\rho})$ has a perfect fuzzy value, that is, $f(T_{\xi_j}^{\rho}) \in (0, 1)$. Consider the static hazard as a malfunction represented by an actual or potential deviation from the intended behavior of the system. We can detect all static hazards of the V-fuzzy function $f(x)$ by considering the following extension of Shannon normal form. Let $f(\bar{x})$, $\bar{x} = (x_1, x_2, ..., x_n)$, be a fuzzy function and denote the vector

$$(x_1, x_2, x_{j-1}, x_{j+1}, ..., x_n) \text{ by } x^j.$$

By successive applications of the rules of Fuzzy Algebra, the function $f(x)$ may be expanded about x_j as follows:

$$f(x) = x_j f_1(x^j) + \bar{x_j} f_2(x^j) + x_j \bar{x_j} f_3(x^j) + f_4(x^j),$$

where f_1, f_2, f_3, and f_4 are fuzzy functions. It is clear that the same expansion holds when the fuzzy functions are replaced by B-fuzzy functions of the same dimension. Let ξ and ρ be two adjacent n-dimensional binary vectors that differ only in their j^{th} component. Treating ξ_j as a perfect fuzzy variable during transition time implies that $T^{\rho}_{\xi_j}$ is a 1-hazard of f iff $f(\xi) = f(\rho) = 1$ and $f(T^{\rho}_{\xi_j}) \in [0, 1)$. We show that the above conditions for the vector $T^{\rho}_{\xi_j}$ to be 1-hazard, yielding the following result:

Theorem 1 ([12]): *The vector $T^{\rho}_{\xi_j}$ is a 1-hazard of the B-fuzzy function $f(x)$ given above iff the binary vector ξ_j is a solution of the following set of Boolean equations:*

$$f_1(x^j) = 1, \qquad f_2(x^j) = 1, \qquad f_4(x^j) = 0.$$

Proof

State 1: $\xi_j = 1$ and $\overline{\xi}_j = 0$ imply $f_1(\xi^j) + f_4(\xi^j) = 1$.
State 2: $\xi_j = 0$ and $\overline{\xi}_j = 1$ imply $f_2(\xi^j) + f_4(\xi^j) = 1$.
Transition state: $\xi_j \in (0, 1)$ [which implies $\overline{\xi}_j \in (0, 1)$], and thus:

$$0 \leq max\{min[\xi_j, f_1(\xi^j)], min[\overline{\xi}_j, f_2(\xi^j)], min[\xi_j, \overline{\xi}, f_3(\xi^j)], f_4(\xi^j)\} < 1.$$

It is clear from the transition state that $f_4(\xi_j)$ cannot be equal to one, and thus:

$$f_4(\xi^j) = 0, \qquad f_1(\xi^j) = f_2(\xi^j) = 1.$$

Several items must be pointed out. The system is not a fuzzy system. It is a Boolean system. The modeling of the system as a fuzzy system is due to the lack of knowledge regarding the behavior of x_j during the transition. It is providing us with a tool to make decisions (regarding the Boolean values of f_1, f_2 and f_4) with no data whatsoever regarding x^j. Thus, we were able to make non-fuzzy decisions in a deterministic environment with no data.

3.2 Fuzzy Expectation and Variance

Ordinarily, imprecision and indeterminacy are considered to be statistical, random characteristics and are taken into account by the methods of the Probability Theory. In real situations, a frequent source of imprecision is not only the presence of random variables, but the impossibility, in principle, of operating with exact data as a result of the complexity of the system, or the imprecision of the constraints and objectives. At the same time, classes of objects that do not have clear boundaries appear in the problems; the imprecision of such classes is expressed in the possibility that an element not only belongs or does not belong to a certain class, but that intermediate grades of membership are also possible. The membership grade is subjective; although it is natural to assign a lower membership grade to an event that have a lower probability of occurrence. The fact that the assignment of a membership function of a fuzzy set is "non-statistical" does not mean that we cannot use probability distribution functions in assigning membership functions. As a matter of fact, a careful examination of the variables of fuzzy sets reveals that they may be classified into two types: statistical and non-statistical.

Definition 1 ([12]): Let B be a Borel field (σ-algebra) of subsets of the real line Ω. A set function $\mu(\cdot)$ defined on B is called a fuzzy measure if it has the following properties:

1. $\mu(\Phi) = 0 \quad$ (Φ is the empty set);
2. $\mu(\Omega) = 1$;
3. If $\alpha, \beta \in B$ and $\alpha \subset \beta$ then $\mu(\alpha) \leq \mu(\beta)$;
4. If $\{\alpha_j \mid 1 \leq j < \infty\}$ is a monotonic sequence, then

$$\lim_{j\to\infty} [\mu(\alpha_j)] = \mu[\lim_{j\to\infty} (\alpha_j)].$$

Clearly, $\Phi, \Omega \in B$; also, if $\{\alpha_j \mid 1 \leq j < \infty, \alpha_j \in B\}$ is a monotonic sequence then $\lim_{j\to\infty} (\alpha_j) \in B$. In the above definition, (1) and (2) mean that the fuzzy measure is bounded and nonnegative, (3) means monotonicity (in a similar way to finite additive measures used in probability), and (4) means continuity. It should be noted that if Ω is a finite set, then the continuity requirement can be deleted. (Ω, B, μ) is called a fuzzy measure space; $\mu(\cdot)$ is the fuzzy measure of (Ω, B). The fuzzy measure μ is defined on subsets of the real line. Clearly, $\mu[\chi_A \geq T]$ is a non-increasing, real-valued function of T when χ_A is the membership function of set A. Throughout our discussion, we use ξ^T to represent $\{x \mid \chi_A(x) \geq T\}$ and $\mu(\xi^T)$ to represent $\mu[\chi_A \geq T]$, assuming that the A set is well specified. Let $\chi^A\colon \Omega \to [0, 1]$ and $\xi^T = \{x \mid \chi_A(x) \geq T\}$. The function χ_A is called a B-measurable function if $\xi^T \in B, \forall T \in [0, 1]$. Definition 2 introduces the fuzzy expected value (FEV) of χ_A when $\chi_A \in [0, 1]$. Extension of this definition when $\chi_A \in [a, b], a < b < \infty$, is presented in [12].

Definition 2 ([12]): Let χ_A be a B-measurable function such that $\chi_A \in [0, 1]$. The fuzzy expected value (FEV) of χ_A over a set A, with respect to the measure $\mu(\cdot)$, is defined as $\left(T \overset{sup}{\in} [0, 1]\right)\{min[T, \mu(\xi^T)]\}$, where $\xi^T = \{x \mid \chi_A(x) \geq T\}$. Now, $\mu\{x \mid \chi_A(x) \geq T\} = f_A(T)$ is a function of the threshold T. The actual calculation of $FEV(\chi_A)$ then consists of finding the intersection of the curves $T = f_A(T)$. The intersection of the two curves will be at a value $T = H$, so that $FEV(\chi_A = H \in [0, 1]$. It should be noted that when dealing with the $FEV(\eta)$ where $\eta \in [0, 1]$, we should not use a fuzzy measure in the evaluation but rather a function of the fuzzy measure, η', which transforms η under the same transformation that χ and T undergo to η and T', respectively. In general the FEV has the promise and the potential to be used as a very powerful tool in developing ECP technologies.

3.3 Fuzzy Relational Databases and Fuzzy Social Network Architecture

The FRDB model which is based on research in the fields of relational databases and theories of fuzzy sets and possibility is designed to allow representation and manipulation of imprecise information. Furthermore, the system provides means for "individualization" of data to reflect the user's perception of the data [42]. As such, the FRDB model is suitable for use in fuzzy expert system and other fields of imprecise information-processing that model human approximate reasoning such as FSNA [15, 19].

The objective of the FRDB model is to provide the capability to handle imprecise information. The FRDB should be able to retrieve information corresponding to natural language statements as well as relations in FSNA. Although most of these situations cannot be solved within the framework of classical database management systems, they are illustrative of the types of problems that human beings are capable of solving through the use of approximate reasoning. The FRDB model and the FSNA model retrieve the desired information by applying the rules of fuzzy linguistics to the fuzzy terms in the query.

The FRDB as well as the FSNA development [15, 19, 42] were influenced by the need for easy-to-use systems with sound theoretical foundations as provided by the relational database model and theories of fuzzy sets and possibility. They address the following issues:

1. representation of imprecise information,
2. derivation of possibility/certainty measures of acceptance,
3. linguistic approximations of fuzzy terms in query languages,
4. development of fuzzy relational operators (IS, AS...AS, GREATER, ...),
5. processing of queries with fuzzy connectors and truth quantifiers,
6. null-value handling using the concept of the possibilities expected value,
7. modification of the fuzzy term definitions to suit the individual user.

The FRDB and the FSNA are collections of fuzzy time-varying relations which may be characterized by tables, graphs, or functions, and manipulated by recognition (retrieval) algorithms or translation rules.

As an example, let us take a look at one of these relations, the similarity relation. Let D_i be a scalar domain, $y \in D_i$. Then $s(x, y) \in [0, 1]$ is a similarity relation with the following properties: Reflexivity: $s(x, x) = 1$; Symmetry: $s(x, y) = s(y, x)$; Θ-transitivity: where Θ is most commonly specified as max-min transitivity. If, $y, z \in U$, then $s(x, z) \geq max(y \in D_i)\, min(s(x, y), s(y, z))$. Another example is the proximity relation defined below. Let D_i be a numerical domain and , $y, z \in D_i$. Here $p(x, y) \in [0, 1]$ is a proximity relation that is reflexive, and symmetric with transitivity of the form $p(x, z) \geq max(y \in D_i) p(x, y) * p(y, z)$.

The generally used form of the proximity relations is $p(x, y) = e^{-\beta|x-y|}$, where $\beta > 0$. This form assigns equal degrees of proximity to equally distant points. For this reason, it is referred to as the absolute proximity in the FRDB and FSNA models. Similarity and proximity are used in evaluation of queries of the general form: "Find X such that $X.A \ominus d$ " Where $X.A$ is an attribute of X, $d \in D$ is a value of attribute A defined on the domain D, and $\ominus$ is a fuzzy relational operator. Clearly, both FRDS and FSNA may have numerous applications in epidemical outbreak prediction.

In many ECP/DPM programs the amount of information is determined by the amount of the uncertainty—or, more exactly, it is determined by the amount by which the uncertainty has been reduced; that is, we can measure information as the decrease of uncertainty. The concept of information itself has been implicit in many ECP models. That is, both as a substantive concept important in its own right and as a consonant concept that is ancillary to the entire structure of ECP.

3.4 Neuro-Fuzzy Systems

The term Neuro-Fuzzy systems refers to combinations of artificial neural networks and Fuzzy logic. Neuro-Fuzzy systems enable modeling human reasoning via fuzzy inference systems along with the modeling of human learning via the learning and connectionist structure of neural networks. Neuro-Fuzzy systems can serve as highly efficient mechanisms for inference and learning under uncertainty. Furthermore, incremental learning techniques can enable observing outliers and the Fuzzy inference can allow these outliers to coexist (with low degrees of membership) with "main-stream" data. As more information about the outliers becomes available, the information, and the derivatives of the rate of information flow, can be used to identify potential epidemical crises that are hidden in the outliers. The classical model of Neuro-Fuzzy systems can be extended to include multidimensional Fuzzy logic and inference systems in numerical domains and in domains characterized by linguistic variables.

Assuming that people form opinions that are fuzzy and that the information exchange between people influences the opinion formation, the opinion formation process is naturally modeled by structures such as fuzzy coupled map networks and fuzzy and neuro-fuzzy networks [35–39]. In these networks of information aggregation and personal and collective opinion formation, interesting dynamic processes that eventually produce self-organization in structured opinion groups is developed [35, 39].

3.5 Complex Fuzzy Systems

Several aspects of the ECP program can utilize the concept of complex fuzzy logic [7, 12, 14, 23, 24, 32–34, 41].

Complex fuzzy logic can be used to represent the two-dimensional information embedded in the description of an epidemical crisis; namely, the severity and uncertainty. In addition, inference based on complex fuzzy logic can be used to exploit the fact that variables related to the uncertainty that is a part of epidemical crises are multi-dimensional and cannot be readily defined via single dimensional clauses connected by single dimensional connectives. Finally, the multi-dimensional fuzzy space defined as a generalization of complex fuzzy logic can serve as a media for clustering of epidemical crisis information in a linguistic variable-based feature space.

Tamir et al. introduced a new interpretation of complex fuzzy membership grade and derived the concept of pure complex fuzzy classes [32]. This section introduces the concept of a pure complex fuzzy grade of membership, the interpretation of this concept as the denotation of a fuzzy class, and the basic operations on fuzzy classes.

To distinguish between classes, sets, and elements of a set we use the following notation: a class is denoted by an upper case Greek letter, a set is denoted by an upper case Latin letter, and a member of a set is denoted by a lower case Latin letter.

The Cartesian representation of the pure complex grade of membership is given in the following way:

$$\mu(V, z) = \mu_r(V) + j\mu_i(z),$$

where $\mu_r(V)$ and $\mu_i(z)$, the real and imaginary components of the pure complex fuzzy grade of membership, are real value fuzzy grades of membership. That is, $\mu_r(V)$ and $\mu_i(z)$ can get any value in the interval [0, 1]. The polar representation of the pure complex grade of membership is given by:

$$\mu(V, x) = r(V)e^{j\sigma\phi(z)},$$

where $r(V)$ and $\phi(z)$, the amplitude and phase components of the pure complex fuzzy grade of membership, are real value fuzzy grades of membership. That is, they can get any value in the interval [0, 1]. The scaling factor σ is in the interval $[0, 2\pi]$. It is used to control the behavior of the phase within the unit circle according to the specific application. Typical values of σ are $\{1, \frac{\pi}{2}, \pi, 2\pi\}$. Without loss of generality, for the rest of the discussion in this section we assume that $\sigma = 2\pi$.

The difference between pure complex fuzzy grades of membership and the complex fuzzy grade of membership proposed by Ramot et al. [23, 24] is that both components of the membership grade are fuzzy functions that convey information about a fuzzy set. This entails different interpretation of the concept as well as a different set of operations and a different set of results obtained when these operations are applied to pure complex grades of membership. This is detailed in the following sections.

3.5.1 Complex Fuzzy Class

A fuzzy class is a finite or infinite collection of objects and fuzzy sets that can be defined in an unambiguous way and comply with the axioms of fuzzy sets given by Tamir and Kandel [33] and the axioms of fuzzy classes given by [3, 6]. While a general fuzzy class can contain individual objects as well as fuzzy sets, a *pure fuzzy class of order one* can contain only fuzzy sets. In other words, individual objects cannot be members of a pure fuzzy class of order one. A pure fuzzy class of order M is a collection of pure fuzzy classes of order $M - 1$. We define a *Complex Fuzzy Class* Γ to be a pure fuzzy class of order one, i.e., a fuzzy set of fuzzy sets. That is, $\Gamma = \{V_i\}_{i=1}^{\infty}$; or $\Gamma = \{V_i\}_{i=1}^{N}$ where V_i is a fuzzy set and N is a finite integer. Note that despite the fact that we use the notation $\Gamma = \{V_i\}_{i=1}^{\infty}$ we do not imply that the set of sets $\{V_i\}$ is enumerable. The set of sets $\{V_i\}$ can be finite, countably infinite, or uncountably infinite. The use of the notation $\{V_i\}_{i=1}^{\infty}$ is just for convenience.

The class Γ is defined over a universe of discourse U. It is characterized by a pure complex membership function $\mu_\Gamma(V, z)$ that assigns a complex-valued grade of membership in Γ to any element $z \in U$. The values that $\mu_\Gamma(V, z)$ may receive lie within the unit square or the unit circle in the complex plane, and are in one of the following forms:

$$\mu_\Gamma(V, z) = \mu_r(V) + j\mu_i(z),$$
$$\mu_\Gamma(z, V) = \mu_r(z) + j\mu_i(V),$$

where $\mu_r(\alpha)$ and $\mu_i(\alpha)$, are real functions with a range of [0, 1].
Alternatively:

$$\mu_\Gamma(V, z) = r(V)e^{j\theta\phi(z)},$$
$$\mu_\Gamma(z, V) = r(z)e^{j\theta\phi(v)},$$

where $r(\alpha)$ and and $\phi(\alpha)$, are real functions with a range of [0, 1] and $\theta \in (0, 2\pi]$.

In order to provide a concrete example we define the following pure fuzzy class. Let the universe of discourse be the set of all the pandemics that hit the U.S. (in any time in the past) along with a set of attributes related to the pandemic, such as spread mechanism, speed of spread, symptoms, etc. Let M_i denote the set of pandemics that hit the U.S. in the last i years. Furthermore, consider a function (f_1) that associates a number between 0 and 1 with each set of pandemics. For example, this function might reflect the severity of the pandemics in terms risk to affected people. In addition, consider a second function (f_2) that associates a number between 0 and 1 with each specific epidemic. For example, this function might denote the incubation time of the relevant micro-organisms. The functions (f_1, f_2) can be used to define a pure fuzzy class of order One. A compound of the two functions in the form of a complex number can represent the degree of membership in the pure fuzzy class of "high risk pandemics"in the set of pandemics that have occurred in the last 10 years.

Formally, let U be a universe of discourse and let 2^U be the powerset of U. Let f_1 be a function from 2^U to [0, 1] and let f_2 be a function that maps elements of U to the interval [0, 1]. For $V \in 2^U$ and $z \in U$ define $\mu_\Gamma(V, z)$ to be:

$$\mu_\Gamma(V, z) = \mu_r(V) + j\mu_i(z) = f_1(V) + jf_2(z)$$

Then, $\mu_\Gamma(V, z)$ defines a pure fuzzy class of order one, where for every $V \in 2^U$, and for every $z \in U$, $\mu_\Gamma(V, z)$, is the degree of membership of z in V and the degree of membership of V in Γ. Hence, a complex fuzzy class Γ can be represented as the set of ordered triples: $\Gamma = \{V, z, \mu_\Gamma(V, z) \mid V \in 2^U, z \in U\}$

Depending on the form of $\mu_\Gamma(\alpha)$ (Cartesian or polar), $\mu_r(\alpha)$, $\grave{\imath}_i(\alpha)$, $r(\alpha)$, and $\phi(\alpha)$ denote the degree of membership of z in V and/or the degree of membership of V in Γ Without loss of generality, however, we assume that $\mu_r(\alpha)$ and $r(\alpha)$ denote the degree of membership of V in Γ for the Cartesian and the polar representations respectively. In addition, we assume that $\mu_i(\alpha)$ and $\phi(\alpha)$ denote the degree of membership of z in V for the Cartesian and the polar representations respectively. Throughout this chapter, the term *complex fuzzy class* refers to a pure fuzzy class with pure complex-valued membership function, while the term *fuzzy class* refers to a traditional fuzzy class such as the one defined by [3].

Degree of Membership of Order ***N***

The traditional fuzzy grade of membership is a scalar that defines a fuzzy set. It can be considered as degree of membership of order 1. The pure complex degree of membership defined in this chapter is a complex number that defines a pure fuzzy class. That is, a fuzzy set of fuzzy sets. This degree of membership can be considered as degree of membership of order 2 and the class defined can be considered as a pure fuzzy class of order 1. Additionally, one can consider the definition of a fuzzy set (a class of order 0) as a mapping into a one-dimensional space and the definition of a pure fuzzy class (a class of order 1) as a mapping into a two-dimensional space. Hence, it is possible to consider a degree of membership of order N as well as a mapping into an N -dimensional space. The following is a recursive definition of a fuzzy class of order. Note that part 2 of the definition is not really necessary, it is given in order to connect the terms pure complex fuzzy grade of membership and the term grade of membership of order 2.

Definition 3 ([32]):

(1) A fuzzy class of order 0 is a fuzzy set; it is characterized by a degree of membership of order 1 and a mapping into a one dimensional space.
(2) A fuzzy class of order 1 is a set of fuzzy sets. It is characterized by a pure complex degree of membership. Alternatively, it can be characterized by a degree of membership of order Two and a mapping into a two-dimensional space.
(3) A fuzzy class of order N is a fuzzy set of fuzzy classes of order $N - 1$; it is characterized by a degree of membership of order $N + 1$ and a mapping into an $(N + 1)$-dimensional space.

Generalized Complex Fuzzy Logic

A general form of a complex fuzzy proposition is: "x...A...B..." where A and B are values assigned to linguistic variables and '...' denotes natural language constants. A complex fuzzy proposition P can get any pair of truth values from the Cartesian interval $[0, 1] \times [0, 1]$ or the unit circle. Formally a fuzzy interpretation of a complex fuzzy proposition P is an assignment of fuzzy truth value of the form $p_r + jp_i$, or of the form $r(p)e^{j\theta(p)}$, to P. In this case, assuming a proposition of the form "x...A...B...," then $p_r(r(p))$ is assigned to the term A and $p_i(\theta(p))$ is assigned to the term B.

For example, under one interpretation, the complex fuzzy truth value associated with the complex proposition:

x is a young person that lives close to the north pole of jupiter

can be $0.1 + j0.5$. Alternatively, in another context, the same proposition can be interpreted as having the complex truth value $0.3e^{j0.2}$. As in the case of traditional propositional fuzzy logic, we use the tight relation between complex fuzzy classes / complex fuzzy membership to determine the interpretation of connectives. For example, let C denote the complex fuzzy set of "young people that live close to the north pole of jupiter," and let $f_c = c_r + jc_i$, be a specific fuzzy membership function of C, then f_c can be used as the basis for interpretations of P. Next we define several connectives along with their interpretation.

Table 1 includes a specific definition of connectives along with their interpretation. In this table, P, Q and S denote complex fuzzy propositions and f_S denotes the complex fuzzy interpretation of S. We use the fuzzy Łukasiewicz logical system as

Table 1 Basic propositional fuzzy logic connectives

Operation	Interpretation
Negation	$f('P) = 1 + j1 - f(P)$
Disjunction	$f(P \oplus Q) = \max(p_R, q_R) + j \times \max(p_I, q_I)$
Conjunction	$f(P \otimes Q) = \min(p_R, q_R) + j \times \min(p_I, q_I)$
Implication	$f(P \longrightarrow Q) = \min(1, 1 - p_R + q_R) + j \times \min(1, 1 - p_I + q_I)$

the basis for the definitions [5, 9]. Hence, the max t-norm is used for conjunction and the min t-conorm is used for disjunction. Nevertheless, other logical systems, such as Gödel fuzzy systems, can be used [5, 20].

The axioms used for fuzzy logic are used for complex fuzzy logic, and Modus ponens is the rule of inference.

Complex Fuzzy Propositions and Connectives Examples

Consider the following propositions(P, Q, and S respectively):
P: "x is a young person that lives close to the north pole of Jupiter".
Q: "x has elevated body temperature with a severe headach".
S: "x is closely monitored due to high risk of acquiring the pandemic".

Let A be the term *"young person,"* and let B denote the term *"close to the north pole of Jupiter."* Furthermore, let C be the term *"elevated body temperature,"* (alternatively, the term *"high fever"* can be used) and let D denote the term *"severe headache"* Hence, P is of the form: "x is a A that B," and Q is of the form " x is C with D." In this case, the terms *"young person," " close to the north pole of Jupiter," "high fever,"* and *"severe headache"* are values assigned to linguistic variables. Furthermore, a term such as *"headache,"* can get fuzzy truth values (between 0 and 1) or fuzzy linguistic values such as *"minor," "mild,"* and *"severe,"* (the terms *"that,"* and *"with,"* are linguistic constants). Assume that the complex fuzzy interpretation (i.e., degree of confidence or complex fuzzy truth value) of P is $p_r + jp_i$, while the complex fuzzy interpretation of Q is $q_r + jq_i$. Thus, the truth value of "x is a *young person*," is p_R, and the truth value assigned to *"x lives close to the north pole of Jupiter,"* is p_i. The truth value of "x has high fever." is q_r , and the truth value of "x has a severe headach," is q_i, Suppose that the term *"old"* stands for *"not young,"* the term *"far,"* stands for *"not close,"* the term *"low,"* stands for *"not high,"* and the term *"no headache"* denotes the negation of *"severe headache."* In a similar way, S is of the form: *"x is E due to F,"* where the complex fuzzy interpretation of S is $s_r + js_i$. Note that this is not the only way to define these linguistic terms and it is used to exemplify the expressive power and the inference power of the logic. Then, the complex fuzzy interpretation of the following composite propositions is:

(1) $f('P) = (1 - p_r) + j(1 - p_I)$
That is, $'P$ denotes the proposition
"x is an old person that lives close to the north pole of Jupiter".
The confidence level in $'P$ is $(1 - p_r) + j(1 - p_i)$; where the fuzzy truth value of the term "x is an old person," is $(1 - p_r)$ and the fuzzy truth value of the term "...lives far ...," is $(1 - p_i)$

(2) $f(P \oplus 'Q) = \max(p_r, 1 - q_r) + j \times \max(p_i, 1 - q_i)$.
That is, $(P \oplus 'Q)$ denotes the proposition
"x is a young person that lives close to the north pole of Jupiter". OR
"x has low fever and no headach". The truth values of individual terms, as well as the truth value of $P \oplus 'Q$ are calculated according to Table 1.

(3) $f('P \oplus Q) = \min(1 - p_r, q_r) + j \times \min(1 - p_i, q_i)$.
That is, $('P \oplus Q)$ denotes the proposition
"x is an old person that lives far from the north pole of Jupiter". AND
"x has high fever and severe headach". The truth values of individual terms, as well as the truth value of $'P \oplus Q$ are calculated according to Table 1.

(4) Let the term R stand for $(P \oplus Q)$, (the complex fuzzy interpretation of R is $r_r + jr_i$.) then, R $\longrightarrow$ S = min $(1, 1 - r_r + s_r) + j \times$min $(1, 1 - r_i + s_r)$

Thus, (R $\longrightarrow$ S) denotes the proposition
IF "x is a young person that lives close to the north pole of Jupiter". AND
"x has high fever and severe headach".
THEN
"x *is* closely monitored due to high risk of acquiring the pandemic disease". The truth values of individual terms, as well as the truth value of are calculated according to Table 1.

Complex Fuzzy Inference Example

Assume that the degree of confidence in the proposition R defined above is $r_r + jr_i$, and assume that the degree of confidence in the fuzzy implication $T = R \longrightarrow S$ is $t_r + jt_i$. Then, using Modus ponens

$$\frac{\begin{array}{c} R \\ \\ R \longrightarrow S \end{array}}{S}$$

one can infer S with a degree of confidence $\min(r_r, t_r) + j \times \min(r_i, t_i)$.
In other words if one is using:
"x is a young person that lives close to the north pole of Jupiter"
AND "x has high fever and severe headach".
IF
"x is a young person that lives close to the north pole of Jupiter"
AND "x has high fever and severe headach".
THEN
"x is closely monitored due to high risk of acquiring the pandemic".
"x *is* closely monitored due to high risk of acquiring the pandemic".
Hence, using Modus ponens one can infer:
"x *is* closely monitored due to high risk of acquiring the pandemic disease". with a degree of confidence of $\min(r_r, t_r) + j \times \min(r_i, t_i)$.

3.6 Fuzzy Graph Theory

Graph theory and in specific fuzzy graph theory can be used for deriving algorithms for early identification of pandemic outbreaks. In this section we provide the basic definitions and list some of the relevant algorithms. A literature search performed has revealed inconsistencies in the definitions. For this reason, we review some of the definitions of non-fuzzy graphs and provide a detailed and precise definition of the basic terms related to fuzzy graphs.

3.6.1 Non-fuzzy Graphs

A directed graph G is a tuple of the form $G = (V, E, \varphi)$, where V is a set referred to as the set of vertices. E is a set of edges, and φ is a function $\varphi: E \rightarrow V \times V$, such that for every $e \in E, \varphi(e) = (u, v)$, where $u \in V$ and $v \in V$. We assume that $V \cap E = \emptyset$ and in general, we use the form $e = (a, b)$ to denote a specific edge that is "said" to connect the vertices a and b. For an undirected graph, both $e_1 = (u, v)$ and $e_2 = (v, u)$ are in the domain/range of φ. A relation $E \subseteq V \times V$ can be used as an implicit definition of an undirected graph with a set of vertices V and a set of edges E. A weighted graph G is a quintuple $G = (V, E, \varphi, w_1, w_2)$, where V, E, and φ are defined before and $w_1 : V \rightarrow R$; $w_2 : E \rightarrow R$; are functions thatmap vertices and or edges to the set of real numbers R (it is possible but less common to assign complex weights to vertices and edges).

We list some of the important terms and algorithms related to non-fuzzy graphs. These terms and algorithms can be found in numerous textbooks [2]. The fundamental terms related to graphs are the *order of the graph*, the *order of a vertex*, and the *connectedness of the graph*. Other fundamental terms related to graphs are *complete graphs, planner graphs* and *simple graphs, sub-graphs, spanning sub-graphs, cliques, paths, cycles, tours, connectivity, Euler tours, Euler cycles, Hamiltonian tours, Hamiltonian cycles, forests*, and *trees*. The fundamental algorithms applied to weighted graphs are: (1) finding the shortest path between vertices, (2) finding the minimum spanning tree, (3) identifying maximal cliques, (4) finding the minimal Euler tour/cycle and (5) finding the minimal Hamiltonian tour/cycle. In this context, *short, max* and *min* might relate to the number of vertices/edges or the sum of weights of the relevant vertices/edges.

3.6.2 Fuzzy Graphs

A fuzzy directed graph G is a quadruple of the form $\hat{G} = (\hat{V}, \sigma, \hat{E}, \varphi)$, where $\hat{V}$ is a set referred to as the set of vertices and $\hat{E} \subseteq \hat{V} \times \hat{V}$ is a set of edges, $\sigma: \hat{V} \rightarrow [0, 1]$ is a mapping (function) from $\hat{V}$ to $[0, 1]$ (i.e., σ is the assignment of degrees of membership to members of $\hat{V}$), and $\varphi: \hat{E} \rightarrow [0, 1]$ is a function that maps elements of the form $e \in \hat{E} = (u, v)$, to $[0, 1]$ (i.e., φ is the assignment of degrees of membership to members of $\hat{E}$), where $u \in \hat{V}$ and $v \in \hat{V}$. We assume that $\hat{V} \cap \hat{E} = \emptyset$ and in general, we use the form $e = (a, b)$ to denote a specific edge that is "said" to connect the vertices a and b. For an undirected graph, both $e_1 = (u, v)$ and $e_2 = (v, u)$ are in the domain of φ. A weighted fuzzy graph G is an sextuple $\hat{G} = (\hat{V}, \sigma, w_1 \hat{E}, \varphi, w_2)$, where $\hat{V}, E, \sigma$, and φ are as defined previously, $w_1: \hat{V} \rightarrow R$; and $w_2: \hat{E} \rightarrow R$ are functions that map vertices and or edges to the set of real numbers R (it is possible but less common to assign complex weights to vertices and edges).

Following these definitions, a fuzzy graph is a special case of a non-fuzzy weighted graph. Hence, many of the algorithms applied to weighted graphs might be of interest when fuzzy graph and their semantics are considered. Moreover, without loss of generality, we can assume that weights represented by real numbers are normalized to the range of $[0, 1]$. In this case the vertices and/or the edges can be represented via a complex number representing the degree of membership of the vertex/edge in the graph. Clearly, a non-weighted fuzzy graph is a special case of a "regular" fuzzy graph. Fuzzy graphs were used in neuro-fuzzy models of information propagation and aggregation, including opinion formation [37, 39]. In the next section we provide important definitions related to complex fuzzy graphs.

3.6.3 Complex Fuzzy Graphs

A complex fuzzy directed graph $\tilde{G}$ is a quadruple of the form $\tilde{G} = (\tilde{V}, \sigma, \tilde{E}, \varphi)$, where $\tilde{V}$ is a complex fuzzy set referred to as the set of vertices and $\tilde{E} \subseteq \tilde{V} \times \tilde{V}$ is a complex fuzzy set of edges, $\sigma : \tilde{V} \rightarrow [0, 1] \times [0, 1]$ is a mapping from $\tilde{V}$ to $[0, 1] \times [0, 1]$ (i.e., σ is the assignment of a complex degrees of membership to members of $\tilde{V}$), and $\varphi : \tilde{E} \rightarrow [0, 1] \times [0, 1]$ is a function that maps elements of the form $e \in \tilde{E} = (u, v)$, to $[0, 1] \times [0, 1]$ (i.e., φ is the assignment of complex degrees of membership to members of $\tilde{E}$), where $u \in \tilde{V}$ and $v \in \tilde{V}$. We assume that $\tilde{V} \cap \tilde{E} = \emptyset$ and in general, we use the form $e = (a, b)$ to denote a specific edge that is "said" to connect the vertices a and b. For an undirected graph, both $e_1 = (u, v)$ and $e_2 = (v, u)$ are in the domain of φ. Note that the use of complex fuzzy logic is a very strong "tool" that enables dealing with the edges/vertices as carrying complex fuzzy membership values. Hence, it enables exploiting the features of complex fuzzy set theory, complex fuzzy set theory, and complex fuzzy inference. In general, one can use the two components of the complex number assigned to vertices/edges as denoting complex fuzzy information. Alternatively one can use on of the two components as a real fuzzy value and the second component as a weight. To illustrate we provide the following example.

Complex Fuzzy Graph Example

Consider a pandemic that adversely affects "young" people that live in the "north" part of Jupiter. The main initial symptoms of the pandemic disease are: (1) "high" fever, (2) "severe" headaches. Many of the affected people are starting to post status and queries to a social network. While they do not clearly disclose infection, their status/queries might be indicative of a pandemic outbreak. Furthermore, assume that Bob, who is 27 years old and lives "close" to the north pole of Jupiter, sends a "Twitter®" type of message to Alice, who is 57 years old and lives in the same area. The message reads "Staying home today." Alice responds with "What's wrong?" Bob response is "I have a headache and a bit of fever". Figure 1 depicts some of this information in a complex fuzzy graph.

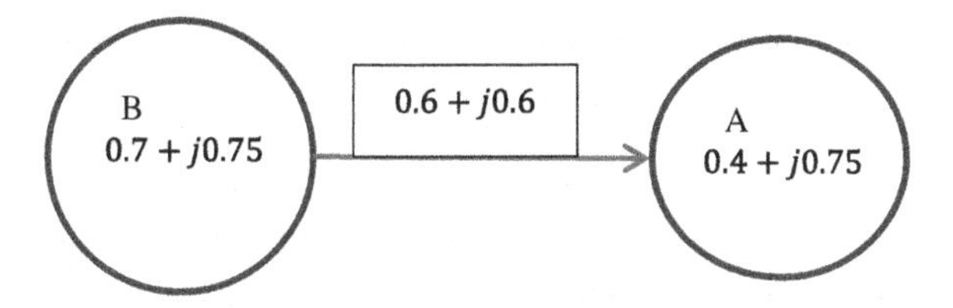

Fig. 1 A complex fuzzy graph representing pandemic related communication

In this graph the vertex 'B' represents Bob along with its degree of membership in a complex fuzzy class. The vertex 'A' represents the same for Alice and the directed edge from Bob to Alice represents the part of the communication between Bob and Alice where Bob reveals his membership in yet another complex fuzzy class. This type of graphs can be used to represent information, actions, and inference.

3.6.4 Complex Fuzzy Graphs' Features and Algorithms

The order of a complex fuzzy graph $\tilde{G}$ is the cardinality of $\tilde{V}$ (denoted as $|\tilde{V}|$). The order of a vertex in $\tilde{G}$ is the number of edges incident to the vertex. Other terms related to complex fuzzy graphs are *complete graphs, planner graphs and simple graphs, sub graphs, spanning sub-graphs, cliques, paths, cycles, tours, Euler tours, Euler cycles, Hamiltonian tours, Hamiltonian cycles, forests,* and *trees*. The fundamental algorithms applied to complex fuzzy graphs are: (1) finding the shortest path between vertices, (2) finding the minimum spanning tree, (3) identifying maximal cliques, (4) finding the minimal Euler tour/cycle, and (5) finding the minimal Hamiltonian tour/cycle. In this context, *short, max* and *min*, might relate to the number of vertices/edges or the sum of weights of the relevant vertices/edges. These terms and algorithms are derived from their definitions in the context of non-fuzzy graphs [2] and fuzzy graphs [22]. We are currently working on the extension of these algorithms to complex fuzzy graphs. One interesting and relevant example is finding the maximal complex fuzzy clique in a complex fuzzy graph. In this example, we expand the work reported in [22] and use a neuro-fuzzy system as a "tool" for addressing the problem.

3.7 Dynamic and Incremental Fuzzy Clustering

Clustering is a widely used mechanism for pattern recognition and classification. Fuzzy clustering (e.g., the Fuzzy C-means) enables patterns to become members of more than one cluster. Additionally, it enables maintaining clusters that represent outliers through low degree of membership. These clusters would be discarded in clustering of hard (vs. Fuzzy) data. Incremental and dynamic clustering (e.g., the incremental Fuzzy ISODATA) enable the clusters' structures to change as information is accumulated. Again, this is a strong mechanism for enabling identification of unlikely events without premature discarding of these events. The clustering can be performed in a traditional feature space composed of numerical measurements of feature values. Alternatively, the clustering can be performed in a multidimensional Fuzzy logic space where the features represent values of linguistic variables The combination of powerful classification capability, adaptive and dynamic mechanisms, as well as the capability to consider uncertain data, maintain data with low likelihood of occurrence, and use a combination of numerical and linguistic values makes this tool, one of the most promising tools for predicting epidemical crisis outbreaks. We currently conduct research on dynamic and incremental fuzzy clustering and it is evident that the methodology can serve as a highly efficient tool for identifying outliers. We plan to report on this research in the near future.

4 Conclusion

In this chapter, we have outlined features of epidemical crises outbreaks. We have shown that an important challenge related to an epidemical crisis is the identification of slow-evolving uncertain events that points to the potential of occurrence of the crisis before it occurs and of fast-evolving data concerning the secondary effect of epidemical crises after the occurrence of primary crisis. We have outlined a set of fuzzy logic-based tools that can be used to address these and other challenges related to ECP.

Recent epidemical crises are showing that there is still a lack of technology-based tools, particularly specific decision-support tools, for addressing epidemical crises, mitigating their adverse impact, and managing crisis response programs.Additional activities that will assist in ECP programs include [17]:

1. Accelerated delivery of technical capabilities for ECP
2. Preparation for an uncertain future
3. Development of world-class science, technology, engineering and mathematics (STEM) capabilities

On top of these important tasks, one should never forget that in the development of ECP programs we do not have the luxury of neglecting human intelligence [16]. In any fuzzy event related to a gray swan, investigation after the fact reveals enough clear data points which had been read correctly but had not been treated properly.

In the future, we intend to investigate the ECP utility of several additional fuzzy logic-based tools including:

1. Value-at-Risk (VaR) under fuzzy uncertainty
2. Non-cooperative fuzzy games
3. Fuzzy logic-driven web crawlers and web-bots
4. Fuzzy Expert Systems and Fuzzy Dynamic Forecasting

Finally, we plan to expand our research on complex fuzzy graphs and their applications, complex fuzzy logic-based neuro-fuzzy systems, and research on incremental and dynamic fuzzy clustering. These research threads are expected to provide significant advancement to our capability to identify and neutralize (as much as possible) primary and secondary adverse effects of epidemical crises.

References

1. Aaron, B., Tamir, D.E., Rishe, N., Kandel, A.: Dynamic incremental K-means clustering, In: The 2014 International Conference on Computational Science and Computational Intelligence, Las Vegas, USA (2014)
2. Balakrishnan, R., Ranganathan, K.: Textbook of Graph Theory. Springer, New York (2001)
3. Běhounek, L., Cintula, P.: Fuzzy class theory. Fuzzy Sets Syst. **154**(1), 34–55 (2005)
4. Bochare, A., Gangopadhyay, A., Yesha, Y., Joshi, A., YeshaYa., Grasso, M.A., Brady, M., Rishe, N.: Integrating domain knowledge in supervised machine learning to assess the risk of breast cancer. Int. J. Med. Eng. Informatics. (In press)
5. Casasnovas, J., Rossell6, F.: Scalar and fuzzy cardinalities of crisp and fuzzy multisets. Int. J. Intell. Syst. **24**(6), 587–623 (2009)
6. Cintula, P.: Advances in LΠ and LΠ1/2 logics. Arch. Math. Logic **42**, 449–468 (2003)
7. Dick, S.: Towards complex fuzzy logic. IEEE Trans. Fuzzy Syst. **13**, 405–414 (2005)
8. Elliott, P., Wartenberg, D.: Spatial epidemiology: current approaches and future challenges. Environ. Health Perspect. **112**, 998–1006 (2004)
9. Fraenkel, A.A., Bar-Hillel, Y., Levy, A., Foundations of Set Theory, 2nd (edn.). Elsevier, Amsterdam (1973)
10. Hájek, P.: Fuzzy logic and arithmetical hierarchy. Fuzzy Sets Syst. **3**(8), 359–363 (1995)
11. Kandel, A., Tamir, D.E., Rishe, N., Fuzzy logic and data mining in disaster mitigation. In: Teodorescu, H.N., (ed.) Improving Disaster Resilience and Mitigation-NewMeans and Tools, Trends, Book Series: NATO Science Series Computer and Systems Sciences, to be published, 2014, Springer, New York (2014)
12. Kandel, A.: Fuzzy Mathematical Techniques with Applications. Addison-Wesley, Reading (1986)
13. Kugaonkar, R., Gangopadhyay, A., Yesha, Y., Joshi, A., YeshaYa., Grasso, M.A., Brady, M., Rishe, N.: Finding associations among SNPs for prostate cancer using collaborative filtering. In: DTMBIO-12: Proceedings of the ACM Sixth International Workshop on Data and Text Mining in Biomedical Informatics, pp. 57–60. Hawaii, USA, 29 Oct 2012
14. Klir, G.J., Tina, A.: Fuzzy Sets, Uncertainty, and Information. Prentice Hall, Upper Saddle River (1988)
15. Last, M., Kandel, A., Bunke, H. (eds.): Data Mining in Time Series Databases, Series in Machine Perception and Artificial Intelligence, vol. 57. World Scientific, Singapore (2004)
16. Last, M., Kandel, A. (eds.): Fighting Terror in Cyberspace, Series in Machine Perception and Artificial Intelligence, vol. 65. World Scientific, Singapore (2005)
17. Lemnios, Z.J., Shaffer, A.: The critical role of science and technology for national defense, Comput. Res. News **21**(5), (2009) (a publication of the CRA)
18. Lu, Y., Zhang, M., Li, T., Guang, Y., Rishe, N.: Online spatial data analysis and visualization system. In: Proceedings of the 19th ACM SIGKDD Conference on Knowledge Discovery and Data Mining (KDD'13), pp. 72–79. Chicago, Illinois USA, 11–14 August 2013
19. Mikhail, R.F., Berndt, D., Kandel, A.: Automated Database Application Testing, Series in Machine Perception and Artificial Intelligence, vol. 76. World Scientific, Singapore (2010)
20. Montagna, F.: On the predicate logics of continuous t-norm BL-algebras. Arch. Math. Logic **44**, 97–114 (2005)

21. Mundici, D., Cignoli, R., D'Ottaviano, I.M.L.: Algebraic foundations of many-valued reasoning. Kluwer Academic Press, Dordrecht (1999)
22. Nair, P.S., Cheng, S.C.: Cliques and fuzzy cliques in fuzzy graphs. In: IFSA World Congress and 20th NAFIPS Joint 9th IEEE International Conference, vol. 4, pp. 2277–2280. (2005)
23. Ramot, D., Milo, R., Friedman, M., Kandel, A.: Complex fuzzy sets. IEEE Trans. Fuzzy Syst. **10**(2), 171–186 (2002)
24. Ramot, D., Friedman, M., Langholz, G., Kandel, A.: Complex fuzzy logic. IEEE Trans. Fuzzy Syst. **11**(4), 450–461 (2003)
25. Ribitzky, R., Yesha, Y., Karnieli, E., Rishe, N.: Knowledge mining & bio-informatics techniques to advance personalized diagnostics & therapeutics. Report to the U.S. National Science Foundation (NSF) on the Outcomes and Consensus Recommendations of the NSF-sponsored International Workshop, February 2012, in Florence, Italy, http://CAKE.fiu.edu/HITpapers/Book_post_NSF_Workshop_Knowledge_Mining_and_Bioinformatics_Techniques_to_Advance_Personalized_Diagnostics_and_Therapeutics.pdf
26. Rishe, N., Sun, Y., Chekmasov, M., Selivonenko, A., Graham, S.: System architecture for 3D terraFly online GIS. In: IEEE Sixth International Symposium on Multimedia Software Engineering (MSE2004), pp. 273–276. Miami, FL, 13–15 December 2004
27. Rishe, N., Espinal, C., Lucic, T., Yesha, Y., YeshaYa., Mathee, K., Marty, A.: Geospatial data for intelligent solutions in public health. In: e-Proceedings of Vaccinology 2012, Rio de Janeiro, 3–7 September 2012
28. Rishe, N., Yesha, Y., YeshaYa., Lucic, T.: Intelligent solutions in public health: models and opportunities. In: Proceedings of the Second Annual International Conference on Tropical Medicine: Intelligent Solutions for Emerging Diseases, p. 45. Miami, Florida, 23–24 February 2012
29. Rishe, N., Yesha, Y., YeshaYa., Lucic, T.: Data mining and querying in electronic health records. In: Proceedings of Up Close and Personalized, International Congress on Personalized Medicine (UPCP 2012), Florence, Italy, p. 40. 2–5 February 2012
30. Taleb, N.N.: Fooled by Randomness. Random House, NY (2004)
31. Taleb, N.N.: The Black Swan. Random House, NY (2007)
32. Tamir, D.E., Lin, J., Kandel, A.: A new interpretation of complex membership grade. Int. J. Intell. Syst. **26**(4), (2011)
33. Tamir, D.E., Kandel, A.: An axiomatic approach to fuzzy set theory. Inf. Sci. **52**, 75–83 (1990)
34. Tamir, D.E., Kandel, A.: Fuzzy semantic analysis and formal specification of conceptual knowledge. Inf. Sci. Intell. Syst. **82**(3–4), 181–196 (1995)
35. Teodorescu, H.N.: Self-organizing uncertainty-based networks. In: Systematic Organisation of Information in Fuzzy Systems. Book Series: NATO Science Series Computer and Systems Sciences, vol. 184, pp. 131–159, 2003, NATO Advanced Research Workshop on Systematic Organisation of Information in Fuzzy Systems, Vila Real, Portugal, 24–26 Oct 2001
36. Teodorescu, H.N., Kandel, A., Schneider, M.: Fuzzy modeling and dynamics. Fuzzy Sets Syst. **106**(1), 1–2 (1999)
37. Teodorescu, H.N.: Information, data, and information aggregation in relation to the user model. NATO Advanced Research Workshop on Systematic Organisation of Information in Fuzzy Vila Real, Portugal, Oct 24–26, 2001. In: Systematic Organisation of Information in Fuzzy Systems Book Series: NATO Science Series Computer and Systems Sciences, vol. 184, pp. 7–10, 2003
38. Teodorescu, H.N.: Pattern formation and stability issues in coupled fuzzy map lattices. Stud. Informatics Control **20**(4), 345–354 (2011)
39. Teodorescu, H.N.: Parallelizing dynamic models based on fuzzy coupled map networks. In: IDAACS 2007: Proceedings of The 4th IEEE Workshop on Intelligent Data Acquisition and Advanced Computing Systems, Book Series: IEEE International Workshop on Intelligent Data Acquisition and Advanced Computing Systems-Technology and Applications-IDAACS, pp. 170–175, 2007
40. Yesha, Y., Rishe, N., YeshaYa., Lucic, T.: Clinical-genomic analysis using machine learning techniques to predict risk of disease. In: Proceedings of Up Close and Personalized, International Congress on Personalized Medicine (UPCP 2012), p. 48. Florence, Italy, 2–5 Feb 2012

41. Zadeh, L.A.: The concept of a linguistic variable and its application to approximate reasoning - Part I. Inf. Sci. **7**, 199–249 (1975)
42. Zemankova-Leech, M., Kandel, A.: Fuzzy Relational Data Bases—A key to Expert Systems. Verlag TUV Rheinland, Koln (1984)
43. Zolotov, S. Ben Yosef, D., Rishe, N., Yesha, Y., Karnieli, E.: Metabolic profiling in personalized medicine: bridging the gap between knowledge and clinical practice in Type 2 diabetes. Personalized Med. **8**(4), 445–456 (2011)

An ACP-Based Approach to Intelligence and Security Informatics

Fei-Yue Wang, Xiaochen Li and Wenji Mao

1 Introduction

The field of Intelligence and security informatics (ISI) is resulted from the integration and development of advanced information technologies, systems, algorithms, and databases for international, national, and homeland security-related applications, through an integrated technological, organizational, and policy-based approach [2]. Traditionally, ISI research and applications have focused on information sharing and data mining, social network analysis, infrastructure protection, and emergency responses for security informatics. Recent years, with the continuous advance of related technologies and the increasing sophistication of national and international security, new directions in ISI research and applications have emerged that address the research challenges with advanced technologies, especially the advancements in social computing. This is the focus of discussion in the current chapter.

As a new paradigm of computing and technology development, social computing can help us understand and analyze individual and organizational behavior and facilitate ISI research and applications in many aspects. To meet the challenges and achieve a methodology shift in ISI research and applications, in this chapter, we shall propose a social computing-based research paradigm consisting of a three-stage modeling, analysis, and control approach that researchers have used successfully to solve many natural and engineering science problems, namely the ACP (Artificial societies, Computational experiments and Parallel execution) approach [10–14].

F.-Y. Wang (✉) · X. Li · W. Mao
The State Key Laboratory of Management and Control for Complex Systems,
Institute of Automation, Chinese Academy of Sciences, Beijing, China
e-mail: feiyue.wang@ia.ac.cn; feiyue@gmail.com

X. Li
e-mail: xiaochen.li@ia.ac.cn

W. Mao
e-mail: wenji.mao@ia.ac.cn

F.-Y. Wang
The Research Center for Computational Experiments and Parallel Systems Technology,
The National University of Defense Technology, Changsha, Hunan, China

R.R. Yager et al. (eds.), *Intelligent Methods for Cyber Warfare*,
Studies in Computational Intelligence 563, DOI 10.1007/978-3-319-08624-8_3

Based on the ACP approach, in this chapter, we shall focus on behavioral modeling, analysis and prediction in security informatics. We shall first present a knowledge extraction approach to acquire behavioral knowledge from open-source intelligence and facilitate behavioral modeling. On the basis of behavioral modeling, we shall then present two approaches to group behavior prediction. The first approach employs plan-based inference and explicitly takes the observed agents preferences into consideration. The second approach employs graph theory and incorporates a graph search algorithm to forecast complex group behavior. We finally provide the results of experimental studies to demonstrate the effectiveness of our proposed methods.

2 The ACP Approach

The ACP approach [10–14] is composed of three interconnected parts: artificial societies for modeling, computational experiments for analysis and parallel execution for control. We shall discuss them in detail below.

2.1 Modeling with Artificial Societies

In the literature, there are no effective formal methods to model complex social-techno systems, especially those heavily involving human and social behavior. The ACP framework posits that agent-based artificial societies are the most suitable modeling approach to social modeling and social computing. An artificial society-based approach has three main components: agents, environments, and rules for interactions. In this modeling approach, how accurately the actual system can be approximated is no longer the only objective of modeling, as it is the case in traditional computer simulations. Instead, the artificial society developed is considered as an actual systemłan alternative possible realization of the target society. Along this line of thinking, the actual society is also considered as one possible realization. As such, the behaviors of the two societies, the actual and the artificial, are different but fit different evaluation and analysis purposes. Note that approximation with high fidelity is still the desired goal for many applications when it is achievable but can be relaxed otherwise, representing a necessary compromise that recognizes intrinsic limits and constraints of dealing with complex social-techno-behavioral systems.

2.2 Analysis with Computational Experiments

Traditional social studies primarily rely on passive observations, small-scale human subject experimental studies, and more recently computer simulations. Repeatable experiments are very difficult to conduct, due to a number of reasons including but not limited to research ethics, resource constraints, uncontrollable conditions, and unobservable factors. Artificial societies can help alleviate some of these problems. Using artificial societies as social laboratories, we can design and conduct controllable experiments that are easy to manipulate and repeat. Through agent and environmental setups and interaction rule designs, one can evaluate and quantitatively analyze various factors and what-if scenarios in social-computing problems. These artificial society-based computational experiments are a natural extension to traditional computer simulation. Basic experimental design issues related to model calibration, analysis, and verification need to be addressed. Furthermore, design principles such as replication, randomization, and blocking, guide these computational experiments just as they would guide experiments in the physical world.

2.3 Control and Management Through Parallel Execution

Parallel execution refers to the fact that long-lived artificial systems can run in parallel and co-evolve with the actual systems they model. This is a generalization of controllers as used in classical automation sciences, which use analytical models to drive targeted physical processes to desired states. This parallel execution idea provides a powerful mechanism for the control and management of complex social systems through co-evolution of actual and artificial systems. The entire system of systems can have three major modes of operations. In the learning and training mode, the actual and artificial systems are disconnected. The artificial systems can be used to train personnel. In the experimenting and evaluating mode, connections or syncing between the actual and artificial systems take place in discrete times. Computational experiments can be conducted between these syncs, evaluating various policies. In the controlling and managing mode, the artificial systems are used as the generalized controllers of the actual systems with two systems constantly connected. Social computing applications, especially those involving security, control and management of social activities, can benefit directly from parallel execution.

3 Modeling Organizational Behavior

Action knowledge has been widely used in modeling and reasoning about agent's behavior. Action knowledge is typically represented using plan representation, which includes domain actions and the states causally associated with the actions (i.e., action preconditions and effects) [3]. Action precondition is the condition that must be made true before action execution. Action effect is the state achieved after action execution. Since action knowledge is the prerequisite of various security-related applications in behavior modeling, explanation, recognition and prediction, in this section, we present a knowledge extraction approach to acquire action knowledge, making use of the massive online data sources. The action extraction procedure includes action data collection, raw action extraction and action refinement [7]. Below we introduce the automatic extraction of action preconditions and effects from online data.

3.1 Extracting Action Knowledge from the Web

Extracting causal relations has been studied in previous related research (e.g., [4, 6, 8]). The focus of our work is different from those of previous research in two aspects. First, instead of finding general causal relations between two clauses or noun phrases, our focus is to find the causal relations between actions and states C action knowledge for behavioral modeling. Second, we need to acquire richer knowledge types C not only causal relations, but also goals, reasons and conditions associated with the actions. Sil et al. [9] propose a SVM-based approach to build classifiers for identifying action preconditions and effects. As their work only tests a small number of actions all selected from one frame in FrameNet, and all the actions are treated as single verbs, the performance of their approach in complex and open domain is unclear.

In extracting action preconditions, we differentiate several types of precondition: necessity/need, permission/possibility and means/tools. We classify the patterns into four categories based on their types and polarities. Tables 1 and 2 shows the linguistic patterns we design for extracting action preconditions and effects. To ensure the quality of the extracted causal knowledge, we prefer rule-based approach which can achieve relatively high precision. On the other hand, as our work is based on the open source data, recall rate could be compensated by the huge volume of online resources.

3.2 Computational Experiment on Terrorist Organization

As a great amount of reports about this group and its historical events are available online, we employ computational methods to automatically extract group actions and causal knowledge from relevant open source textual data. The textual data we use are the news about Al-Qaeda reported in *The Times*, *BBC*, *USA TODAY*, *The New York Times* and *The Guardian*, with totally 953,663 sentences.

Among the official investigation reports, 13 real attacks perpetrated by $Al-Qaeda$ have relatively complete descriptions. Intelligence analyst helped us manually compose the action knowledge of each attack based on these descriptions, and these form the basis of our experiment. We evaluate the performance of our method by checking how many actions and states specified in each attack example are already covered by the domain actions and causal knowledge we extract. Table 3 shows the results of the experimental study. The average coverage rates of the actions, preconditions, effects and states (preconditions plus effects) are 85.8, 74.1, 78.7 and 75.6 %, respectively. In general, the experimental results verify the effectiveness of our approach.

After action knowledge acquisition, we collect organizational behavior knowledge with quality. Based on the action knowledge we collect, we further employ planning algorithm to generate attack plans about this group and construct the plan library

Table 1 Patterns for extracting action preconditions

<table>
<tr><td>Precondition</td><td>Necessity/need</td><td><action (verb-ing+object/verb-ing) set> require | demand | need <precondition set>
<node-name> need <precondition set> to <action (verb+object/verb) set></td></tr>
<tr><td></td><td>Permission/possibility</td><td><precondition set> allow <node-name> to <action (verb+object/verb) set>
<precondition set> enable | create the possibility for <nodename> to <action (verb+object/verb) set></td></tr>
<tr><td></td><td>Means/tools</td><td><node-name> use <precondition set> to <action (verb+object/verb) set>
provide | supply | offer <precondition set> for <node-name> to <action (verb+object/verb) set>
provide | supply | offer <node-name> with <precondition> to <action (verb+object/verb) set></td></tr>
<tr><td></td><td>Negative patterns</td><td><¬precondition set> prevent | stop <node-name> from <action (verb-ing+object/verb-ing) set>
<¬precondition set> disable | undermine <node-name> to <action (verb+object/verb) set>
lack of <precondition set> prevent | stop <node-name> from <action (verb-ing+object/verb-ing) set>
the shortage of <precondition set> disable | undermine <action (verb-ing+object/verb-ing) set>
cannot <action (verb+object/verb) set> without | unless <precondition set></td></tr>
</table>

Table 2 Patterns for extracting action effects

<table>
<tr><td rowspan="2">Effect</td><td>Causation</td><td><action (verb-ing+object/verb-ing) set> *bring about* | *lead to* | *result in* | *trigger* | *cause* | *produce* | *give rise to* <effect set>
<effect set> *be caused* | *produced* | *triggered* | *brought about by* <action (verb-ing+object/verb-ing) set>
<node-name> <action (verb+object/verb) set> *to cause* | *bring about* | *produce* | *trigger* <effect set>
<node-name> <action (verb+object/verb) set>, *causing* | *producing* | *triggering* | *resulting in* | *leading to* <effect set>
<node-name> <action (verb+object/verb) set>, *which* | *that bring about* | *lead to* | *result in* | *trigger* | *cause* | *produce* <effect set> <effect set> caused | produced by <action (verb-ing+object/verb-ing) set>
What bring about | *lead to* | *result in* | *trigger* | *cause* | *produce* | *give rise to* <effect set> be <action (verb-ing+object/verb-ing) set></td></tr>
<tr><td>Reason/goal</td><td>*the reason of* | *reason for* | *cause of* <action (verb-ing+object/verbing) set > *be* <effect set>
<node-name> <action (verb+object/verb) set> *because of* | *on account of* | *due to* <effect set>
<action (verb-ing+object/verb-ing) set> *be due to* <effect set> <node-name> <action (verb+object/verb) set> *for the purpose of* | *in an attempt to* | *in an effort to* | *in order to* | *so as to* | *in the cause of* <effect set></td></tr>
</table>

Table 3 Experimental results on causal knowledge and action extraction

Attack example	Action converge	Precondition converge	Effect converge	State converge
1	0.833	0.818	0.833	0.824
2	0.800	0.778	0.800	0.786
3	0.900	0.727	0.900	0.781
4	0.889	0.737	0.778	0.750
5	0.857	0.733	0.714	0.727
6	0.833	0.727	0.833	0.765
7	0.875	0.765	0.750	0.760
8	0.833	0.667	0.833	0.722
9	0.889	0.684	0.889	0.750
10	0.900	0.800	0.800	0.800
11	0.857	0.750	0.714	0.739
12	0.857	0.692	0.714	0.700
13	0.833	0.750	0.667	0.722
Average	0.858	0.741	0.787	0.756

[7]. Plan library represents the groups strategic plans and behavioral patterns, which are the key of organizational behavior modeling. Below is an example plan in this groups plan library (The rectangles denote actions and the rounded rectangles denote preconditions and effects).

4 Forecasting Group Behavior via Plan Inference

Group behavior prediction is an emergent research and application field in intelligence and security informatics, which studies computational methods for the automated prediction of what a group might do. As many security-related applications could benefit from forecasting an entitys behavior for decision making, assessment and training, it is gaining increasing attention in recent years. Recent progress has made it possible to automatically extract plan knowledge (i.e., actions, their preconditions and effects) from online raw textual data and construct group plans by means of planning algorithm, albeit in the restrictive security informatics domain (see Sect. 3). On the basis of this, we present two plan-based approaches to group behavior forecasting in this section. The first approach is based on probabilistic plan inference, and the second approach is aimed at forecasting complex group behavior via multiple plan recognition (Fig. 1).

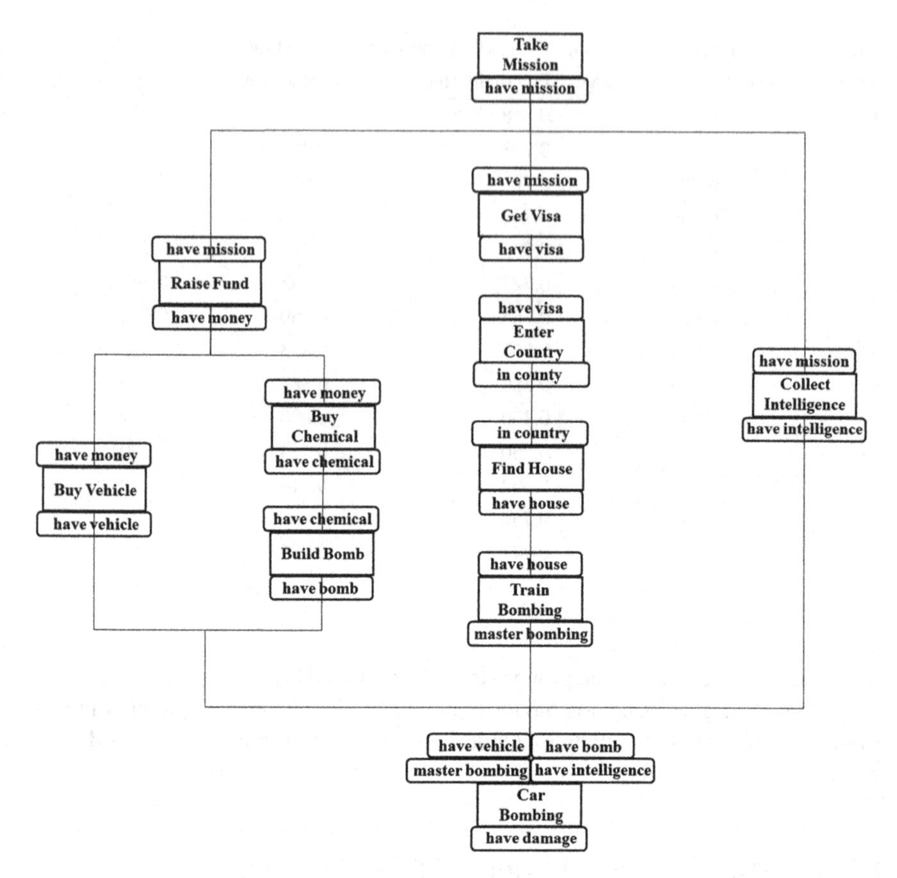

Fig. 1 An example strategic plan of the group

4.1 The Probabilistic Plan Inference Approach

Plan representations are used by many intelligent systems. In a probabilistic plan representation, the likelihood of states is represented by probability values. To represent the success and failure of action execution, we use execution probability $P_{execution}$ to represent the likelihood of successful action execution given action preconditions are true. An action effect can be nondeterministic and/or conditional nondeterministic. We use effect probability P_{effect} to represent the likelihood of the occurrence of an action effect given the corresponding action is successfully executed, and conditional probability $P_{conditional}$ to represent the likelihood of the occurrence of its consequent given a conditional effect and its antecedents are true. The desirability of action effects (i.e., their positive/negative significance to an agent) is represented by utility values. Outcomes are those action effects with non-zero utility values. We use expected utility (EU) to represent the overall benefit or disadvantage of a plan.

Our approach is based on the fundamental MEU (maximum expected utility) principle underlying decision theory, which assumes that a rational agent will adopt a plan maximizing the expected utility. The computation of expected plan utility captures two important factors. One is the desirability of plan outcomes. The other is the likelihood of outcome occurrence, represented as outcome probability. We use the observed evidence to incrementally update state probabilities and the probabilities of action execution. The computation process is realized through recursively using plan knowledge represented in plans.

4.1.1 Probability of States

Let E be the evidence. If state x is observed, the probability of x given E is 1.0. Observations of actions change the probabilities of states. If action A is observed executing, the probability of each precondition of A should be 1.0, and the probability of each effect of A is the multiplication of its execution probability and effect probability. If A has conditional effects, the probability of a consequent of a conditional effect of A is the product of its execution probability, conditional probability and the probabilities of each antecedent of the conditional effect.

- IF $x \in precondition(A)$, $P(x|E) = 1.0$
- IF $x \in effect(A)$, $P(x|E) = P_{execution}(A|precondition(A)) \times P_{effect}(x|A)$
- IF $x \in consequent(e) \wedge e \in conditional - effect(A)$,
  $P(x|E) = P_{execution}(A|precondition(A)) \times P_{conditional}(x|antecedent(e), e) \times \prod_{e' \in antecedent(e)} P(e'|E)$

Otherwise, the probability of x given E is equal to the prior probability of x.

4.1.2 Probability of Action Execution

If an action A is observed executed, the probability of successful execution of A given E is 1.0, that is, $P(A|E) = 1.0$. If A is observed executing, $P(A|E)$ equals to its execution probability. Otherwise, the probability of successful execution of A given E is computed by multiplying the execution probability of A and the probabilities of each action precondition.

$$P(A|E) = P_{execution}(A|precondition(A)) \times \prod_{e \in precondition(A)} P(e|E)$$

4.1.3 Outcome Probability and Expected Utility of Actions

The probability changes of action execution impact the calculation of outcome probabilities and expected utilities of actions. Let O_A be the outcome set of action A, and outcome $o_i \in O_A$. The probability of o_i given E is computed by multiplying the probability of executing A and the effect probability of o_i.

$$P_{action}(o_i|E) = P(A|E) \times P_{effect}(o_i|A)$$

4.1.4 Outcome Probability of Plans and Expected Plan Utility

Let O_P be the outcome set of plan P, and outcome $o_j \in O_P$. Let $\{A_1, ..., A_k\}$ be the partially ordered action set in P leading to o_j, where o_j is an action effect of A_k. The probability of o_j given E is computed by multiplying the probabilities of executing each action leading to o_j and the effect probability of o_j (Note that $P(A_i|E)$ is computed according to the partial order of A_i in P).

$$P_{plan}(o_j|E) = (\prod_{i=1,...,k} P(A_i|E)) \times P_{effect}(o_j|A_k)$$

The expected utility of P given E is computed using the utilities of each plan outcome in P and the probabilities with which each outcome occurs.

$$EU(P|E) = \sum_{o_j \in O_P} (P_{plan}(o_j|E) \times Utility(o_j))$$

4.2 The Multiple Plan Recognition Approach

In real-world situations, a group often engages in complex behavior and may pursue multiple plans/goals simultaneously. These complex group behaviors can hardly be captured by conventional plan inference approaches as they often assume that an agent only commits to one plan at a time. To achieve complex group behavior forecasting, we propose a novel multiple plan recognition approach in this section.

From a computational perspective, multiple plan recognition poses great challenge. For observed group actions, the hypothesis space of multiple plan recognition turns out to be rather huge and the computational complexity is extremely high. To address the challenge, our approach consider using searching techniques to efficiently find the best explanation. Intuitively, if we view the actions of plans as vertexes and links between actions as edges, we can convert plans into a graph. We intend to map multiple plan recognition into a graph theory problem and adopt graph search techniques to find a near best explanation.

Below we first give the problem definition and represent the hypothesis space of input observations as a directed graph (i.e. explanation graph). We then describe how to compute the probability of an explanation. We finally present an algorithm for finding the best explanation.

4.2.1 Problem Definition

Our approach adopts the hierarchical plan representation. A hierarchical plan library, PL, is a set of hierarchical partial plans. Each partial plan is composed of abstract and/or primitive actions. The actions in the partial plan form a tree-like structure, where an abstract action corresponds to an AND node (i.e., there exists only one way of decomposition) or an OR node (i.e., multiple ways of decomposition exist) in the plan structure. At an AND node, each child is decomposed from its parent with decomposition probability 1. At an OR node, each child is a specialization of its parent. The sum of specialization probabilities of each child is 1.

For each observation, it can be either a primitive action or a state. Given a plan library, an explanation, SE_i, for a single observation, O_i, is an action sequence starting from a top-level goal, G_0, to O_i: $SE_i = \{G_0, SG_1, SG_2, ..., SG_m, O_i\}$, where SG_1, SG_2,..., and SG_m are a set of abstract actions. There can be multiple explanations for a single observation. An explanation, E_j, for an observation set $O = \{O_1, O_2, ..., O_n\}$ is defined as $E_j = SE_1 \cup SE_2 \cup ... \cup SE_n$, where SE_i is an explanation for the single observation O_i. If $SE_1 = SE_2 = ... = SE_n$, the explanation E_j corresponds to a single plan. Otherwise it corresponds to multiple plans.

We define the multiple plan recognition problem as follows. Given a hierarchical plan library PL and an observation set O, the task of multiple plan recognition is to find the most likely explanation (best explanation), E_{max}, from the explanation set, E, for O

$$E_{max} = \underset{E_i \in E}{argmax}\ P(E_i|O)$$

4.2.2 Constructing Explanation Graph

Given a set of observed actions and a plan library, the procedure of constructing explanation graph is as follows:

Step 1. Construct an explanation graph EG which is an empty graph and add all observations to the bottom level of EG.

Step 2. Expand the parents of each observation following a breadth-first strategy and add these parents to EG. Decomposition/specialization links between actions are treated as directed edges and are also added to EG. The direction of the edges denotes decomposition or specialization relation. A decomposition/specialization probability is attached to each edge. Duplicate actions and edges are combined during expansion.

Step 3. Apply this breadth-first expansion strategy on EG until all the actions in EG are expanded.

Step 4. Then add a dummy node on the top of the graph and connect the dummy node to all the top-level goals. The edges from the dummy node to top-level goals are associated with the prior probabilities of each top-level goal.

Now our approach constructs an explanation graph which contains all the possible explanations for the given observed actions.

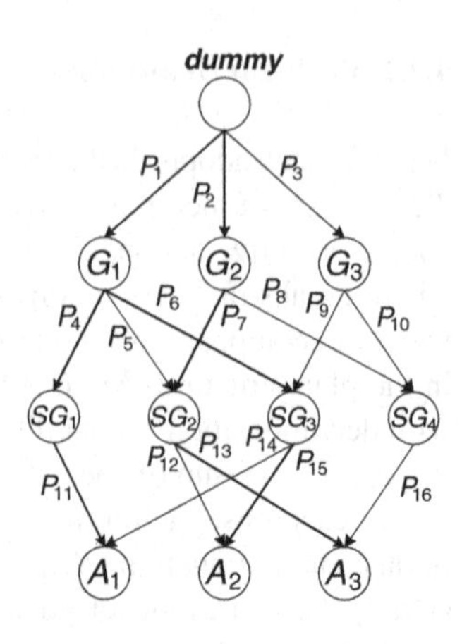

Fig. 2 Illustration of an explanation graph

Figure 2 is an explanation graph for the observed actions A_1, A_2 and A_3. It is a directed graph with bold lines denoting an explanation. The symbols P_1, P_2,..., P_{16} denote the decomposition/specialization probabilities associated with edges. An explanation corresponds to a connected sub-graph in the explanation graph containing the dummy node, top-level goals, sub-goals, and all the observations. In the explanation, the nodes with input degree 1 correspond to observations.

Here we define an explanation for an observation set as a tree in an explanation graph, in which the root is a dummy node and the leaves are all the observations. The tree exactly specifies an explanation for each observation.

4.2.3 Computing the Probability of an Explanation

Let $O_{1:i} = \{O_1, O_2, ..., O_i\}$ be observed actions, the probability of an explanation E_j is computed as

$$P(E_j|O_{1:i}) = P(E_j, O_{1:i})|P(O_{1:i}) = P(O_{1:i}|E_j)P(E_j)|P(O_{1:i})$$

As $1/P(O_{1:i})$ is a constant for each explanation, we denote it as K. $P(O_{1:i}|E_j)$ is the probability that $O_{1:i}$ occurs given the explanation E_j and is 1 for all hypotheses. $P(E_j)$ is the prior probability of TH explanation, i.e., the probability of entire tree of the explanation graph. For explanation E_j, let $G_{1:m} = \{G_1, ..., G_m\}$ be top-level goals and $SG_{1:n} = \{SG_1, ..., SG_n\}$ be sub-goals. We denote the vertex set of the tree E_j as $V = dummy \cup G_{1:m} \cup SG_{1:n} \cup O_{1:i}$. Let $E = \{e_1 = dummy \rightarrow G_1, ..., e_s = SG_x \rightarrow SG_y, ..., e_t = SG_z \rightarrow O_i\}$ be the set of edges of E_j, where $1 \leq x, y, z \leq n$ and $1 \leq s \leq t$. Here we assume the decomposition of each action is directly influenced by its parent node. The prior probability of the explanation E_j is

$$
\begin{aligned}
P(E_j) &= P(V|E) \\
&= P(O_i, e_t|V/O_i, E/e_t) * P(V/O_i, E/e_t) \\
&= P(e_t) * P(V/O_i, E/e_t) \\
&= ... = P(dummy) * \prod_{edge \in E} P(edge)
\end{aligned}
$$

$P(O_i, e_t|V/O_i, E/e_t)$ is the conditional probability that the decomposition rule, e_t, activates and O_i is decomposed given the tree $(V/O_i, E/e_t)$. This is equal to $P(e_t = SG_z \to O_i)$ according to our decomposition assumption. In addition, $P(edge)$ is the probability of the edge in the explanation and $P(dummy)$ is the prior probability that an observed agent will pursue goals. This is constant for each explanation.

4.2.4 Finding the Best Explanation

Now the problem of finding the best explanation can be formulated as

$$
\begin{aligned}
E_{max} &= argmax P(O_{1:i}|E_j)P(E_j)|P(O_{1:i}) \\
&= \underset{E_j \in E}{argmax} \prod_{edge_i \in E_j} P(edge_i) \\
&= \underset{E_j \in E}{argmax} \sum_{edge_i \in E_j} ln(P(edge_i)) \\
&= \underset{E_j \in E}{argmax} \sum_{edge_i \in E_j} ln(P(\frac{1}{edge_i}))
\end{aligned}
$$

where $P(edge_i)$ is the decomposition probability associated with $edge_i$. We denote $ln(1/P(edge_i))$ as the weight of $edge_i$. As $0 < P(edgei) < 1$, we get $ln(1/P(edge_i)) > 0$. For explanation graph EG, we attach the weight $ln(1/P(e))$ to each edge $e \in EG$ (where $P(e)$ is the probability on edge e) and then we can convert an explanation graph to a directed weighted graph. Now the problem of finding the most likely explanation is reformulated as finding a minimum weighted tree in the explanation graph with the dummy node as the root and observations as leaf nodes.

Finding a minimum weight tree in a directed graph is known as the directed Steiner tree problem in graph theory [1, 15]. It is defined as follows: given a directed graph, $G = (V, E)$, with weights, $w(w0)$, on the edges, a set of terminals, $S \subseteq V$, and a root vertex, r, find a minimum weight tree, T, rooted at r, such that all the vertices in S are included in T. A number of algorithms have been developed to solve this problem. In our approach, we employ an approximation algorithm proposed by Charikar et al. [1].

4.3 Computational Experiments

4.3.1 Experimental Study 1

We still choose $Al - Qaeda$ as the representative realistic group for our study. Among the official investigation reports, 13 real attacks perpetrated by $Al - Qaeda$ have relatively complete descriptions. Based on our automatically generated plans, intelligence analyst helped choose 13 plans that match the reported real attacks. These plans form the plan library for our experimental study. We randomly generate a set of evidence using the combination of actions and initial world states in the plan library, and collect 95 lines of evidence. Each line contains either two observations (constituting 49 % of the evidence set) or three observations (constituting 51 % of the evidence set).

Four human raters experienced in security informatics participate in the experiment. According to the plan library we construct, each rater examined the evidence set line by line and predicted the most likely plans based on each line of evidence. The test set is composed of each raters predictions together with the corresponding evidence, with inter-rater agreement ($Kappa$) 0.764. The prior state probabilities, action execution probabilities and effect probabilities used by our approach (less than 100 items in total) were assigned by intelligence analyst. The intelligence analyst also assigned prior and conditional probabilities for Bayesian reasoning. Mapping plans to Bayesian networks is based on the generic method provided in [5].

Table 4 shows the experimental results using our approach and Bayesian reasoning. We measure the agreement of the probabilistic plan inference approach and each rater using the $Kappa$ statistic. The average agreements between our approach and human raters are 0.664 (for two observations) and 0.773 (for three observations), which significantly outperform the average agreements between Bayesian reasoning and the raters. As $0.6 < k < 0.8$ indicates substantial agreement, the empirical results show good consistency between the predictions generated by our approach and those of human raters.

4.3.2 Experimental Study 2

We still choose $Al - Qaeda$ as a representative group. Based on our previous work [7], we automatically extract group actions and construct group plans from relevant open source News (e.g., $Times\ Online$ and $USATODAY$). Domain experts helped connect the hierarchical partial plans in the plan library. The plan library we use for this experiment includes 10 top-level goals and 35 primitive and abstract actions (we allow primitive/abstract actions to appear in multiple plans). Although large numbers of plans are computationally feasible by our approach, we prefer a relatively small and realistic plan library so that it is tractable by human raters in the experiment. Figure 3 illustrates the plan structure for a top-level goal in the plan library.

Table 4 Kappa agreements between algorithms and human raters

Rater	Plan inference						Bayesian reasoning					
	Two observations			Three observations			Two observations			Three observations		
	P(A)	*P(E)*	*K*	*P(A)*	*P(E)*	*K*	*P(A)*	*P(E)*	*K*	*P(A)*	*P(E)*	*K*
1	0.826	0.108	0.805	0.878	0.106	0.864	0.436	0.078	0.388	0.469	0.106	0.406
2	0.696	0.090	0.666	0.837	0.105	0.818	0.435	0.086	0.382	0.469	0.107	0.405
3	0.609	0.096	0.567	0.776	0.097	0.752	0.435	0.098	0.374	0.490	0.102	0.432
4	0.652	0.088	0.618	0.694	0.101	0.660	0.370	0.089	0.308	0.388	0.092	0.326
Avg			0.664			0.773			0.363			0.392

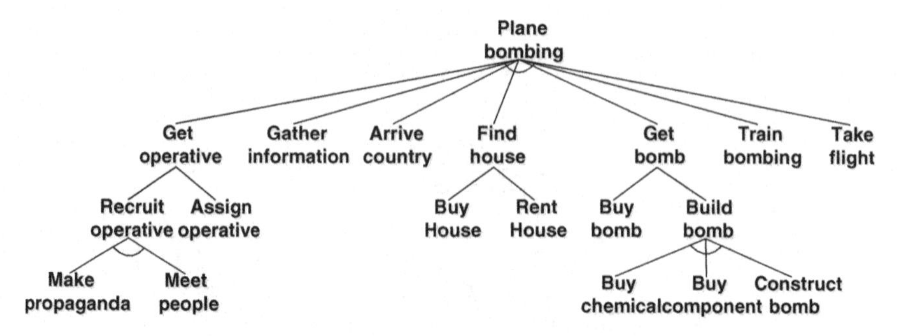

Fig. 3 Plan structure for a top-level goal in the plan library

Table 5 Agreements between MPR algorithm and human raters for various observations (obs.)

Rater	$P(A)$			K		
	2 obs.	3 obs.	4 obs.	2 obs.	3 obs.	4 obs.
1	0.967	0.932	0.899	0.961	0.908	0.874
2	0.978	0.913	0.869	0.974	0.887	0.837
3	0.956	0.917	0.859	0.949	0.891	0.830
4	0.906	0.924	0.902	0.856	0.899	0.877
5	0.838	0.895	0.878	0.765	0.866	0.847

We randomly generate a number of observation sets using the combination of primitive actions in the plan library. We collect 90 lines of observation sets in total, each line corresponding to one observation set. Among them, 30 observation sets contain two observations each, 30 contain three observations each and 30 contain four observations each. Five human raters who have at least 3 years experience in the security informatics domain participated in the experiment. Based on the constructed plan library, each rater examined the observation sets one by one and predicted the most likely plans (single plan or multiple plans) based on each observation set. The test set is composed of each raters predictions together with corresponding observations (with inter-rater agreement of 0.88).

Table 5 shows the experimental results between the multiple plan recognition approach and each human rater. We measure the agreement of the results generated by our approach and those of the raters for two, three and four observations using precision, $P(A)$, and Kappa statistics, K. The agreements between our approach and human raters for two observations, three observations, and four observations are all above 0.8, thus the empirical results show good consistency between the predictions generated by our MPR approach and those of human raters.

5 Conclusion

In this chapter, we propose an ACP-based approach for behavioral modeling, analysis and prediction in security informatics. To facilitate behavioral modeling, we present a knowledge extraction approach to acquire behavioral knowledge from open-source intelligence. On the basis of behavioral modeling, we propose two plan inference approaches for group behavior forecasting. The first explicitly takes the observed agents preferences into consideration to infer the most likely plan of groups. The second employs a graph search algorithm to discover multiple intentions underlying complex group behavior. Experimental results to demonstrate the effectiveness of these computational methods we propose as well as the underlying ACP approach.

References

1. Charikar, M., Chekuri, C., yat Cheung, T., Dai, Z., Goel, A., Guha, S., Li, M.: Approximation algorithms for directed steiner problems. In: Proceedings of the Ninth Annual ACM-SIAM Symposium on Discrete Algorithms, pp. 192–200 (1998)
2. Chen, H., Wang, F.Y., Zeng, D.: Intelligence and security informatics for homeland security: information, communication, and transportation. IEEE Trans. Intell. Transp. Syst. **5**(4), 329–341 (2004)
3. Fikes, R.E., Nilsson, N.J.: Strips: a new approach to the application of theorem proving to problem solving. Artif. Intell. **2**(3), 189–208 (1971)
4. Girju, R.: Automatic detection of causal relations for question answering. In: Proceedings of the ACL 2003 Workshop on Multilingual Summarization and Question Answering, pp. 76–83 (2003)
5. Huber, M.J., Durfee, E.H., Wellman, M.P.: The automated mapping of plans for plan recognition. In: Proceedings of the Tenth International Conference on Uncertainty in Artificial Intelligence, pp. 344–351 (1994)
6. Khoo, C.S.G., Chan, S., Niu, Y.: Extracting causal knowledge from a medical database using graphical patterns. In: Proceedings of the 38th Annual Meeting on Association for Computational Linguistics, pp. 336–343 (2000)
7. Li, X., Mao, W., Zeng, D., Wang, F.Y.: Automatic construction of domain theory for attack planning. In: IEEE International Conference on Intelligence and Security Informatics, pp. 65–70 (2010)
8. Persing, I., Ng, V.: Semi-supervised cause identification from aviation safety reports. In: Proceedings of the Joint Conference of the 47th Annual Meeting on Association for Computational Linguistics, pp. 843–851 (2009)
9. Sil, A., Huang, F., Yates, A.: Extracting action and event semantics from web text. In: AAAI Fall Symposium on Common-Sense Knowledge, vol. 40 (2010)
10. Wang, F.Y.: Computational experiments for behavior analysis and decision evaluation of complex systems. Acta Simulata Systematica Sinica **5**, 008 (2004)
11. Wang, F.Y.: Social computing: concepts, contents, and methods. Int. J. Intell. Control Syst. **9**(2), 91–96 (2004)
12. Wang, F.Y.: A computational framework for decision analysis and support in isi: Artificial societies, computational experiments, and parallel systems. In: Intelligence and Security Informatics, pp. 183–184. Springer, Berlin (2006)
13. Wang, F.Y.: Parallel management systems: concepts and methods. J Complex Syst. Complex. Sci. **3**(2), 26–32 (2006)

14. Wang, F.Y.: Toward a paradigm shift in social computing: the acp approach. IEEE Intell. Syst. **22**(5), 65–67 (2007)
15. Zosin, L., Khuller, S.: On directed steiner trees. In: Proceedings of the Thirteenth Annual ACM-SIAM Symposium on Discrete Algorithms, pp. 59–63 (2002)

Microfiles as a Potential Source of Confidential Information Leakage

Oleg Chertov and Dan Tavrov

Abstract Cyber warfares, as well as conventional ones, do not only comprise direct military conflicts involving weapons like DDoS attacks. Throughout their history, intelligence and counterintelligence played a major role as well. Information sources for intelligence can be closed (obtained during espionage) or open. In this chapter, we show that such open information sources as microfiles can be considered a potentially important additional source of information during cyber warfare. We illustrate by using real data based example that ignoring issues concerning providing group anonymity can lead to leakage of confidential information. We show that it is possible to define fuzzy groups of respondents and obtain their distribution using appropriate fuzzy inference system. We conclude the chapter with discussing methods for protecting distributions of crisp as well as fuzzy groups of respondents, and illustrate them by solving the task of providing group anonymity of a fuzzy group of "respondents who can be considered military enlisted members with the high level of confidence."

1 Introduction

With the development of appropriate information technologies, the role of open information sources as a way of obtaining confidential information becomes more and more significant. Such technologies include means of processing very large amounts of data, text and data mining methods, hardware and software based ways of obtaining and analyzing information from different sources, to name just a few.

According to the research conducted by the International Data Corporation [1], about 30 % of digital information in the world need protection, and this number will rise to roughly 40 % by 2020.

O. Chertov (✉) · D. Tavrov
National Technical University of Ukraine, Kyiv Polytechnic Institute,
37 Peremohy Prospekt, 03056 Kyiv, Ukraine
e-mail: chertov@i.ua

D. Tavrov
e-mail: dan.tavrov@i.ua

R.R. Yager et al. (eds.), *Intelligent Methods for Cyber Warfare*,
Studies in Computational Intelligence 563, DOI 10.1007/978-3-319-08624-8_4

A *microfile* is a collection of primary data with information about a sample respondent set. Microfiles are constructed using census or other statistical and sociological surveys data, marketing research data, social networks analysis data etc. With the help of the primary microfile data, as opposed to aggregated ones, one can try to obtain answers to questions not foreseen by the microfile creators.

Microfiles can be considered a potentially important source of information during cyber warfare. With their help, it is possible to violate individual or group anonymity. *Anonymity* of an object means that this object is unidentifiable among the set of certain objects [2]. *Individual* data anonymity is a property of information on a single respondent to be unidentifiable within the data set [3, p. 1]. *Group* data anonymity is a condition, under which [4, p. 11] data features that cannot be distinguished while considering individual records only are protected.

Individual anonymity can be violated even when attributes that uniquely identify microfile respondents are removed. For instance, as L. Sweeney showed experimentally in 2001, 97 % of the voters in the state of Massachusetts possess unique combination of birth date (day, month, and year) and nine-digit ZIP code [5]. Appropriate methods for providing individual anonymity were introduced, such as randomization [6], microaggregation [7], data swapping [8], data matrix factorization [9] and singular value decomposition [10], wavelet transforms (WT) [11], etc.

Group anonymity can be violated by analyzing distributions of the microfile data over certain attribute values. For example, Fig. 1 presents the regional distribution of power engineering specialists obtained from the microfile containing results of the 1999 population census in France [12]. The higher the cylinder, the more specialists live in a particular region. Since the French energy sector primarily consists of nuclear stations (78 % of all energy produced in 2011 [13]), the highest number of power engineering specialists occurred exactly in those regions where nuclear power plants are situated (black cylinders in Fig. 1). Therefore, to conceal the site of any secret nuclear research center, one should distort the real regional distribution of French power engineering specialists.

In the literature, several classes of the task of providing group anonymity (TPGA) are distinguished. The quantity TPGA defined as the task of providing anonymity of a respondent group quantity distribution over the set of values (e.g. military personnel regional distribution) was introduced in [14]. In terms of quantity task, it is impossible to solve the task of concealing concentration distribution of respondents. Such tasks are called concentration group anonymity tasks [15]. One of them is the concentration difference task [16], which implies concealing the distribution of the difference between two concentration distributions. The problems of providing group anonymity are most elaborately covered in [4]. The general methodology of providing group anonymity is presented in [3].

Most of existing methods of solving the TPGA deal with the so called *crisp* groups of respondents, i.e. those ones, to which a particular respondent either belongs or not. The membership in such a group can be determined by analyzing values of one or several specific attributes, e.g., "Occupation," as in the case of French power engineering specialists. To protect anonymity, one can use existing methods, or, in the crudest case, remove appropriate attributes from the microfile.

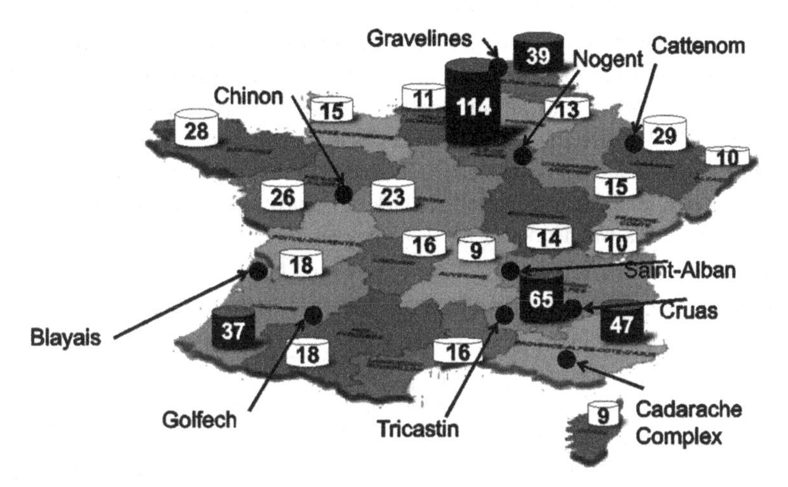

Fig. 1 Regional power engineering specialists distributed according to the microfile containing results of the 1999 French population census

In some cases, however, it is possible to violate group anonymity for a *fuzzy* group of respondents, i.e. the one, to which a respondent can belong only to a certain degree. Whether a respondent belongs to a group, is determined by analyzing values not of some special attributes, but of one or several rather general ones, such as "Age," "Sex," etc. In this chapter, for instance, we discuss a real data based example, in which the fuzzy group consists of people who can be considered military enlisted members with the high level of confidence. The membership in such a group can be deduced from analyzing values of such general purpose attributes as "Age," "Sex," "Black or African American," "Marital Status," "Educational Attainment," and "Hours per Week in 1999." We show that even if group anonymity is provided for a crisp group of military personnel, it might still be possible to retrieve sensitive information from the microfile using the concept of a fuzzy respondent group.

Importance of easily accessed data for retrieving hidden information should not be underestimated. E.g., the famous Russian chemist D. Mendeleyev was able to find out the secret composition of the French powder [17, pp. 353–354] by analyzing annual shipment report of the railroad company that supplied the factory.

Since it is obviously not an option to remove important attributes like "Age," "Sex," etc. from the microfile, appropriate anonymity-providing methods should be developed.

2 General Approach to Violating Group Anonymity

2.1 Group Anonymity Basics

Microdata are the data about respondents (people, households, enterprises etc.). Let **M** denote a (depersonalized) *microfile* with microdata collected in a file of attributive records, which can be viewed as a matrix with rows $u_i, i = \overline{1, \mu}$, corresponding to respondents, and columns $w_j, j = \overline{1, \eta}$, corresponding to attributes.

Let $\mathbf{w}_j$ denote the set of all attribute w_j values. The *vital* set is a subset $\mathbf{V} = \{V_1, V_2, \ldots, V_{l_v}\}$ of the Cartesian product of *vital* attributes. The elements of $\mathbf{V}$ are the *vital value combinations*. $\mathbf{V}$ enables us to define the respondent group.

We can define linguistic variables [18] L_i corresponding to each microfile attribute. Universes of discourse for L_i consist of the ith microfile attribute values. Values of L_i belong to its term-set $T(L_i)$. The *generalized vital set* $\tilde{\mathbf{V}} = \left\{\tilde{V}_1, \tilde{V}_2, \ldots, \tilde{V}_{l_{\tilde{v}}}\right\}$ is a subset of the Cartesian product of term-sets of all the linguistic variables corresponding to the vital microfile attributes. The elements of $\tilde{\mathbf{V}}$ are the *generalized vital value combinations*.

Let the *parameter set* $\mathbf{P} = \{P_1, P_2, \ldots, P_{l_p}\}$ be a subset of values corresponding to the *parameter* microfile attribute, which is not vital. The elements of $\mathbf{P}$ are *parameter values*. They enable us to split the microfile $\mathbf{M}$ into *parameter submicrofiles* $\mathbf{M}_1, \ldots, \mathbf{M}_{l_p}$ with $\mu_j, j = \overline{1, l_p}$, records in them.

We will denote by $G(\mathbf{V}, \mathbf{P})$ the *group*, i.e. the set consisting of $\mathbf{V}$ and $\mathbf{P}$. We will denote by $\tilde{G}\left(\tilde{\mathbf{V}}, \mathbf{P}\right)$ the *fuzzy group*, i.e. the set consisting of $\tilde{\mathbf{V}}$ and $\mathbf{P}$.

We can determine the membership grade $\mu_{\tilde{G}}(u_i)$ of every respondent $u_i, i = \overline{1, \mu}$, in $\tilde{G}$. We denote the set of all grades by $\tilde{\mathbf{M}}_{\tilde{G}} = \left\{\mu_{\tilde{G}1}, \mu_{\tilde{G}2}, \ldots, \mu_{\tilde{G}q}\right\}$.

By *goal representation* $\Omega\left(\mathbf{M}, \tilde{G}\right)$ of $\mathbf{M}$ with respect to $\tilde{G}$ we define a dataset of arbitrary structure representing features of $\tilde{G}$ in a way proper for analyzing.

2.2 An Overview of Goal Representations

2.2.1 Goal Signals

The goal representation which is frequently used in the literature is the *goal signal* $\theta = \left(\theta_1, \theta_2, \ldots, \theta_{l_p}\right)$, which reflects such potentially sensitive properties of a group as [4, p. 77] extreme values, statistical features, etc. For the sake of simplicity, we assume that each goal signal value corresponds to one parameter submicrofile $\mathbf{M}_k$, $k = \overline{1, l_p}$. The goal signal may be treated as a function $\theta = \theta(\mathbf{P}, \mathbf{V})$ of parameter values $\mathbf{P}$ and a term $\mathbf{V}$ defining the set of vital value combinations, with each $\theta_k = \theta(P_k, \mathbf{V})$.

In the literature, there are distinguished several kinds of goal signals. Among the more popular ones is the *quantity signal* $\mathbf{q} = \left(q_1, q_2, \ldots, q_{l_p}\right)$ introduced in [14]. The elements $q_k, k = \overline{1, l_p}$, stand for the quantities of respondents with a particular parameter value P_k and values of vital attributes belonging to $\mathbf{V}$.

In many cases, absolute quantities are not representative, and should be replaced with the relative ratios. In these cases, the *concentration signal* $\mathbf{c} = \left(c_1, c_2, \ldots, c_{l_p}\right)$ introduced in [19] is used instead of the quantity one. The elements $c_k, k = \overline{1, l_p}$, are obtained by dividing q_k by the overall number of respondents in a specified parameter submicrofile:

$$c_k = \frac{q_k}{\mu_k}, \quad k = \overline{1, l_p}\,. \tag{1}$$

Vital attributes enable us to split each parameter submicrofile $\mathbf{M}_j$ into *vital submicrofiles* $\mathbf{M}_k^{(G)}$, $k = \overline{1, l_p}$, which contain all microfile records with a parameter value P_k and values of vital attributes belonging to $\mathbf{V}$, and *non-vital submicrofiles* $\mathbf{M}_k^{\overline{(G)}}$, $k = \overline{1, l_p}$, which contain the microfile records with a parameter value P_k and values of vital attributes not belonging to $\mathbf{V}$. Each submicrofile $\mathbf{M}_k^{(G)}$ contains q_k records, each submicrofile $\mathbf{M}_k^{\overline{(G)}}$ contains $(\mu_k - q_k)$ records.

2.2.2 Goal Surfaces

When we need to deal with the anonymity of fuzzy groups, the goal signal is not sufficient to embrace all the information about the microfile respondents. We need to introduce the generalization of the goal signal called the *goal surface* Θ. It can be treated as a function $\Theta = \Theta\left(\mathbf{P}, \tilde{\mathbf{M}}_{\tilde{G}}, \tilde{\mathbf{V}}\right)$ of parameter values $\mathbf{P}$, membership grades of a particular respondent in the fuzzy group $\tilde{\mathbf{M}}_{\tilde{G}}$, and a term $\tilde{\mathbf{V}}$ defining the set of generalized vital value combinations, with each $\Theta_{jk} = \Theta\left(P_k, \mu_{\tilde{G}j}, \tilde{\mathbf{V}}\right)$.

There can be distinguished two kinds of goal surfaces, a *quantity surface* $\mathbf{Q}$ and a *concentration surface* $\mathbf{C}$. To build $\mathbf{Q}$, one needs to calculate the membership grades $\mu_{\tilde{G}}(u_i)$ in the fuzzy group $\tilde{G}$ for every microfile respondent $u_i \in \tilde{G}$, that is, every respondent whose vital attribute values belong to the universes of discourse of appropriate linguistic variables. This can be carried out by applying a properly designed fuzzy inference system (FIS). In this chapter, we will use the Mamdani FIS [20], which typically consists of several input and output variables, the fuzzification module, the fuzzy inference engine, the fuzzy rule base, and the defuzzification module [21]. Each rule j in the rule base is in the form

$$\text{if } x_1 \text{ is } A_{1j}, \ldots, x_n \text{ is } A_{nj}, \text{ then } y \text{ is } B_j,$$

where $x_1, \ldots, x_n$, and y are input and output variables, respectively; $A_{1j}, \ldots, A_{nj}$, and B_j are values (fuzzy sets) of input and output variables. The inference engine works on the basis of compositional rule of inference (CRI) [22].

To build a FIS, one needs to accomplish the following steps:

1. Choose input, output variables, define appropriate membership functions (MFs).
2. Construct the fuzzy rule base.
3. Choose methods of fuzzy intersection, implication, and aggregation.
4. Choose appropriate defuzzification algorithm.

For the FIS for building **Q**, one needs to take linguistic variables L_j as the input ones. The output variable should represent the membership grade in $\tilde{G}$ of a certain respondent. The generalized vital value combinations should represent the antecedents of the fuzzy rules. In some cases, the problem at hand can impose crisp restrictions to be considered aside from those in the fuzzy rule base.

Having calculated membership grades using FIS, one needs to count the number of respondents with particular parameter values and membership grades in $\tilde{G}$:

$$Q_{jk} = \left| \left\{ u_i \left| z_{iw_p} = P_k, \mu_{\tilde{G}}(u_i) = \mu_{\tilde{G}j} \right. \right\} \right| . \quad (2)$$

However, in most cases exact values of membership grades are irrelevant and do not shed much light on the distribution to be analyzed. The numbers of membership grades belonging to certain intervals can provide information that is much more useful. We need to split $\tilde{\mathbf{M}}_{\tilde{G}}$ into intervals $\Delta_{\tilde{\mathbf{M}}s}$, $s = \overline{1,r}$, and count the number of respondents with membership grades belonging to them:

$$Q_{sk} = \left| \left\{ u_i \left| z_{iw_p} = P_k, \mu_{\tilde{G}}(u_i) \in \Delta_{\tilde{\mathbf{M}}s} \right. \right\} \right| . \quad (3)$$

It is wise to build (3) only for intervals with high values of membership grades.

Using (3), each parameter submicrofile $\mathbf{M}_k$ may be split into *vital submicrofiles of grade* $\Delta_{\tilde{\mathbf{M}}s}$ $\mathbf{M}_k^{\left(\tilde{G}_{\Delta_{\tilde{\mathbf{M}}s}}\right)}$ and *non-vital submicrofiles* $\mathbf{M}_k^{\left(\tilde{G}_{\Delta_{\tilde{\mathbf{M}}0}}\right)}$, $k = \overline{1,l_p}$, $s = \overline{1,r}$. Vital submicrofiles contain the microfile records u_i with the parameter value P_k and $\mu_{\tilde{G}}(u_i) \in \Delta_{\tilde{\mathbf{M}}s}$. Non-vital submicrofiles contain the microfile records u_i with the parameter value P_k and the membership grade in $\tilde{G}$ not belonging to any interval. Each vital submicrofile of a certain grade $\mathbf{M}_k^{\left(\tilde{G}_{\Delta_{\tilde{\mathbf{M}}s}}\right)}$ contains Q_{sk} records, each non-vital submicrofile $\mathbf{M}_k^{\left(\tilde{G}_{\Delta_{\tilde{\mathbf{M}}0}}\right)}$ contains $\left(\mu_k - \sum_{s=1}^{r} Q_{sk}\right)$ records.

In cases when the absolute numbers of respondents are not representative, it is better to use the concentration surface **C** with the elements

$$C_{sk} = \frac{Q_{sk}}{\mu_k}, \quad k = \overline{1,l_p} . \quad (4)$$

2.3 The General Approach to Creating the FIS for Violating Anonymity of a Fuzzy Group

In Sect. 2.2.2, we outlined several steps that need to be accomplished to create a FIS for building a quantity surface **Q** for a fuzzy respondent group $\tilde{G}$. However, when the task of violating anonymity of a fuzzy group is concerned, the group whose anonymity needs to be violated is not precisely defined. For example, when the task is to violate anonymity of a fuzzy group of "respondents who can be considered military with the high level of confidence," it is not clear what vital attributes should be taken, and what values the corresponding linguistic variables have.

In general, to build a FIS for classifying respondents as belonging to a given fuzzy group with a certain grade, one needs to proceed according to such steps:

1. According to external statistical data and/or expert judgment, determine the microfile attributes, which can be used in combination to describe respondents belonging to the fuzzy group with a *high* membership grade.
2. Split, if necessary, the values of these attributes into meaningful intervals, and obtain the distributions over the values of each attribute for the respondents belonging to the fuzzy group with a high membership grade.
3. Define the ranges of the values of these attributes, outside which respondents are considered (in a crisp way) as not belonging to the group.
4. Exclude from the set of the attributes defined on step 1 those ones, distribution over which is sufficiently close to the uniform one.
5. Exclude from the set of the attributes defined on step 4 those ones, distribution over which for the respondents belonging to the fuzzy group with a high membership grade is sufficiently close to the distribution for the respondents at large.
6. According to external statistical data and/or expert judgment, determine the microfile attributes which can be used in combination to describe respondents belonging to the fuzzy group with a *low* membership grade, and add them to the set defined on step 5.
7. Split, if necessary, the values of the newly added attributes into meaningful intervals, and obtain the distributions over the values of each attribute for the respondents belonging to the fuzzy group with a low membership grade.
8. Define the ranges of the values of the newly added attributes inside which respondents are considered (in a crisp way) as not belonging to the group.
9. Define the values of all the input variables of the FIS. Variables correspond to some or all of the attributes from the set defined on step 6, 10. Judging from external statistical data and/or expert judgment, define values of the output linguistic variable, and construct a meaningful set of fuzzy rules.
10. Choose appropriate methods of fuzzy union, intersection, implication, aggregation, and defuzzification.

3 Practical Results

3.1 Violating Anonymity of the Crisp Group of Military Personnel

In this section, we will show how anonymity of the crisp group of respondents can be violated using publicly available microfile data. In particular, we want to show how the potential sites of the military bases can be determined using the regional distribution of the military personnel. For our purpose, we used the 5-Percent Public

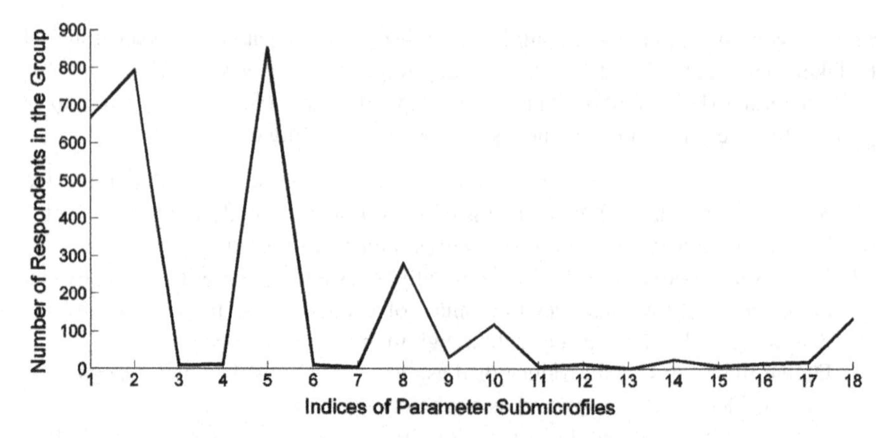

Fig. 2 The quantity signal obtained for the crisp group of active duty military personnel

Use Microdata Sample Files from the U. S. Census Bureau [23] corresponding to the 2000 U. S. Census microfile data on the state of Florida.

In accordance with Sect. 2.1, we took "Place of Work Super-PUMA" (where PUMA stands for "Public Use Microdata Area") as the parameter attribute. We took codes of all the statistical areas of Florida, i. e. each 10th value in the range 12010–12180, as the parameter values. With the help of these parameter values, the microfile can be split into 18 parameter submicrofiles, $\mathbf{M}_1, \ldots, \mathbf{M}_{18}$, with the total number of respondents in each of them, $\mu_1, \ldots, \mu_{18}$, given as follows:

$$\boldsymbol{\mu} = (\mu_1, \ldots, \mu_{18}) = (8375, 10759, 9683, 10860, 25753, 10153, 6916, 50680, 39892, 10453, 9392, 9016, 8784, 11523, 11158, 24124, 30666, 46177) \ . \quad (5)$$

We took "Military Service" as a vital attribute. Its value "1," standing for "Active Duty," was chosen as the only vital value. Thus, we have defined the *group G* of active duty military personnel distributed over statistical areas of Florida. The quantity signal $\mathbf{q}$ is shown in Fig. 2.

As we see, there are three extreme values in the quantity signal. More precisely, above 75 % of all the active duty military personnel work in the first, second, and fifth statistical areas. Such disproportionate quantities may point to the sites of military bases. Thus, anonymity can be violated relatively easily for a crisp group.

3.2 Violating Anonymity of the Fuzzy Group of Military Enlisted Members

In the previous section, we showed that anonymity of a crisp group of military personnel can be violated by analyzing extreme values of an appropriate quantity signal. One of the crudest ways to prevent such violation is to remove completely from the microfile the "Military Service" attribute. However, as we show in this section, anonymity can also be violated for a fuzzy group $\tilde{G}$ of "respondents who can be considered military enlisted members with the high level of confidence."

To construct a quantity surface, we need to build appropriate FIS. We decided to use the demographic analysis of the military personnel conducted by the Office of the Deputy under Secretary of Defense [24] in 2011 and updated in November 2012 as our main source of relevant statistical data. We also used certain expert judgments, e.g. that the military enlisted members in majority tend to work more than 40 h per week. We then followed along the steps outlined in Sect. 2.3:

1. We chose microfile attributes "Age," "Sex," "Black or African American," "Marital Status," "Educational Attainment," and "Hours per Week in 1999" as the ones that can be used in combination to describe respondents belonging to our fuzzy group with a high membership grade.
2. According to [24], the distributions of the active duty enlisted members over the values of the chosen attributes are as follows:
   - 49.3 % are 25 years of age or younger, 22.8 % are 26–30 years of age, 13.1 % are 31 to 35 years of age, 9.2 % are 36–40 years of age, 5.5 % are 41 years of age or older;
   - 85.8 % are male, and 14.2 % are female;
   - 16.9 % are Black or African American, whereas 83.1 % are not;
   - 54.0 % are married, 41.3 % never married, and 4.6 % are divorced;
   - 93.4 % have less than Bachelor's Degree, 5.3 % have Bachelor's or Advanced Degree (other 1.3 % either have no High School diploma, or their educational level is unknown).
3. Having analyzed information presented in [24], we decided to consider respondents whose are younger than 18 years of age or older than 45 years of age as those ones who do not belong to our fuzzy group in a crisp sense.
4. We excluded from the set of supposedly vital attributes "Marital Status" because it provides the distribution, which is very close to the uniform one.
5. We decided to skip this step since all attributes provide significant information.
6. Using expert judgment that every enlisted member has to exhibit a certain level of English, we added the attribute "English Ability" to our set of attributes.
7. We decided to skip this step as not necessary.
8. We decided to choose "English Ability" values "3" and "4" (standing for "Not well" and "Not at all," respectively) as those ones which correspond to respondents who do not belong to the fuzzy group in the crisp sense.
9. We decided to take five input variables for the FIS, namely, "Age," "Sex," "Black or African American," "Educational Attainment," and "Hours per Week." Values of "Age," "Sex," and "Hours per Week" are presented in Figs. 3–5, respectively (codes for the "Educational Attainment" variable are given in Table 1). Variable "Sex" has two values, "Male" and "Female," with the MFs

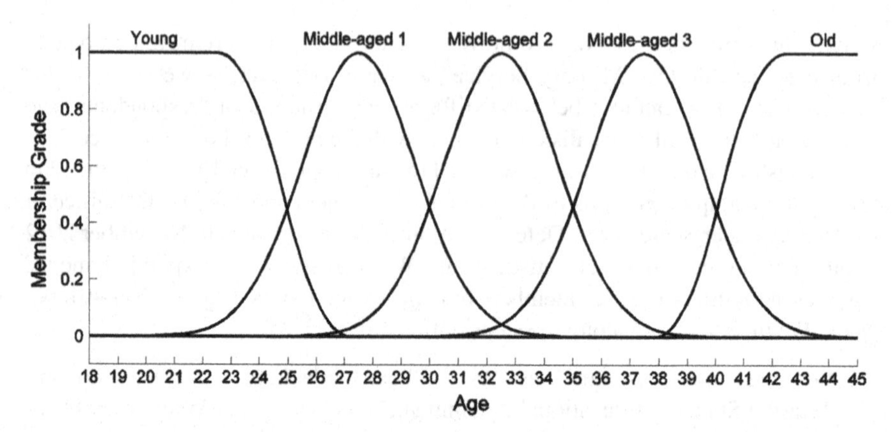

Fig. 3 Membership functions for the "Age" variable

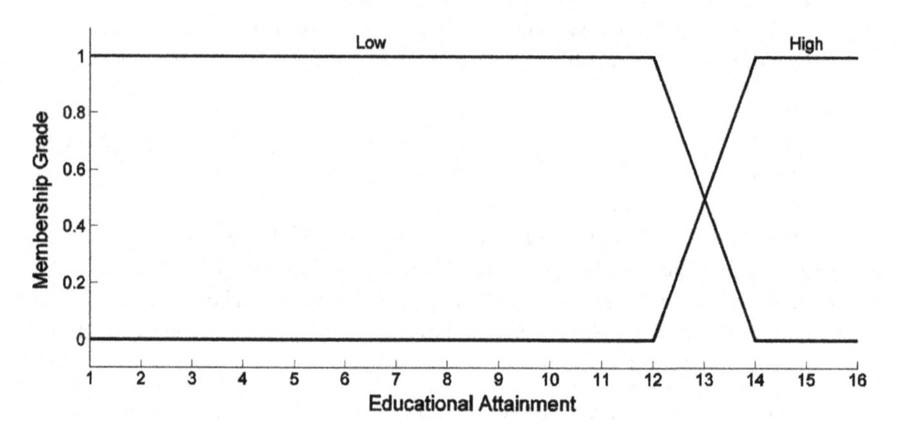

Fig. 4 Membership functions for the "Educational Attainment" variable

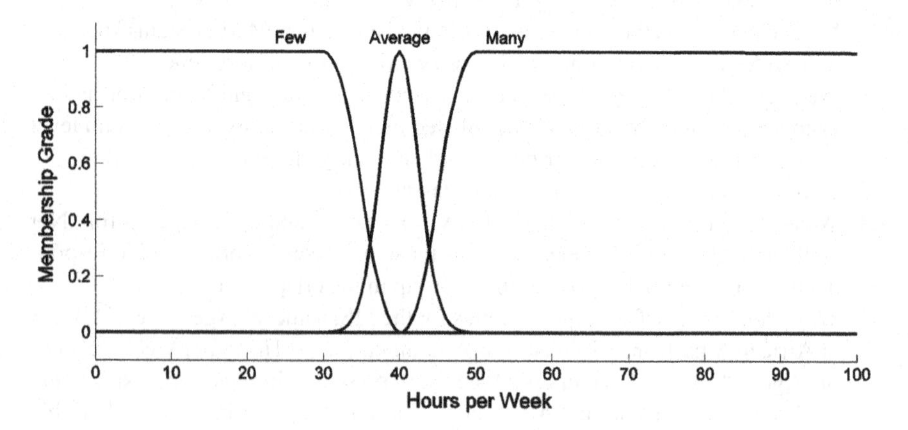

Fig. 5 Membership functions for the "Hours per Week" variable

Table 1 Codes for the "Educational Attainment" variable

Code	Description	Code	Description
1	No schooling completed	9	High school graduate
2	Nursery school to 4th grade	10	Some college, but less than 1 year
3	5th grade or 6th grade	11	One or more years of college, no degree
4	7th grade or 8th grade	12	Associate degree
5	9th grade	13	Bachelor's degree
6	10th grade	14	Master's degree
7	11th grade	15	Professional degree
8	12th grade, no diploma	16	Doctorate degree

$$\mu_{\text{Male}}(x) = \begin{cases} 1, & x = 1 \\ 0, & x \neq 1 \end{cases}, \quad \mu_{\text{Female}}(x) = \begin{cases} 1, & x = 2 \\ 0, & x \neq 2 \end{cases},$$

where "1" is the microfile attribute value standing for "Male," and "2" is the value standing for "Female." Variable "Black or African American" has two values, "No" and "Yes," with the MFs

$$\mu_{\text{No}}(x) = \begin{cases} 1, & x = 0 \\ 0, & x \neq 0 \end{cases}, \quad \mu_{\text{Yes}}(x) = \begin{cases} 1, & x = 1 \\ 0, & x \neq 1 \end{cases},$$

where "0" is the value standing for "Not Black," and "1" is the value standing for "Black."

10. Values of the output variable "Membership in a Fuzzy Group" are presented in Fig. 6. The set of rules is presented in Table 2. These rules were largely determined by analyzing [24]. For instance, if almost half of all enlisted members are young, many of them work more than 40 h per week, and the absolute majority are "White," "Male," and "Lowly educated," then respondents with such characteristics can be considered enlisted members with "high" membership grade. For less obvious vital value combinations we used expert judgment.
11. We decided to take *maximum* as fuzzy union and aggregation, *minimum* as fuzzy intersection and implication, and *centroid method* for defuzzification.

Using the FIS constructed in accordance with these 11 steps, we calculated membership grades for all the respondents in the microfile that belong to the group in a crisp sense. We decided to choose the following intervals to construct the quantity surface **Q** (3): $\Delta_{\tilde{\mathbf{M}}1} = (0.5; 0.6]$, $\Delta_{\tilde{\mathbf{M}}2} = (0.6; 0.7]$, $\Delta_{\tilde{\mathbf{M}}3} = (0.7; 0.8]$, $\Delta_{\tilde{\mathbf{M}}4} = (0.8; 0.9]$.

The quantity surface does not provide necessary information for violating anonymity of the fuzzy group. To determine potential sites of military bases, it is better to use the concentration surface **C** (4) obtained using (5). Surfaces **Q** and **C** are given below (we present all the results with three decimal numbers; the calculations throughout the chapter had been carried out with higher precision):

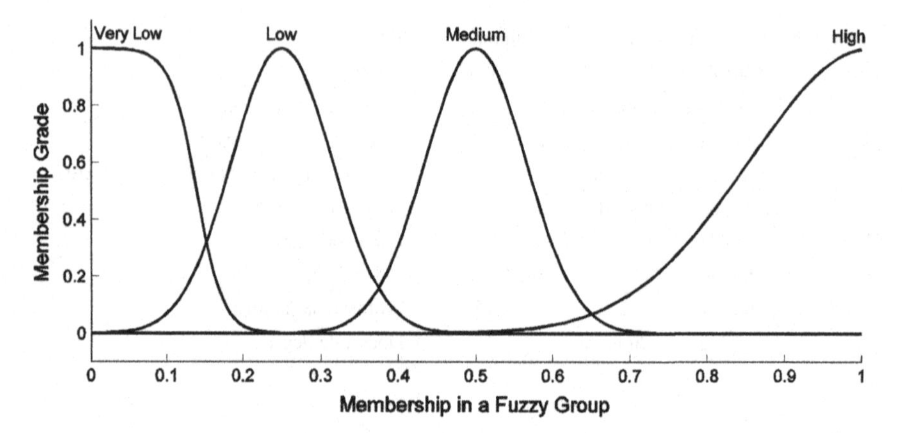

Fig. 6 Membership functions for the "Membership in a Fuzzy Group" variable

$$
\mathbf{Q}^T = \begin{pmatrix}
204 & 23 & 56 & 328 \\
218 & 54 & 94 & 377 \\
159 & 12 & 46 & 183 \\
179 & 29 & 53 & 284 \\
438 & 97 & 151 & 730 \\
160 & 28 & 41 & 211 \\
116 & 25 & 34 & 170 \\
834 & 159 & 238 & 1099 \\
745 & 142 & 226 & 900 \\
144 & 23 & 45 & 192 \\
155 & 35 & 49 & 276 \\
144 & 31 & 48 & 191 \\
150 & 33 & 46 & 183 \\
194 & 30 & 53 & 245 \\
176 & 31 & 60 & 238 \\
330 & 57 & 95 & 374 \\
433 & 62 & 122 & 439 \\
619 & 86 & 192 & 738
\end{pmatrix}, \quad
\mathbf{C}^T = \begin{pmatrix}
0.024 & 0.003 & 0.007 & 0.039 \\
0.020 & 0.005 & 0.009 & 0.035 \\
0.016 & 0.001 & 0.005 & 0.019 \\
0.016 & 0.003 & 0.005 & 0.026 \\
0.017 & 0.004 & 0.006 & 0.028 \\
0.016 & 0.003 & 0.004 & 0.021 \\
0.017 & 0.004 & 0.005 & 0.025 \\
0.016 & 0.003 & 0.005 & 0.022 \\
0.019 & 0.004 & 0.006 & 0.023 \\
0.014 & 0.002 & 0.004 & 0.018 \\
0.017 & 0.004 & 0.005 & 0.029 \\
0.016 & 0.003 & 0.005 & 0.021 \\
0.017 & 0.004 & 0.005 & 0.021 \\
0.017 & 0.003 & 0.005 & 0.021 \\
0.016 & 0.003 & 0.005 & 0.021 \\
0.014 & 0.002 & 0.004 & 0.016 \\
0.014 & 0.002 & 0.004 & 0.014 \\
0.013 & 0.002 & 0.004 & 0.016
\end{pmatrix}.
$$

The sum of rows of **C** is shown in Fig. 7 along with the superimposed quantity signal **q** obtained in Sect. 3.1 (to fit the scale, we normalized both vectors by dividing them by their maximal values). By analyzing extreme values obtained from the concentration surface **C**, we can determine the same statistical areas we determined in Sect. 3.1. It is worth noting that extreme value in the element 11 was not present in **q**, however, all the extremes that actually *were* present in **q** have been successfully determined, even though the attribute "Military Service" was removed from the microfile.

Thus, we successfully managed to violate anonymity for the fuzzy group of "respondents who can be considered military enlisted members with the high level

Table 2 Fuzzy rule base for the FIS in example

Hours per week	Educational attainment	Sex	Black or Afr. Amer.	Age				
				Young	Mid. -aged 1	Mid. -aged 2	Mid. -aged 3	Old
Few	Low	Male	Yes	VL	VL	VL	VL	VL
			No	L	VL	VL	VL	VL
		Female	Yes	VL	VL	VL	VL	VL
			No	VL	VL	VL	VL	VL
	High	Male	Yes	VL	VL	VL	VL	VL
			No	VL	VL	VL	VL	VL
		Female	Yes	VL	VL	VL	VL	VL
			No	VL	VL	VL	VL	VL
Average	Low	Male	Yes	L	VL	VL	VL	VL
			No	H	L	L	L	L
		Female	Yes	VL	VL	VL	VL	VL
			No	L	VL	VL	VL	VL
	High	Male	Yes	VL	VL	VL	VL	VL
			No	VL	VL	VL	VL	VL
		Female	Yes	VL	VL	VL	VL	VL
			No	VL	VL	VL	VL	VL
Many	Low	Male	Yes	L	VL	VL	VL	VL
			No	H	M	M	M	L
		Female	Yes	VL	VL	VL	VL	VL
			No	L	VL	VL	VL	VL
	High	Male	Yes	VL	VL	VL	VL	VL
			No	VL	L	L	L	L
		Female	Yes	VL	VL	VL	VL	VL
			No	VL	VL	VL	VL	VL

of confidence." In other words, even if group anonymity is provided for a crisp group of military personnel (Sect. 3.1), it is still possible to retrieve sensitive information from the microfile using the concept of a fuzzy respondent group.

4 Providing Anonymity for Crisp and Fuzzy Respondent Groups

4.1 The Generic Scheme of Providing Group Anonymity

The *task of providing group anonymity* in a microfile is the task of modifying it for a group $\tilde{G}(\tilde{\mathbf{V}}, \mathbf{P})$, so that sensitive (for the task solved) data features become confided. The generic scheme of providing group anonymity goes as follows:

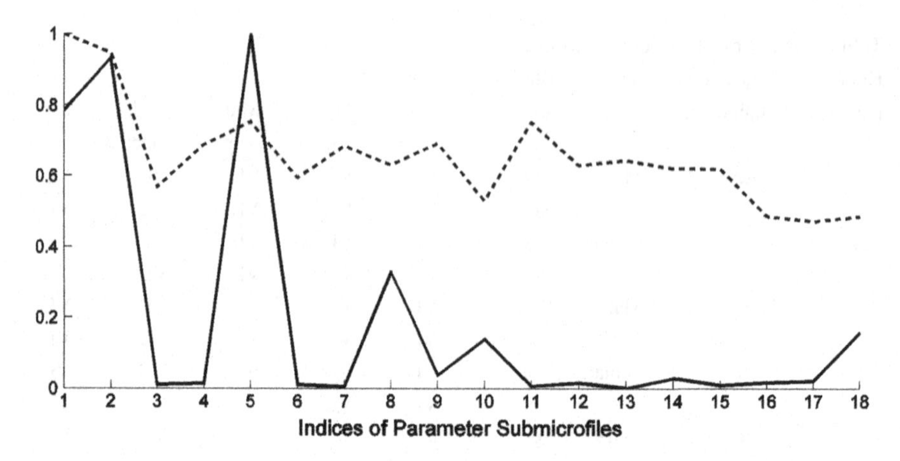

Fig. 7 The quantity signal (*solid line*) and the sum of the rows of the concentration surface (*dashed line*) for the example

1. Prepare a depersonalized microfile $\mathbf{M}$.
2. Define groups $\tilde{G}_i(\tilde{\mathbf{V}}_i, \mathbf{P})$, $i = \overline{1, k}$, representing respondents to be protected.
3. For each i from 1 to k:
   - choose data *goal representation* $\Omega_i(\mathbf{M}, \tilde{G}_i)$ representing particular features of the group in a way appropriate for its further modification;
   - define the *goal mapping function* $\Upsilon_i : \mathbf{M} \to \Omega_i(\mathbf{M}, \tilde{G}_i)$ and obtain the goal representation;
   - define the *modifying functional* $\Xi_i : \Omega_i(\mathbf{M}, \tilde{G}_i) \to \Omega_i^*(\mathbf{M}, \tilde{G}_i)$ and obtain the *modified goal representation*;
   - define the *inverse goal mapping function* $\Upsilon_i^{-1} : \Omega_i^*(\mathbf{M}, \tilde{G}_i) \to \mathbf{M}^*$ and obtain the modified microfile.
4. Prepare the modified microfile $\mathbf{M}^*$ for publishing.

The first three operations at step 3 constitute the *first stage* of solving the TPGA. Obtaining the modified microfile using the inverse goal mapping function at step 3 is the only operation constituting the *second stage* of solving the TPGA.

4.2 Wavelet Transforms as the Modifying Functional

4.2.1 One-Dimensional Wavelet Transforms as the Modifying Functional for the Goal Signals

We will introduce wavelet transforms (WT) to the extent necessary for applying them to modifying the goal signal. For more information on wavelets, consult [25]. For a detailed discussion of applying WT to solving the TPGA, refer to [4].

Let $\mathbf{h} = (h_1, h_2, \ldots, h_n)$ and $\mathbf{l} = (l_1, l_2, \ldots, l_n)$ denote the *high-frequency* and *low-frequency wavelet filter*, respectively. To perform the goal signal one-level wavelet decomposition, we need to perform the following operations:

$$\mathbf{a}_1 = \theta *_{\downarrow 2} \mathbf{l}\,, \quad \mathbf{d}_1 = \theta *_{\downarrow 2} \mathbf{h}\,, \tag{6}$$

where $*_{\downarrow 2}$ denotes the convolution with the follow-up dyadic downsampling, array $\mathbf{a}_1$ ($\mathbf{d}_1$) consists of level one approximation (detail) coefficients.

To simplify the notation, let us introduce the following operations:

$$\mathbf{z} = \left(\left(\theta *_{\downarrow 2} \mathbf{f}\right) \underbrace{*_{\downarrow 2} \mathbf{f}\right) \ldots *_{\downarrow 2} \mathbf{f}}_{k-1 \text{ times}} = \prod_{i=1}^{k} \left(\theta *_{\downarrow 2} \mathbf{f}\right)\,, \tag{7}$$

$$\mathbf{z} = \left(\left(\theta *_{\uparrow 2} \mathbf{f}\right) \underbrace{*_{\uparrow 2} \mathbf{f}\right) \ldots *_{\uparrow 2} \mathbf{f}}_{k-1 \text{ times}} = \prod_{i=1}^{k} \left(\theta *_{\uparrow 2} \mathbf{f}\right)\,, \tag{8}$$

where $*_{\uparrow 2}$ denotes the dyadic upsampling with the follow-up convolution.

To obtain decomposition coefficients of arbitrary level k, we need to perform (6) with the goal signal replaced by the approximation coefficients of level $k-1$:

$$\mathbf{a}_k = \prod_{i=1}^{k} \left(\theta *_{\downarrow 2} \mathbf{l}\right)\,, \quad \mathbf{d}_k = \left(\prod_{i=1}^{k-1} \left(\theta *_{\downarrow 2} \mathbf{l}\right)\right) *_{\downarrow 2} \mathbf{h}\,. \tag{9}$$

To obtain the goal signal approximation and details of level k, we need to perform the following operations:

$$\mathbf{A}_k = \prod_{i=1}^{k} \left(\mathbf{a}_k *_{\uparrow 2} \mathbf{l}\right)\,, \quad \mathbf{D}_k = \prod_{i=1}^{k-1} \left(\left(\mathbf{d}_k *_{\uparrow 2} \mathbf{h}\right) *_{\uparrow 2} \mathbf{l}\right)\,. \tag{10}$$

The goal signal can be decomposed into the following sum:

$$\theta = \mathbf{A}_k + \sum_{i=1}^{k} \mathbf{D}_i\,. \tag{11}$$

Wavelet approximation $\mathbf{A}_k$ of the signal represents its smoothed version. Wavelet details of all levels $\mathbf{D}_i$, $i = \overline{1,k}$, represent high-frequency fluctuations in it.

To protect such properties of the goal signal as its extreme values, two different approaches may be proposed [4]. According to the *extremum transition approach*, the goal signal has to be modified in such way that its new extreme values differ from the initial ones. The other approach called the *Ali Baba's wife approach* implies not eliminating existing extreme values but adding several new alleged ones, which makes it impossible to discriminate between real and fake extreme values.

Aside from protecting signal properties, it is important to guarantee that the overall data utility is not reduced very much. WT can be successfully applied in order to achieve both goals. To mask extreme values, we can modify the goal signal approximation, whereas leaving the signal details intact (or modifying them at most proportionally) preserves important properties of the initial data.

However, mere modifying the approximation will not do much good, because internal structure of the signal will be tampered with. The better way of modifying the signal approximation is to modify its approximation coefficients. To do this, we need to know the explicit dependence of the approximation values on the approximation coefficients. This dependence can be retrieved from the so called wavelet reconstruction matrix (WRM) $\mathbf{M}_{rec}$ introduced in [14]:

$$\mathbf{A}_k = \mathbf{M}_{rec} \cdot \mathbf{a}_k \,. \tag{12}$$

With the help of the WRM, we can represent each approximation element as the linear combination of approximation coefficients and $\mathbf{M}_{rec}$ elements. The latter ones are dependent on wavelet filter elements and the size of the goal signal. Using (12), we can construct restrictions for the linear programming problem, whose solution yields modified approximation coefficients $\tilde{\mathbf{a}}_k$. These coefficients can be used to obtain modified approximation $\tilde{\mathbf{A}}_k$ according to (10).

Using (11), we can obtain the signal $\breve{\theta} = \tilde{\mathbf{A}}_k + \sum_{i=1}^{k} \mathbf{D}_i$. If any of its elements are negative, we need to add to the signal a sufficiently great number γ to make all the signal entries non-negative. To preserve the mean value of the goal signal after this operation, we need to multiply it by an appropriate coefficient:

$$\theta^* = \left(\breve{\theta} + \gamma\right) \cdot \frac{\sum\limits_{k=1}^{l_p} \theta_k}{\sum\limits_{k=1}^{l_p} \left(\breve{\theta}_k + \gamma\right)} \,. \tag{13}$$

When the goal signal is the concentration signal, it is necessary to apply (13) not only to the signal itself, but to the corresponding quantity signal as well, so that the overall number of respondents in the microfile does not change.

4.2.2 Separable Two-Dimensional Wavelet Transforms as the Modifying Functional for the Goal Surfaces

To perform one-level wavelet decomposition of the goal surface Θ, we need to carry out the following calculations:

$$\mathbf{a}_1 = \overbrace{\underbrace{\left(\Theta *_{\downarrow 2} \mathbf{l}\right)}_{\text{row-wise}} *_{\downarrow 2} \mathbf{l}}^{\text{column-wise}} , \mathbf{d}_{h1} = \overbrace{\underbrace{\left(\Theta *_{\downarrow 2} \mathbf{l}\right)}_{\text{row-wise}} *_{\downarrow 2} \mathbf{h}}^{\text{column-wise}} ,$$
$$\mathbf{d}_{v1} = \overbrace{\underbrace{\left(\Theta *_{\downarrow 2} \mathbf{h}\right)}_{\text{row-wise}} *_{\downarrow 2} \mathbf{l}}^{\text{column-wise}} , \mathbf{d}_{d1} = \overbrace{\underbrace{\left(\Theta *_{\downarrow 2} \mathbf{h}\right)}_{\text{row-wise}} *_{\downarrow 2} \mathbf{h}}^{\text{column-wise}} . \quad (14)$$

These operations are the generalized versions of (6). However, instead of one array of detail coefficients, we obtain three of them, i.e. *horizontal detail coefficients* $\mathbf{d}_{h1}$, *vertical detail coefficients* $\mathbf{d}_{v1}$, and *diagonal detail coefficients* $\mathbf{d}_{d1}$.

The goal surface can be decomposed into the sum of its approximation and three types of details:

$$\Theta = \mathbf{A}_k + \sum_{i=1}^{k} \mathbf{D}_{hi} + \sum_{i=1}^{k} \mathbf{D}_{vi} + \sum_{i=1}^{k} \mathbf{D}_{di} . \quad (15)$$

To modify the goal surface using WT, we can use the method similar to the one described in Sect. 4.2.1. Each element of the two-dimensional approximation can be presented as the linear combination of the approximation coefficients and some values dependent on the wavelet filter elements and the size of the goal surface. This representation is useful for constructing restrictions of a linear programming problem, whose solution yields modified approximation coefficients $\tilde{\mathbf{a}}_k$.

Applying (15), we can obtain the surface $\breve{\Theta} = \tilde{\mathbf{A}}_k + \sum_{i=1}^{k} \mathbf{D}_{hi} + \sum_{i=1}^{k} \mathbf{D}_{vi} + \sum_{i=1}^{k} \mathbf{D}_{di}$, which can be amended if necessary using the procedure described in Sect. 4.2.1 yielding the modified goal surface Θ^*:

$$\Theta^* = \left(\breve{\Theta} + \gamma\right) \cdot \frac{\sum_{s=1}^{r} \sum_{k=1}^{l_p} \Theta_{sk}}{\sum_{s=1}^{r} \sum_{k=1}^{l_p} \left(\breve{\Theta}_{sk} + \gamma\right)} . \quad (16)$$

When the goal surface is the concentration surface, it is necessary to apply (16) not only to the surface itself, but to the corresponding quantity surface as well. In the latter case, the surface needs to be rounded afterwards. If the sum of all the surface elements differs from the initial one after such rounding by a small number ϵ, it is permissible to add ϵ to the greatest element of the rounded surface.

4.3 Inverse Goal Mapping Functions for Minimizing Microfile Distortion

4.3.1 Inverse Goal Mapping Functions for Crisp Respondent Groups

Modifying the microfile in order to adjust it to the modified goal representation by applying inverse goal mapping function implies introducing into it a certain level of distortion, whose overall amount has to be minimized. In general, it is a good practice to modify the microfile by applying the inverse goal mapping function to the modified quantity signal (or surface), even when the goal representation is the concentration signal (or surface). In this section, we will assume that the inverse goal mapping function is applied to the modified quantity signal $\mathbf{q}^*$.

To modify the microfile in order to adjust it to the modified quantity signal $\mathbf{q}^*$, one needs to alter values of the parameter attribute for certain respondents. To make sure that the number of respondents in each parameter submicrofile remains the same, respondents should be altered in pairs. One of the respondents in a pair has to belong to the group G, whereas the other one has to lie outside the group. We call this operation the *swapping of the respondents between the submicrofiles* (SRBS).

Let *influential attributes* [3] be the ones, whose distribution plays a great role for researchers. To minimize overall microfile distortion, one needs to search for pairs of respondents to swap between submicrofiles that are close to each other. To determine how "close" respondents are, one can use the *influential metric* [3]:

$$\mathrm{InfM}\left(u_1, u_2\right) = \sum_{l=1}^{n_{ord}} \omega_l \left(\frac{u_1\left(I_l\right) - u_2\left(I_l\right)}{u_1\left(I_l\right) + u_2\left(I_l\right)}\right)^2 + \sum_{k=1}^{n_{nom}} \gamma_k \chi^2\left(u_1\left(J_k\right), u_2\left(J_k\right)\right) , \quad (17)$$

where I_l stands for the lth ordinal influential attribute (their total number is n_{ord}); J_k stands for the kth nominal influential attribute (their total number is n_{nom}); $u\left(\cdot\right)$ returns respondent u's specified attribute value; $\chi\left(v_1, v_2\right)$ is equal to χ_1 if values v_1 and v_2 fall into one category, and χ_2 otherwise; ω_l and γ_k are non-negative weighting coefficients to be taken judging from the importance of a certain attribute (the more important is the attribute, the greater is the coefficient).

To organize the process of the pairwise SRBS, let us introduce the notion of the *valence* δ_k^i of the submicrofile $\mathbf{M}_k^i$ as a number, whose absolute value determines how many respondents need to be added to or removed from the submicrofile, and whose sign shows whether the respondents need to be added (negative valence) or removed (positive valence) from the submicrofile. The valences of the vital submicrofiles $\mathbf{M}_k^{(G)}$, $k = \overline{1, l_p}$, are equal to the values of the so called *difference signal*

$$\delta^{(G)} = \mathbf{q} - \mathbf{q}^* . \quad (18)$$

The valences of the non-vital submicrofiles $\mathbf{M}_k^{\left(\overline{G}\right)}$, $k = \overline{1, l_p}$, are determined to ensure that the number of respondents in each parameter submicrofile is the same:

Table 3 The valence matrix for anonymizing crisp respondent groups

	P_1	P_2	...	P_{l_p}
G	$\delta_1^{(G)}$	$\delta_2^{(G)}$	...	$\delta_{l_p}^{(G)}$
$\overline{G}$	$\delta_1^{(\overline{G})}$	$\delta_2^{(\overline{G})}$	...	$\delta_{l_p}^{(\overline{G})}$

$$\delta^{(\overline{G})} = (\mu_k - q_k) - \left(\mu_k - q_k^*\right) = -\delta^{(G)}, \quad k = \overline{1, l_p}. \tag{19}$$

Valences of submicrofiles can be arranged into the *valence matrix* Δ (Table 3). Performing the swapping is expressed with the help of the *swapping cycle*:

$$C = ((i_1, j_1), (i_1, j_2), (i_2, j_2), (i_2, j_1)), \tag{20}$$

where (i_1, j_1) determines the positive valence of the vital submicrofile: $\Delta_{i_1 j_1} > 0$, $i_1 = 1$; (i_1, j_2) determines the negative valence of the vital submicrofile: $\Delta_{i_1 j_2} < 0$, $i_1 = 1$; (i_2, j_2) determines the positive valence of the non-vital submicrofile: $\Delta_{i_2 j_2} > 0$, $i_2 = 2$; (i_2, j_1) determines the negative valence of the non-vital submicrofile: $\Delta_{i_2 j_1} < 0$, $i_2 = 2$; $i_1 \neq i_2$, $j_1 \neq j_2$. Cycle entries are called *cycle vertices*.

To define the swapping cycle, it is sufficient to specify its first two vertices.

Respondents to be swapped over C have to belong to the submicrofiles with positive valences and be close with respect to (17). The *swap* is a triplet

$$S = \langle C, I_1, I_2 \rangle, \tag{21}$$

where C is the cycle (20); I_1 is the index of the respondent (the *first candidate to be swapped*, FCS) in the vital submicrofile with the valence defined by the first C vertex; I_2 is the index of the respondent (the *second candidate to be swapped*, SCS) in the non-vital submicrofile with the valence defined by the third C vertex.

The SRBS over C is interpreted as the transferring of the FCS from the submicrofile defined by the vertex 1 to the one defined by the vertex 2, and the simultaneous transferring of the SCS from the submicrofile defined by the vertex 3 to the one defined by the vertex 4. After performing the SRBS according to S, one needs to reduce by one $\Delta_{i_1 j_1}$ and $\Delta_{i_2 j_2}$, and add one to $\Delta_{i_1 j_2}$ and $\Delta_{i_2 j_1}$.

The *cost* of the swap $c(S)$ is a value of (17) calculated for the FCS and SCS. The task of modifying the microfile at the second stage of solving the TPGA lies in determining such an ordered sequence of swaps called the *swapping plan* $\mathbf{S} = \left(S_1, \ldots, S_{n_{swap}}\right)$ that satisfies two conditions:

1. After performing all the swaps, $\Delta_{ik} = 0 \; \forall i = \overline{1, r} \; \forall k = \overline{1, l_p}$.
2. The overall cost of the swapping plan $c(\mathbf{S}) = \sum_{i=1}^{n_{swap}} c(S_i)$ has to be minimal.

This task is the one that can be solved using only exhaustive search, so heuristic strategies need to be developed for constructing the swapping plan that yields results acceptable from both the computational complexity and the minimal swap-

Table 4 The valence matrix for anonymizing fuzzy respondent groups

	P_1	P_2	...	P_{l_p}
$\tilde{G}_{\Delta_{\tilde{M}1}}$	$\delta_1^{\left(\tilde{G}_{\Delta_{\tilde{M}1}}\right)}$	$\delta_2^{\left(\tilde{G}_{\Delta_{\tilde{M}1}}\right)}$	...	$\delta_{l_p}^{\left(\tilde{G}_{\Delta_{\tilde{M}1}}\right)}$
...	...	...	...	...
$\tilde{G}_{\Delta_{\tilde{M}r}}$	$\delta_1^{\left(\tilde{G}_{\Delta_{\tilde{M}r}}\right)}$	$\delta_2^{\left(\tilde{G}_{\Delta_{\tilde{M}r}}\right)}$	...	$\delta_{l_p}^{\left(\tilde{G}_{\Delta_{\tilde{M}r}}\right)}$
$\tilde{G}_{\Delta_{\tilde{M}0}}$	$\delta_1^{\left(\tilde{G}_{\Delta_{\tilde{M}0}}\right)}$	$\delta_2^{\left(\tilde{G}_{\Delta_{\tilde{M}0}}\right)}$	...	$\delta_{l_p}^{\left(\tilde{G}_{\Delta_{\tilde{M}0}}\right)}$

ping plan cost points of view. Several strategies that meet these requirements have been proposed in [26].

4.3.2 Inverse Goal Mapping Functions for Fuzzy Respondent Groups

In this section, we will assume that the inverse goal mapping function is applied to the modified quantity surface $\mathbf{Q}^*$.

The valences of the vital submicrofiles of different grades $\mathbf{M}_k^{\left(\tilde{G}_{\Delta_{\tilde{M}s}}\right)}$, $k = \overline{1, l_p}$, $s = \overline{1, r}$, are equal to the values of the so-called *difference surface*

$$\delta^{(G)} = \mathbf{Q} - \mathbf{Q}^* . \tag{22}$$

Valences of non-vital submicrofiles $\mathbf{M}_k^{\left(\tilde{G}_{\Delta_{\tilde{M}0}}\right)}$, $k = \overline{1, l_p}$, are determined to ensure that the number of respondents in each parameter submicrofile is the same:

$$\delta_k^{\left(\tilde{G}_{\Delta_{\tilde{M}0}}\right)} = \left(\mu_k - \sum_{i=1}^{r} Q_{ik}\right) - \left(\mu_k - \sum_{i=1}^{r} Q_{ik}^*\right), \quad k = \overline{1, l_p} . \tag{23}$$

Valences of submicrofiles can be arranged into the valence matrix Δ (Table 4).

Because of the procedure for obtaining $\mathbf{Q}^*$ using the two-dimensional WT (Sect. 4.2.2), it is impossible to modify the microfile by performing only the SRBS. However, it is possible to modify $\mathbf{M}$ by performing the *transferring of the respondents from one submicrofile* $\mathbf{M}_k^{\left(\tilde{G}_{\Delta_{\tilde{M}s_1}}\right)}$, $s_1 \geq 0$, *to another* $\mathbf{M}_k^{\left(\tilde{G}_{\Delta_{\tilde{M}s_2}}\right)}$, $s_2 \geq 0$, $s_1 \neq s_2$. To *transfer* respondent u from $\mathbf{M}_k^{\left(\tilde{G}_{\Delta_{\tilde{M}s_1}}\right)}$ to $\mathbf{M}_k^{\left(\tilde{G}_{\Delta_{\tilde{M}s_2}}\right)}$ means to modify its vital attributes values so that its membership grade in $\tilde{G}$ $\mu_{\tilde{G}}(u)$ belongs to the interval $\Delta_{\tilde{M}s_2}$. The interval $\Delta_{\tilde{M}0}$ may be viewed as the interval containing all the values from [0, 1], which don't belong to any other interval $\Delta_{\tilde{M}s}$, $s = \overline{1, r}$.

Performing the transferring is expressed with the help of the *transferring cycle*:

$$C_T = ((i_1, j_1), (i_2, j_1)) , \tag{24}$$

where (i_1, j_1) determines the negative valence of the submicrofile: $\Delta_{i_1 j_1} < 0$; (i_2, j_1) determines the positive valence of the submicrofile: $\Delta_{i_2 j_1} > 0$; $i_1 \neq i_2$.

We propose to determine the respondent to be transferred in the following way:

1. Randomly choose a respondent from the submicrofile defined by the first cycle vertex (we will call this record the *representative respondent*, RR).
2. Choose the respondent from the submicrofile defined by the second vertex closest to the RR with respect to (17) (we will call this respondent the *candidate to be transferred*, CT).

We will perform the transferring of the CT by equating its vital attribute values to the ones taken from the RR. The *transfer* can be represented as a triplet

$$T = \langle C_T, I_1, I_2 \rangle , \tag{25}$$

where C_T is the cycle (24); I_1 is the index of the RR; I_2 is the index of the CT. After performing the transferring according to T, one needs to reduce by one the absolute values of $\Delta_{i_1 j_1}$ and $\Delta_{i_2 j_1}$.

The *cost* of the transfer $c(T)$ is a value of (17) calculated for the RR and CT. The task of modifying the microfile at the second stage of solving the TPGA can be reduced to determining such an ordered sequence of transfers called the *transferring plan* $\mathbf{T} = \left(T_1, \ldots, T_{n_{trans}}\right)$ that satisfies two conditions:

1. After performing all the transfers, $\Delta_{ij} = 0 \; \forall i = \overline{1, r} \; \forall j = \overline{1, l_p}$.
2. The overall cost of the transferring plan $c(\mathbf{T}) = \sum_{i=1}^{n_{trans}} c(T_i)$ has to be minimal.

It is possible to solve the TPGA by performing only transfers, but such approach is not acceptable since it implies perturbing microfile records. We propose to reduce the overall number of the transfers by performing the SRBS beforehand.

The overall number of the transfers to perform in the microfile $\mathbf{M}$ is equal to

$$N_{trans} = \sum_{j=1}^{l_p} \left(\frac{1}{2} \sum_{i=1}^{r} |\Delta_{ij}| \right) . \tag{26}$$

After performing the SRBS over the cycle with vertices 1 and 3 corresponding to the positive valences in Δ, and the vertices 2 and 4 corresponding to the negative ones, N_{trans} is reduced by two. Such cycles are called the *full swapping cycles* (FSC). We will denote them by C_F. FSCs are analogous to the ones defined by (20). After performing the SRBS over the cycle with vertices 1, 3, and 4 corresponding to the positive valences in Δ, and the vertex 2 corresponding to the negative one, N_{trans} is reduced by one. Such cycles are called the *partial swapping cycles* (PSC):

$$C_P = ((i_1, j_1), (i_1, j_2), (i_2, j_2), (i_2, j_1)) , \tag{27}$$

where (i_1, j_1), (i_2, j_2), and (i_2, j_1) determine the positive valences of the submicrofile: $\Delta_{i_1 j_1} > 0$, $\Delta_{i_2 j_2} > 0$, $\Delta_{i_2 j_1} > 0$; (i_1, j_2) determines the negative valence of the submicrofile: $\Delta_{i_1 j_2} < 0$; $i_1 \neq i_2$, $j_1 \neq j_2$.

To define the swapping cycle, it is sufficient to specify its first three vertices.

Respondents to be swapped over FSC or PSC have to belong to the submicrofiles with the positive valences and be close with respect to (17). The *full swap* can be represented as a triplet

$$S_F = \langle C_F, I_{1F}, I_{2F} \rangle \, , \tag{28}$$

where C_F is the cycle (20); I_{1F} is the index of the respondent (the *first candidate to be fully swapped*, FCFS) from the vital submicrofile with the positive valence defined by the first C_F vertex; I_{2F} is the index of the respondent (the *second candidate to be fully swapped*, SCFS) from the non-vital submicrofile with the positive valence defined by the third C_F vertex.

The *partial swap* can be represented as a triplet

$$S_P = \langle C_P, I_{1P}, I_{2P} \rangle \, , \tag{29}$$

where C_P is the cycle (27); I_{1P} is the index of the respondent (the *first candidate to be partially swapped*, FCPS) from the submicrofile with the positive valence defined by the first C_P vertex; I_{2P} is the index of the respondent (the *second candidate to be partially swapped*, SCPS) from the submicrofile with the positive valence defined by the third C_P vertex.

The *cost* of the swap $c(S_F)$ $(c(S_P))$ is a value of (17) calculated for the FCFS (FCPS) and SCFS (SCPS) from appropriate submicrofiles. The task of modifying the microfile at the second stage of solving the TPGA for fuzzy respondent groups lies in determining three ordered sequences:

1. The sequence of full swaps called the *full swapping plan* $\mathbf{S}_F = (S_{1F}, \ldots, S_{n_{swapF}})$. The overall cost of the plan $c(\mathbf{S}_F) = \sum_{i=1}^{n_{swapF}} c(S_{iF})$ has to be minimal. After performing all the swaps from $\mathbf{S}_F$ it is impossible to build full swapping cycles.
2. The sequence of partial swaps called the *partial swapping plan* $\mathbf{S}_P = (S_{1P}, \ldots, S_{n_{swapP}})$. The overall cost of the plan $c(\mathbf{S}_P) = \sum_{i=1}^{n_{swapP}} c(S_{iP})$ has to be minimal. After performing all the swaps from $\mathbf{S}_P$ it is impossible to build partial swapping cycles.
3. The transferring plan $\mathbf{T} = (T_1, \ldots, T_{n_{trans}})$ that has to satisfy two conditions expressed earlier.

The tasks of determining each of three plans are the ones that can be solved using only exhaustive search, so heuristic strategies need to be developed for constructing plans that yield results acceptable from both the computational complexity and the minimal swapping plan cost points of view.

Let $\Delta^{(0)}$ denote the *initial valence matrix*, which is obtained according to Table 4. The generic scheme of all the heuristic strategies for determining the full swapping plan boils down to performing the following steps:

1. Equate the *current valence matrix* to the initial one; set $i = 1$. Perform steps 2–8 while it is possible to build full swapping cycles.
2. Assign $\Delta^{\text{temp}} = \Delta^{(i)}$.
3. By analyzing Δ^{temp}, choose the first vertex of C_{iF}; if it is impossible, stop.
4. Choose the FCFS from the submicrofile defined by the first C_{iF} vertex.
5. By analyzing Δ^{temp}, choose the second vertex of C_{iF}; if it is impossible, equate Δ^{temp} element corresponding to the first C_{iF} vertex to zero and go to 3.
6. By analyzing Δ^{temp}, choose the third vertex of C_{iF}; if it is impossible, equate Δ^{temp} element corresponding to the second C_{iF} vertex to zero and go to 5; otherwise, finish the cycle.
7. Choose the SCFS from the submicrofile defined by the third vertex, which is closest to the first one with respect to (17).
8. Perform the swapping; obtain the current valence matrix $\Delta^{(i)}$ by reducing by one the absolute values of the valences from $\Delta^{(i-1)}$ corresponding to the submicrofiles defined by C_{iF}; set $i = i + 1$; go to 2.

All the strategies differ in particular implementations of steps 3, 4, 5, and 6.

Heuristic strategies for determining the partial swapping plans have the same generic scheme, with several slight differences. Firstly, the first cycle vertex in the case of the partial swapping plans does not necessarily represent the vital microfiles. Secondly, analysis on steps 3, 5, and 6 is carried out using $\Delta^{(i-1)}$, not Δ^{temp}. In addition, the initial valence matrix should be taken as the last current matrix obtained after applying heuristic strategies for determining the full swapping plans.

Since the transferring of the respondents in a parameter submicrofile $\mathbf{M}_k$, $k \in \{1, 2, \ldots, l_p\}$, does not depend on the transferring in any other parameter submicrofile $\mathbf{M}_l$, $l \neq k$, let us discuss the strategies for determining the part of the transferring plan $\mathbf{T}$ corresponding to the kth parameter submicrofile, $k \in \{1, 2, \ldots, l_p\}$.

Let $\Delta_{:k}^{(0)}$ denote the *initial valence matrix column k*, which is the kth column of the valence matrix obtained after performing all swaps. The scheme of the strategies for determining the transferring plan boils down to performing such steps:

1. Equate the *current valence matrix column k* to the initial one; set $i = 1$. Perform steps 2–6 while $\exists l \; \Delta_{lk}^{(i)} \neq \mathbf{0}$.
2. Choose the first vertex of the cycle C_T.
3. Randomly choose the RR from the submicrofile defined by the first C_T vertex.
4. Choose the second vertex of the cycle C_T.
5. Choose the CT closest to the RR with respect to (17).
6. Perform the transferring of the CT; obtain the current valence matrix column k $\Delta_{:k}^{(i)}$ by reducing by one the absolute values of the valences from $\Delta_{:k}^{(i-1)}$ corresponding to the submicrofiles defined by C_T; set $i = i + 1$; go to 2.

All the strategies differ in particular implementations of steps 2 and 4. In this chapter, we decided to use four heuristic strategies for determining swapping cycles by choosing the following implementations of the steps 3, 4, 5, and 6:

1. On step 3, for strategies No. 1 and No. 2 we choose the microfile with the greatest valence, for strategies No. 3 and No. 4—with the smallest one.

2. On step 4, for all strategies we try out all the possible candidates, and choose the one that guarantees the minimum values of (17) on step 8.
3. On step 5, for all strategies we try out all the possible vertices, and choose the one that guarantees the minimum values of (17) on step 8.
4. On step 6, for strategies No. 1 and No. 3 we choose the third vertex from the valence matrix row closest to the row with the first vertex, for strategies No. 2 and No. 4—the third vertex from the last valence matrix row, if possible, or from the row closest to the row with the first vertex, otherwise.

We also chose the following implementations of the steps 2 and 4 of heuristic strategies for determining transferring cycles:

1. On step 2, for strategies No. 1 and No. 2 we choose the microfile with the greatest negative valence, for strategies No. 3 and No. 4—with the smallest negative one.
2. On step 4, for all strategies we try out all the possible vertices, and choose the one that guarantees the minimum values of (17) on step 6.

4.4 Practical Results of Providing Anonymity for the Fuzzy Group of Military Enlisted Members

To solve the TPGA for the group of military enlisted members (Sect. 3.2) at the first stage, we need to obtain the modified concentration surface according to the procedure described in Sect. 4.2.2. We chose the Daubechies tenth-order wavelet decomposition filters [27] to perform WT. Applying (14) to $\mathbf{C}$ from Sect. 3.2, we obtain the following approximation coefficients of the first decomposition level:

$$\mathbf{a}_1 = \begin{pmatrix} 0.016 & 0.019 & 0.007 & 0.022 & 0.021 & 0.019 & 0.018 & 0.015 & 0.019 \\ 0.029 & 0.033 & 0.018 & 0.037 & 0.035 & 0.031 & 0.031 & 0.030 & 0.030 \end{pmatrix}.$$

As we recall from Sect. 3.2, there are extreme values in the first, second, and fifth columns of $\mathbf{C}$. One of the ways to mask them is to use such modified coefficients (we present them with two decimal points due to space limitations):

$$\tilde{\mathbf{a}}_1 = \begin{pmatrix} 54.72 & -134.57 & 85.97 & 118.03 & 213.19 & -106.42 & -7.61 & 42.90 & 253.79 \\ -7.71 & 113.88 & 227.45 & -83.60 & -15.03 & 28.54 & 280.46 & 28.82 & -106.33 \end{pmatrix}.$$

Applying the generalized version of (10), we obtain the new surface approximation $\tilde{\mathbf{A}}_1$. By adding this approximation to the old surface details $\mathbf{D}_{h1}$, $\mathbf{D}_{v1}$, and $\mathbf{D}_{d1}$ according to (15), we obtain the surface $\breve{\mathbf{C}}$. Since this surface contains negative values, we apply to it (16) ($\gamma = \frac{\sum_{i=1}^{4}\sum_{j=1}^{18} c_{ij}}{(4\times 18)} - \min\left(\breve{\mathbf{C}}\right)$), and obtain the modified concentration surface $\mathbf{C}^*$

$$
(\mathbf{C}^*)^T = \begin{pmatrix}
0.001 & 0.008 & 0.017 & 0.010 \\
0.004 & 0.010 & 0.019 & 0.012 \\
0.011 & 0.014 & 0.018 & 0.015 \\
0.013 & 0.011 & 0.008 & 0.011 \\
0.015 & 0.008 & 0.000 & 0.007 \\
0.021 & 0.013 & 0.003 & 0.011 \\
0.024 & 0.018 & 0.011 & 0.017 \\
0.015 & 0.014 & 0.013 & 0.014 \\
0.004 & 0.009 & 0.014 & 0.010 \\
0.001 & 0.010 & 0.021 & 0.012 \\
0.004 & 0.013 & 0.024 & 0.015 \\
0.008 & 0.012 & 0.016 & 0.012 \\
0.012 & 0.009 & 0.006 & 0.009 \\
0.017 & 0.010 & 0.001 & 0.008 \\
0.019 & 0.012 & 0.004 & 0.011 \\
0.020 & 0.015 & 0.009 & 0.014 \\
0.017 & 0.016 & 0.016 & 0.016 \\
0.008 & 0.013 & 0.018 & 0.014
\end{pmatrix}.
$$

Its details are equal to the details of the initial surface $\mathbf{C}$ multiplied by the factor of $11,736.620$, i.e. are modified proportionally, which totally suits our purposes of preserving data utility.

Using the inverse of (4) with (5), we obtain the surface $\breve{\mathbf{Q}}$, applying (16) (with $\gamma = 0$) with the subsequent rounding to which yields the modified surface $\mathbf{Q}^*$:

$$
(\mathbf{Q}^*)^T = \begin{pmatrix}
7 & 64 & 134 & 77 \\
35 & 103 & 186 & 118 \\
98 & 125 & 159 & 131 \\
136 & 113 & 85 & 108 \\
358 & 197 & 0 & 161 \\
198 & 122 & 28 & 104 \\
152 & 115 & 70 & 107 \\
714 & 673 & 622 & 664 \\
143 & 317 & 531 & 356 \\
7 & 93 & 199 & 113 \\
34 & 112 & 207 & 129 \\
64 & 97 & 137 & 104 \\
100 & 75 & 46 & 70 \\
177 & 105 & 16 & 88 \\
195 & 125 & 40 & 110 \\
439 & 337 & 212 & 314 \\
479 & 466 & 450 & 463 \\
351 & 541 & 772 & 584
\end{pmatrix}.
$$

The sum of rows of $\mathbf{C}^*$ is shown in Fig. 8 along with the superimposed quantity signal $\mathbf{q}$ from Sect. 3.1 (to fit the scale, we once again normalized each of two vectors

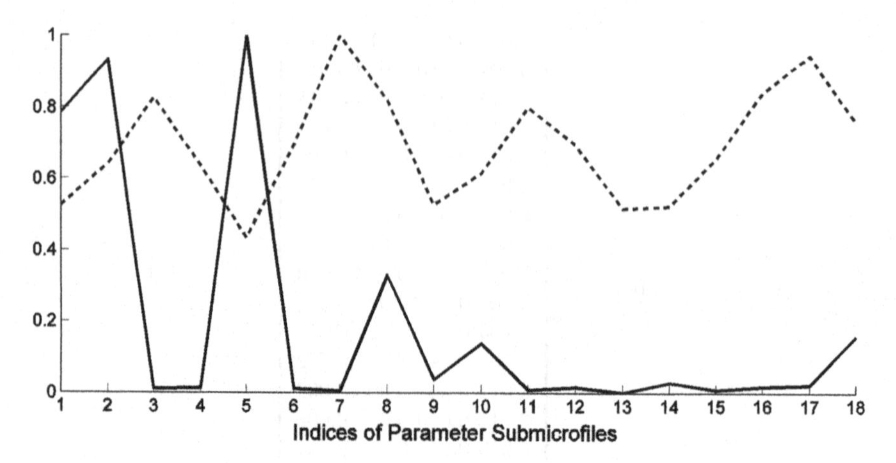

Fig. 8 The initial quantity signal (*solid line*) and the sum of the rows of the modified concentration surface (*dashed line*) for the example

Table 5 Results of applying heuristic strategies to modifying the microfile

Strategy number	Cost of full and partial swapping plans	Cost of transferring plan
1	1,931	14,466
2	2,112	14,306
3	1,931	14,535
4	2,116	14,347

by dividing them by their maximal values). As we can see, extreme values in the first, second, and fifth signal elements have been successfully masked.

Now we need to modify the microfile in order to adjust it to the modified quantity surface $\mathbf{Q}^*$. To perform microfile modification according to Sect. 4.3.2, we took microfile attributes "Sex," "Age," "Black of African American," "Marital Status," "Educational Attainment," "Citizenship Status," "Person's Total Income in 1999," and "Hours per Week in 1999" as the influential ones. For the sake of simplicity, we considered every attribute to be nominal, and we assumed $\gamma_k = 1\ \forall k = \overline{1, 8}$, $\chi_1 = 1$, $\chi_2 = 0$. In this case, (17) shows the overall number of attribute values to be changed in order to provide group anonymity.

The results of applying strategies No. 1–4 to the modified quantity surface are presented in Table 5. Since there are 278,337 respondents that have a positive grade of membership in the fuzzy group of the military enlisted members, and we took 8 influential attributes, we see that to provide anonymity we need to alter at most only $\frac{(1931+14535)}{(8\times278337)} = 0.007$ of all microfile attribute values.

5 Conclusion and Future Research

In the chapter, we showed that microfiles could be considered an important source of information during cyber warfare. We proposed a generic approach to violating anonymity of crisp and fuzzy groups of respondents, and illustrated the importance of such problems with the real data based example concerning violating anonymity of the fuzzy group of "respondents who can be considered military enlisted members with the high level of confidence." We showed that the group anonymity in this case could be provided by modifying values of about 0.7 % of all the microfile attribute values, which is an acceptable cost in most practical situations.

We believe the research can be continued in the direction of developing efficient algorithms for the second stage of solving the TPGA, including evolutionary computation methods. In addition, it is important to enhance the proposed method for constructing FIS for defining fuzzy respondent groups by applying neural network technologies for defining parameters of membership functions.

References

1. Gantz, J., Reinsel, D.: Big data, bigger digital shadows, and biggest growth in the Far East. http://www.emc.com/leadership/digital-universe/iview/executive-summary-a-universe-of.htm (2012)
2. Pfitzmann, A., Hansen, M.: A terminology for talking about privacy by data minimization: anonymity, unlinkability, undetectability, unobservability, pseudonymity, and identity management, Version v0.34, http://dud.inf.tu-dresden.de/Anon_Terminology.shtml (2010)
3. Chertov, O., Tavrov, D.: Data group anonymity: general approach. Int. J. Comput. Sci. Inf. Secur. **8**(7), 1–8 (2010)
4. Chertov, O. (ed.): Group Methods of Data Processing. Lulu.com, Raleigh (2010)
5. Sweeney, L.: Computational Disclosure Control: A Primer on Data Privacy. Ph.D. thesis, Massachusetts Institute of Technology, Cambridge (2001)
6. Evfimievski, A.: Randomization in privacy preserving data mining. ACM SIGKDD Explor. Newslett. **4**(2), 43–48 (2002)
7. Domingo-Ferrer, J., Mateo-Sanz, J.M.: Practical data-oriented microaggregation for statistical disclosure control. IEEE Trans. Knowl. Data Eng. **14**(1), 189–201 (2002)
8. Fienberg, S.E., McIntyre, J.: Data swapping: variations on a theme by Dalenius and Reiss. In: Domingo-Ferrer, J., Torra, V. (eds.) Privacy in Statistical Databases, PSD 2004. LNCS, vol. 3050, pp. 14–29. Springer, Berlin (2004)
9. Wang, J., Zhong, W., Zhang, J.: NNMF-based factorization techniques for high-accuracy privacy protection on non-negative-valued datasets. The 6th IEEE International Conference on Data Mining Workshops. ICDM Workshops 2006, Hong Kong, December 2006, pp. 513–517. IEEE Computer Society Press, Washington (2006)
10. Xu, S., Zhang, J., Han, D., Wang, J.: Singular value decomposition based data distortion strategy for privacy protection. Knowl. Inf. Syst. **10**(3), 383–397 (2006)
11. Liu, L., Wang, J., Zhang, J.: Wavelet-based data perturbation for simultaneous privacy-preserving and statistics-preserving. In: 2008 IEEE International Conference on Data Mining Workshops, Pisa, December 2008, pp. 27–35. IEEE Computer Society Press (2008)
12. National Institute of Statistics and Economic Studies. Minnesota Population Center. Integrated Public Use Microdata Series, International: Version 6.2 [Machine-readable database]. University of Minnesota, Minneapolis, https://international.ipums.org/international/ (2013)

13. Nuclear Power in France, World Nuclear Association, http://www.world-nuclear.org/info/inf40.html
14. Chertov, O., Tavrov, D.: Group anonymity. In: Hllermeier, E., Kruse, R., Hoffmann, F. (eds.) Information Processing and Management of Uncertainty in Knowledge-Based Systems. Applications. CCIS, vol. 81, pp. 592–601. Springer, Berlin (2010)
15. Chertov, O., Tavrov, D.: Group anonymity: problems and solutions. Lviv Polytechnic Natl. Univ. J. Info. Syst. Netw. **673**, 3–15 (2010)
16. Chertov, O., Tavrov, D.: Providing data group anonymity using concentration differences. Mathe. Mach. Syst. **3**, 34–44 (2010)
17. Tishchenko, V., Mladientsev, M.: Dmitrii Ivanovich Miendielieiev, yego zhizn i dieiatielnost. Univiersitietskii pieriod 1861–1890 gg. Nauka, Moskva (1993) (In Russian)
18. Zadeh, L.A.: The concept of a linguistic variable and its application to approximate reasoning. Inf. Sci. **8**, 199–249 (1975)
19. Chertov, O., Tavrov, D.: Providing Group Anonymity Using Wavelet Transform. In: MacKinnon, L.M. (ed.) Data Security and Security Data. LNCS, vol. 6121, pp. 25–36. Springer, Berlin (2012)
20. Mamdani, E.H., Assilian, S.: An experiment in linguistic synthesis with a fuzzy logic controller. Int. J. Man-Mach. Stud. **7**(1), 1–13 (1975)
21. Klir, G.J., Yuan, B.: Fuzzy Sets and Fuzzy Logic: Theory and Applications. Prentice Hall, Upper Saddle River (1995)
22. Zadeh, L.A.: Outline of a new approach to the analysis of complex systems and decision processes. IEEE Trans. Syst. Man Cybern. SMC-**3**(1), 28–44 (1973)
23. U. S. Census 2000. 5-Percent Public Use Microdata Sample Files, http://www.census.gov/main/www/cen2000.html
24. Demographics. Profile of the Military Community. Office of the Deputy under Secretary of Defense (Military Community and Family Policy), http://www.militaryonesource.mil/12038/MOS/Reports/2011_Demographics_Report.pdf (2012)
25. Mallat, S.: A Wavelet Tour of Signal Processing. Academic Press, New York (1999)
26. Chertov, O.R.: Minimizatsiia spotvoren pry formuvanni mikrofailu z zamaskovanymy danymy. Visnyk Skhid-noukrainskoho Natsionalnoho Universytetu imeni Volodymyra Dalia, **8**(179), 256–262 (2012) (In Ukrainian)
27. Daubechies, I.: Ten lectures on wavelets. Soc. Ind. Appl. Math. (1992)

Decision Support in Open Source Intelligence

Daniel Ortiz-Arroyo

Abstract This chapter describes a decision support system specially designed for applications in open source intelligence. The decision support system was developed within the framework of the FP7 VIRTUOSO project. Firstly, we describe the overall scope and architecture of the VIRTUOSO platform. Secondly, we describe with detail some of most representative components of the DSS. The components employ computational intelligence techniques such as knowledge representation, soft-fusion and fuzzy logic. The DSS together with other tools developed for the VIRTUOSO platform will help intelligence analysts to integrate diverse sources of information, visualize them and have access to the knowledge extracted from these sources. Finally, we describe some applications of decision support systems in cyber-warfare.

1 Introduction

Decision support systems (DSS) comprise a set of computational tools whose purpose is to support better decision making processes within organizations.

DSS allow decision makers to visualize, analyze, process and mine data of various types. Data is collected from a diversity of sources and integrated to create a knowledge base repository. The knowledge base is then used by decision makers to help them in reasoning about some possible scenarios.

DSS are increasingly used in a variety of fields such as business intelligence and criminal intelligence. In the area of business intelligence, DSSs help companies to asses their competitor's position in the market, additionally to determining market trends, and in planning future investments.

In criminal intelligence, DSS help intelligence agencies to tackle organized crime, detect crime patterns and analyze the structure of criminal organizations.

D. Ortiz-Arroyo (✉)
Department of Electronic Systems,
Computational Intelligence and Security Laboratory,
Aalborg University Esbjerg, Esbjerg, Denmark
e-mail: do@es.aau.dk

R.R. Yager et al. (eds.), *Intelligent Methods for Cyber Warfare*,
Studies in Computational Intelligence 563, DOI 10.1007/978-3-319-08624-8_5

DSS have become more sophisticated in the internet era. The internet has eased the communication and sharing of information by people, organizations and the IT systems of companies. Moreover, social media is increasingly used by hundreds of millions of people, including government, media and business organizations, but also by organized crime.

The internet is used by traditional mass media for distributing news and information in digital format. For instance, some of the most common sources are RSS feeds, blogs, and specialized web portals that include video, audio and text about events or persons.

The internet has also eased the distribution of specialized technologies. Computer code be found on the internet that may be used to exploit the weaknesses of IT systems and/or to perform malicious or criminal activities. The easy access to specialized code has allowed hackers and organized crime to carry out cyber-attacks against specific servers of government institutions or to the whole computer network of a country. These attacks are carried out using a botnet of "zombies" computers distributed around the world [1].

In the case of intelligence applications, traditionally, the data sources used to track down organized crime were *classified*. Classified, secret data about individuals or organizations is kept in secure data repositories isolated from the internet.

However, intelligence agencies have recognized the value that the information publicly available on the web has in their investigations. In many criminal cases the usage of social media by organized crime and its associates may leave intended or unintended traces that can be collected for analysis [2].

Information collected from open sources on the web is being used not only in criminal investigations but also in a variety of areas such as, situation monitoring and assessment, and to produce early warnings of possible crisis.

The collection of methods used in collecting, managing and analyzing publicly available data is called *Open Source Intelligence* (OSINT).

The data sources employed in OSINT are varied. Some of these data sources are electronic media such as newspapers and magazines, web-based social media such as social networks, or specialized web portals and blogs, public data from government sources, professional and academic literature, geospatial data, scanned documents, video and data streams, among the most common sources.

OSINT creates important technical, legal and ethical challenges. One of the main technical challenges is to collect relevant meaningful information from reliable sources, among the huge amount of data sources available on the web. This is a critical issue since information may be of low reliability or bogus, obsolete, duplicated and/or available only in certain languages. Information may be of different types and being available in different formats.

Once reliable data sources are found, the relevant information contained in them must be identified and extracted. The extraction of relevant information from a corpora of documents is also technically challenging, because entities expressed in natural language such as events, individuals, organization and places must be recognized and disambiguated.

After the extraction process, information should be stored in knowledge bases, a step that coverts raw information into knowledge. The knowledge stored in knowledge bases is commonly represented in the form of ontologies and semantic networks. These ontologies contain specialized and general domain knowledge about entities such as type of crimes, individuals, organizations and events.

The knowledge bases allow analysts to reasoning about entities and their relationships.

One important feature of DSS is their visualization tools. These tools allow the analyst to look at summaries of data from different perspectives, using for instance dashboards, graph viewers, maps, and plots of different types.

All the technical aspects that have been briefly described in previous paragraphs are very important in DSSs for OSINT applications, but they are not the only challenge that must be addressed when such systems are developed.

There other non-technical challenges in OSINT are related to legal and ethical aspects. In OSINT, relevant information must be collected in a way that respects the privacy of individuals and a the same time, that does not violate the existing national and international laws in this regard.

The collection of personal information about individuals and organizations from the numerous open (and proprietary) sources on the internet has opened up the possibility for using such data not only to target specific organizations but also for mass surveillance purposes.[1]

This issue has created a debate on the ethical and legal aspects involved in the use of OSINT-based technologies.

In the case of the VIRTUOSO project, these issues were addressed since the inception of the project. One of the tasks continuously performed during the whole development process of VIRTUOSO, was to make sure that privacy was respected and that no laws (for instance copyright) were violated when collecting and storing data.

The VIRTUOSO platform was developed and implemented, addressing all the technical and ethical challenges described above.

As is described later in this chapter, VIRTUSO consists of several software components The DSS in VIRTUOSO is one of the its key components. The DSS employs computational intelligence techniques that allow analysts to reason under uncertainty, represent and fuse knowledge, among other tasks.

This chapter describes the overall architecture of the VIRTUOSO system and the main components that comprise VIRTUOSO's DSS.

This chapter is organized as follows. Section 2 describes the architecture of the VIRTUOSO platform. Section 3 describes some representative components of the decision support system in VIRTUOSO. Section 4 describes some possible applications of decision support systems in cyberwarfare. Finally, Sect. 5 presents some conclusions.

---

[1] The recent revelations by E. Snowden about the mass surveillance programs deployed by US intelligence agencies has revived the debate within Europe and in other parts of the world about this issue.

2 VIRTUOSO's Architecture

In summary, the goal of the VIRTUOSO project is to retrieving unstructured data from open sources available on the web and converting it automatically into structured actionable knowledge. To achieve this goal a flexible architecture for the whole system was designed.

The architecture of VIRTUOSO is based on a Service Oriented Architecture (SOA). SOA is a recommendation for how to structure component systems based on web services. SOA was proposed to ease communicating, synchronizing and integrating diverse software components that implement these services. This feature was an extremely important issue in VIRTUOSO, because software components could be developed by different partners participating in the project, using different languages and technologies.

In VIRTUOSO the SOA model was implemented using the Weblab[2] platform. Weblab is a platform, whose main purpose is to build software systems specifically for OSINT applications based on the SOA specification.

The architecture of VIRTUOSO consists of three main processing stages: (a) data acquisition, (b) data processing, and (c) decision support. A special portal allows users to configure and monitor the different processing stages.

In the data acquisition stage, data from unstructured open sources is retrieved using web crawling techniques. Web crawlers acquire different types of data from a wide diversity of sites on the web i.e. electronic-text data, multimedia content, and even from scanned papers. These multiple types of data come from web sites, blogs, tweets, RSS feeds, trends, video streaming sites, and paper documents.

Regarding electronic-text data, at the current state of the project more than 500,000 documents are processed every day, written in 39 different languages from 188 countries. These documents are retrieved from 28,000 open sources.

The data acquisition stage is continuously connected to the Internet to retrieve all relevant data. Additionally, at this stage some pre-processing is performed. For instance, normalization, object recognition, entity naming, event extraction, image and video classification, source assessment, and speech recognition.

Normalization of different types of media and documents is performed by representing them in a single XML based format that contains pointers to the real location of data. The source assessment stage attempts to evaluate the reliability of a data source.

The number of pre-processing steps performed by VIRTUOSO platform can be configured.

After the pre-processing stage, a special data repository is created, containing all the results of all pre-processing steps that may have been performed.

Both, the data acquisition and preprocessing stages were implemented by integrating all its components on the SOA model. Figure 1 shows the SOA platform with the tree main processing stages of VIRTUOSO together with the crawlers required to download data from open sources on the web.

---

[2] Weblab is available at weblab-project.org.

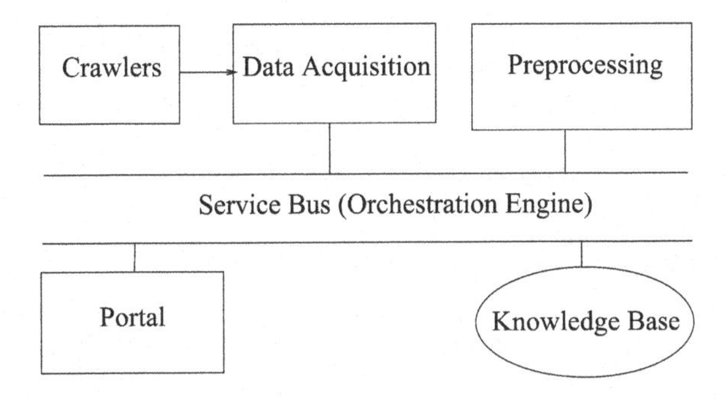

Fig. 1 Data acquisition, preprocessing and data repository

Contrarily, the data processing stage of VIRTUOSO is not connected directly to the internet. This is mainly done for security reasons.

The data processing stage contains several components, among which are: a full text and multimedia search engine, a summarization component, automatic translation of documents, determination of document similarity, and query translation.

The knowledge base is one of the key components in VIRTUOSO. The knowledge base is created apriori with general domain knowledge and is updated with knowledge extracted in the data pre-processing stage.

To being able to use the data repository created in the pre-processing stage, an import/export component is available at the processing stage. During importing, data may be manually or semi-manually validated to ensure that no irrelevant or dubious data is introduced in the knowledge base.

3 The Decision Support System

The decision support system of VIRTUOSO is one of its key components. The purpose of the DSS is to provide intelligence analysts with a set of software components that can be used to extract and store knowledge and to visualize, analyze, process and mine data of various types.

One of the main benefits of using DSS in applications such as VIRTUOSO is to improve the decision making process of analysts in making more informed and effective decisions. For instance, DSS provide analysts with apriori knowledge about certain types of crimes and organizations. Using this knowledge, and the data available for a current situation or event, analysts can look a different views of the data and look for patterns and trends that may help them to asses the importance of the situation or event. Analysts can also look at how participants in an event are related to each other (i.e. their social network) to determine how important these individuals are within the social network.

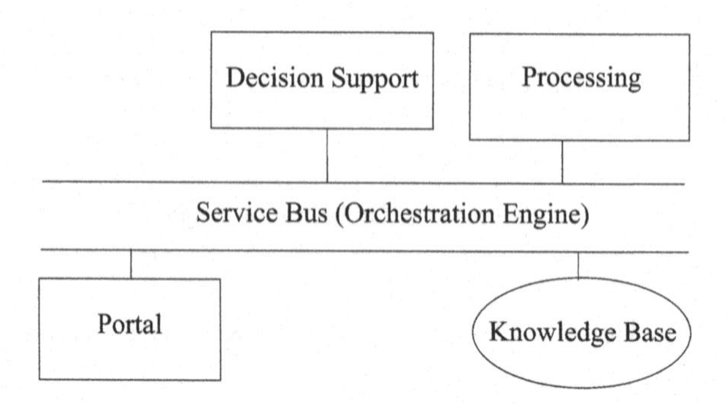

Fig. 2 Data processing, knowledge base and decision support system

However, given that domain scenarios for criminal investigations are different, expert analysts must decide which components of a DSS should be applied in each particular case.

VIRTUOSO's decision support system consists of a variety of software components. Figure 2 shows a high level view of the data processing and decision support components of VIRTUOSO on the SOA platform.

The portal allows users to interact with the DSS. The knowledge base stores apriori knowledge in the form of ontologies and knowledge extracted from open source documents. The knowledge base is also part of the decision support system contained in VIRTUOSO. The data processing stage and the decision support system, share the same SOA-based Weblab infrastructure.

Some of the components that are part of the processing stage in the DSS, can be applied to process documents, directly to data or to both data and documents. For instance, the components that can be applied to documents are: metadata viewer, source assessment, geographical search, multimedia and text search, trend analysis, social media topic, sentiment monitor and semantic search.

The components that can be applied to data are: graph viewer, tabular viewer, graphical SPARQL querying, rule editor, entity editor, similarity of entities, similarity of strings, semantic analysis, and social network analysis component.

Finally, the components that can be applied to both data and documents are: dashboard visualization, knowledge browser, centipede and traceability component.

All these components allow the analyst to perform a wide variety of tasks, from querying semantic knowledge bases, to the visualization and processing of different types of graphs, data and documents. Some of the software components available in VIRTUOSO's decision support system are not used directly by the analyst, but instead, they provide services to other components in the system.

All the components work seamlessly together on Weblab's SOA platform.

Due to the wide variety of components that are available in the DSS of VIRTUOSO, it is not possible to describe all of them in detail. Thus, in the rest of the chapter we will describe a few of the most representative components in the DSS.

The documentation of the VIRTUOSO project contains a detailed description of the rest of the components [3] that were briefly mentioned in previous paragraphs.

3.1 Knowledge Base, Fusion and Uncertainty Management

One of the key components in VIRTUOSO's DSS is its knowledge base. The knowledge base comprises ontological knowledge (conceptual and geographic) and operational knowledge (factual).

The ontological knowledge consists of the known existing relationships among the concepts employed in the intelligence domain. The knowledge base contains several ontologies about general knowledge and specialized knowledge about the domains of criminal and intelligence analysis.

Ontological knowledge is represented as triples in the form (predicate, subject, object) or (p,s,o) for short, as specified in the resource description framework (RDF) schema. Internally the knowledge base employs a slightly different format based on RDF.

Semantic knowledge in VIRTUOSO may be introduced in the knowledge base either manually for highly specialized domains or in an automatic or semiautomatic way for other types of domains. For instance, part of the ontological knowledge included in VIRTUOSO is imported from existing ontology resources. However, to use these or other existing ontology resources a process of semantic disambiguation and fusion of information is performed.

The fusion component in VIRTUOSO merges two graph structures, using an operation called "maximal joint". This method was originally proposed to fuse conceptual graphs in [4]. However, in VIRTUOSO the maximal joint heuristic was applied to semantic graphs.

The joint operation was divided into two parts. First, the compatibility of the two elements to fuse is evaluated. Two entity nodes are considered compatible if the type of entity is the same (e.g. person, location) and if a high proportion of entities' properties is similar. The similarity measure that will be applied depends on the type of properties that entities may have.

In VIRTUOSO the nodes in the graph structures correspond to entities that have properties defined as strings of characters or numbers.

The similarity of string properties, like names for instance, was evaluated using Levensthein string edit distance. This distance basically evaluates how many insertions or deletions of characters are needed to convert one string into the other.

For numerical properties, the similarity was calculated using several techniques. For instance, in the case of date properties the number of days between the two dates was used as the distance. For other types of numerical properties the similarity was evaluated using the following equation:

$$sim_{num}(\beta, x, y) = e^{\frac{\beta(x-y)^2}{\beta-1}} \quad (1)$$

where β represents the sensibility of the measure to the distance between two similar numerical values x and y.

Two entities were considered compatible if the similarity value of the numerical and string properties was above certain threshold value.

Once the similarity between entities was determined, the maximal join operation was used to fused the sub graphs of two distinct but compatible graphs using the following method. Nodes determined as being compatible were fused, creating an extended graph that included the sub-graphs of the two compatible nodes. This procedure was repeated recursively in each node of the subgraphs until incompatibilities were found.

The operational knowledge contained in the knowledge base consists of information extracted from the open sources. The basic entities available in the knowledge base are physical entities (e.g., persons, vehicles), legal entities (e.g., organizations), non-physical entities (e.g., phone number), and event entities (meeting, travel).

The knowledge base contains also various types of metadata, such as time dependencies, validity (or certainty), sensitivity, confidentiality, and provenance information (to being able to trace back to the sources).

To manage uncertainty the RDF triples in the knowledge base were extended by adding an extra parameter β that depending on the entity and type of uncertainty may represent a probability distribution or a possibility distribution. RDF triplets stored in the knowledge base were represented as $\{(predicate, subject, object), \beta\}$, using that information.

3.2 Social Network Analysis Component

Social Network Analysis (SNA) comprises the study of relations, ties, patterns of communication and behavioral performance within social groups. In SNA, a social network is commonly modeled by a graph composed of nodes and edges. The nodes in the graph represent social actors and the links the relationship or ties between them [5].

Since criminal organizations are also a form of social network, they can be represented as graphs in the user interface of VIRTUOSO. The nodes in the graph are the individual members of an organization and the links represent their known relationships. In general, the relationships may be the known connections between individuals (e.g. friendship) or they may represent the structure of command within an organization. These connections may be manually introduced in the network or extracted autoamatically from a document collection and stored in the knowledge base.

In SNA, multiple metrics have been proposed that aim at evaluating the importance of each of the nodes within a social network. One of the most important metrics in SNA is *centrality* [6, 7]. Centrality describes a member's relative position or importance within the context of his/her social network.

One of the applications of the centrality measures that are commonly used in SNA is to discover *key players* [8]. Key players are these nodes in the network that are considered “important” in regard to some criteria, such as the number of its connections (i.e. the degree of a node), their importance regarding the diffusion of information, their influence on the network, etc.

To process social networks, the SNA component in VIRTUOSO employs some of the most popular centrality measures used in SNA such as degree, betweenness, closeness, and eigenvector centrality [9].

In each particular case, the expert analyst must decide which centrality measure should be applied. However, it is also possible that the analyst may be interested in evaluating the overall importance of a group of nodes according to several centrality criteria. In this case, VIRTUOSO will be able to calculate an aggregated centrality value using all or some of the centrality measures available in the SNA component.

The aggregation of centrality measures may be also useful when the analyst is not sure about which centrality measure should be used.

To perform the aggregation, the SNA component in VIRTUOSO employs an ordered weighted aggregation (OWA) operator [10]. The OWA operator is defined as:

$$h_w(a_1, a_2, \ldots, a_n) = w_1 b_1 + w_2 b_2, \ldots + w_n b_n \tag{2}$$

where $w_i \in [0, 1]$ are the weights and $\sum_{i=1}^{n} w_i = 1$, $(b_1, b_2, \ldots, b_n)$ is a permutation of the vector $(a_1, a_2, \ldots, a_n)$ in which the elements are ordered $b_i \geq b_j$ if $i < j$ for any i, j.

The weights used in the OWA operator are normally calculated by stating first what is *andness* value that is expected from the operator. The *andness* value of the OWA operator is defined in terms of the weights of the operator as:

$$Andness(\mathbf{w}) = 1 - \alpha = 1 - \frac{1}{n-1} \sum_{i=1}^{n} w_i (n - i), \alpha \in [0, 1] \tag{3}$$

where α is the *orness* value. This *andness* value represents how close the OWA operator behaves as a fuzzy *and* operator i.e. how close the resulting aggregation value is to the fuzzy AND (minimum) value produced by all the centrality measures. For instance, with *andness* value of 1 the weights of the OWA will be (1,0, ... ,0), which will produce the minimum value in the aggregation. With an *andness* value of 0.5 all weights will be $1/n$ and the OWA will calculate the average value of all its inputs.

When OWA operators are used, expert knowledge about a problem domain is used to decide the most appropriate *andness* value. In the prototype of the SNA component, weight values of (0.1, 0.15, 0.25, 0.5) were assigned by default to the OWA operator. These weights produced an *andness* value of 0.71. Hence, this default centrality measure produced by the OWA operator will be a value that lies between the average centrality of all the measures and the minimum value produced by all of them.

One of the issues of OWA operators, is that very different values in the weights may produce the same *andness* value. This is an important issue that must be considered when OWA operators are used in decision making systems and other applications. In decision making problems we normally want to aggregate all the input values in such a way that all of them contribute to the final decision and not just a few of them or in extreme cases only one. Therefore, one desirable feature of an OWA operator is to get maximum dispersion in the weight values. The weight's dispersion measures the degree with which every input contributes to produce the output of an OWA operator, and is defined as:

$$disp(\mathbf{w}) = -\sum_{i=1}^{n} w_i ln(w_i) \quad (4)$$

In the case of the SNA component the weight dispersion obtained by using the default weight values was 1.20. In general, finding the weight values for an OWA operator is considered as an optimization problem, in which we want to get maximum weight dispersion for a specific *andness* value of interest between [0, 1], subjected to the restriction that the sum of all weights should we 1. This optimization problem has been addressed by other aggregation operators. An example is the maximum entropy OWA (MEOWA) operator [11] that employs Lagrange multipliers to solve the constrained optimization problem.

Other operators like andness-directed multiplicative or implicative weighted aggregation (AMWA or AIWA) operators [12] attempt to aggregate the importance that each input has. All these operators have been also implemented in the SNA component in VIRTUOSO. Thus, the analyst could experiment by aggregating the different centrality measures with different operators to see what is their effect.

It must be noted that for some specific constant values of *andness*, it is possible to use some of the analytical expressions described in [13] to calculate the weights used in the OWA operator. As is described in [13] these expressions provide good dispersion values in the weights' distribution. For instance for an *andness* $=2/3$ $=0.66$ we could use the following equation to calculate each of the n weights used in the OWA operator:

$$w_i = \frac{2i}{n(n+1)} \quad (5)$$

using this expression, the OWA weights will be $(2 * 1/(4 * 5) = 0.1, 2 * 2/20 = 0.2, 2 * 3/20 = 0.3, 2 * 4/20 = 0.4) = (0.1, 0.2, 0.3, 0.4)$, which have a weight dispersion of 1.28 and whose values are close to the values produced by the *andness* of 0.71 used by default in the SNA component.

The application of the OWA, MEOWA or AMWA operators allows the analyst to use all the centrality measures available at once, in such a way that each one contributes partially to the overall calculation of the centrality of every node in the network.

The SNA component can be used used in two different ways in VIRTUOSO. One way is as a REST web service that receives HTTP/POST requests containing the description of the social network to be analyzed. This description is provided in a standard format such as graphml.[3] The output of the service is the calculation of the desired centralities for each node in the network. These values are returned in JSON format encoding.

It is also possible to use the SNA component functionality integrated as a portlet[4] within the weblab platform.

4 Decision Support Systems and Cyber-Warfare

The service oriented-based architecture (SOA) of VIRTUOSO allows to reuse some of its software components to create decision support systems that could be applied in other domains such as cyber-warfare.

Recent studies have found that some cyber attacks have been performed when the computer passwords of certain employees that work in companies or government agencies have been guessed correctly by attackers. To do this, attackers analyze personal data posted by these employees in social media, to learn about the employee's social network [14] connections.

The employee's social network is then used to gain access to some of the acquaintances' computers that are less secure. Once this is done, attackers may use a *spearpishing* attack, which consists in sending emails to the employee from one of his/her colleagues or friends' email accounts. The email sent may include computer code hidden in apparently harmless attachments that is used to guess the employee's password in other computer systems or to determine the answer to certain security questions asked by systems with restricted access.

In this scenario, it may be possible to use VIRTUOSO's DSS to gather data from social media about an employee and its social network. The social network obtained in this way can be analyzed to determine how much public data may be available about an employee and his/her acquaintances. Such data may be used to determine how vulnerable certain employees may be to *spearphising* attacks.

The other area where some of VIRTUOSO's DSS components could be applied, is in the area of cognitive models of decision making for cyber defense. Specifically in designing and applying cyber security ontologies and in scenario ontologies, in a way similar as it is described in [14]. The knowledge base in VIRTUOSO may be used to store these ontologies, together with knowledge extracted from closed or open sources. This knowledge could help analysts to reason about possible cyber attacks.

---

[3] graphml is an XML-based format used to represent graphs or networks.

[4] portlets are pluggable user interface components that are managed by web portals.

5 Conclusion

VIRTUOSO provides a large collection of software components that help analysts to process and visualize a large collection of open source data of various types.

These components together with the decision support system and its knowledge base, will help analysts to reason more easily about a particular scenario.

VIRTUOSO is a complex system consisting of many software components. We have described just a few of them and its features.

At the current stage, the decision support system and all the tools developed in VIRTUOSO have been tested on a few scenarios and a final presentation on the results of the project has been performed for the reviewers of the European commission in charge of assessing the final results of the project.

At the time of writing this chapter, most of the software components available in VIRTUOSO are at pre-release state.

One of the challenges of complex software systems like VIRTUOSO, is to use them in the most effective way. VIRTUOSO requires a group of IT specialists, administrators and analysts that could manage the data sources, maintain the knowledge base, define the most relevant scenarios, and analyze the results provided by the system.

The final report on the VIRTUOSO project included some recommendations in this regard. However, the procedures employed and the type organization that may be needed to use effectively the whole system must be tailored to fit each specific.

As part of the project, it is planned that a demonstrator of the VIRTUOSO platform will be installed at Aalborg University campus Esbjerg in Denmark. When this happens, interested parties within the European Union will be allowed to use the demonstrator and experiment with the system to asses its functionality.

Acknowledgments The author was member of the team from Aalborg University that participated in VIRTUOSO. The research leading to these results has received funding from the European Union's Seventh Framework Programme (FP7/2007-2013) under grant agreement no. 242352.

References

1. Ortiz-Arroyo, D.: Information security threats and policies in europe. In: Laundon, K., Laundon, J., (eds.) Management Information Systems: Managing the Digital firm. Global Edition: Managing the Digital Firm, pp. 357–358. Pearson Longman (2011)
2. Frank, R., Cheng, C., Pun, V.: Social media sites: New fora for criminal, communication, and investigation opportunities. Technical Report 21, Simon Fraser University (2011)
3. Virtuoso Consortium: Decision support and visualization modules report. Technical report, Virtuoso Project Consortium (2013)
4. Laudy, C., Deparis, E., Lortal, G., Mattioli, J.: Multi-granular fusion for social data analysis for a decision and intelligence application. In: Proceedings of 16th International Conference on Information Fusion. IEEE, (2013)
5. Scott, J.: Social Network Analysis: A Handbook. SAGE (2000)

6. Freeman, L.C.: A set of measures of centrality based on betweenness. Sociometry **40**(1), 35–41 (1977)
7. Friedkin, N.E.: Theoretical foundations for centrality measures. The Am. J. Soc. **96**(6), 1478–1504 (1991)
8. Ortiz-Arroyo, D.: Discovering sets of key players in social networks. In: Abraham, A., Hassanien, A., Snasel, V., (eds.) Computational Social Networks Analysis: Trends, Tools and Research Advances, pp. 27–47. Springer, Berlin (2010)
9. Borgatti, S.P., Everett, M.G.: A graph-theoretic framework for classifying centrality measures. Soc. Netw. **28**(4), 466–484 (2006)
10. Yager, R.R.: On ordered weighted averaging aggregation operators in multi-criteria decision making. IEEE Trans. Syst. Man Cybern. **18**, 183–190 (1988)
11. Filev, D., Yager, R.R.: Analytic properties of maximum entropy owa operators. Inform. Sci. **85**, 11–27 (1995)
12. Larsen, L.H.: Multiplicative and implicative importance weighted averaging aggregation operators with accurate andness direction. In: Proceedings of the Joint 2009 International Fuzzy Systems Association World Congress and 2009 European Society of Fuzzy Logic and Technology Conference, pp. 402–407. IEEE (2009)
13. Ahn, B.S.: On the properties of owa operator weights functions with constant level of orness. IEEE Trans. Fuzzy Syst. **14**(4), 511–515 (2006)
14. Oltramari, A., Lebiere, C., Vizenor, L., Zhu, W., Dipert, R.: Towards a cognitive system for decision support in cyber operations. In: Laskey, KB., Emmons, L., Cesar, P., da Costa, G., (eds) STIDS, volume 1097 of CEUR Workshop Proceedings, pp. 94–100. CEUR-WS.org (2013)

Information Fusion Process Design Issues for Hard and Soft Information: Developing an Initial Prototype

James Llinas

1 Introduction

The Data and Information Fusion (DIF) process can be argued to have three main functions: Common Referencing (CR) (also known as "Alignment"), Data Association (DA), and State Estimation, as shown in Fig. 1:

It can be argued that any DIF process can be architected as a network of such nodes (see [1]). In Fig. 1, we have either data or estimates (from a prior Fusion Node) entering this process. Data Alignment or CR is a function that transforms all input data to a common format and also, importantly, a common semantic framework. DA is a function that associates the evidence from multiple sources to asserted entities in the domain of interest; such entities can be not only physical objects but events, behaviors, situational substructures, etc. DA involves accounting for sensor and estimation errors, and also for semantic differences and similarities; the idea is to assemble and partition evidential sets of information so that subsequent inferencing and estimation processes are applied to the most robust collections of evidence about any such entity. DA comprises the three operations shown of Hypothesis Generation (defining feasible associations), Hypothesis Evaluation (a strategy for scoring inter-entity associability of the collective evidence), and Hypothesis Selection (typically some type of optimization scheme to define the best associations of all that are feasible and of higher scores). At this point, there is thus a set of entity-evidence groupings (evidence "assigned" to a given entity), and the assigned evidence is passed to whatever estimation process is at work on the given entity for next-time-increment processing. State Estimation (SE) follows, with various inferencing or estimation methods operating on these aligned and associated evidential sets. By a far greater proportion, the DIF research and development community has applied such techniques to evidential data coming from modern electromechanical sensors, i.e., what some call "physics-based" sensors. In the military/defense domains, these are the modern sensors used for Intelligence, Surveillance, and Reconnaissance (ISR) applications, ranging from satellite-based sensor systems to embedded, Unattended Ground Sensors (UGS's). In

J. Llinas (✉)
Center for Multisource Information Fusion,
University at Buffalo, Buffalo, NY, USA
e-mail: llinas@buffalo.edu

R.R. Yager et al. (eds.), *Intelligent Methods for Cyber Warfare*,
Studies in Computational Intelligence 563, DOI 10.1007/978-3-319-08624-8_6

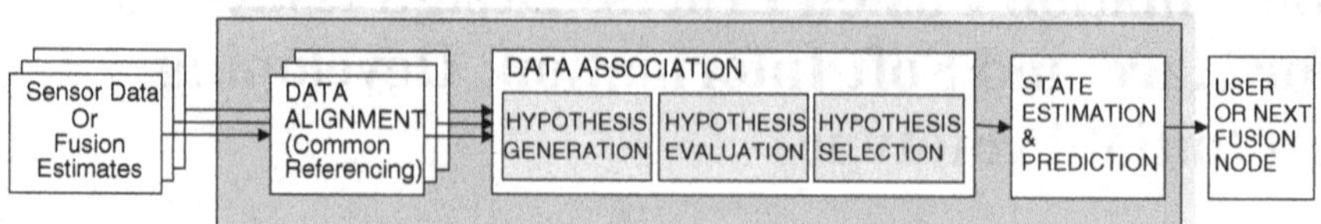

Fig. 1 Nominal fusion node processes

these cases, the designer of automated CR-DA-SE DIF operations enjoys the benefits of dealing with input sources that are well-calibrated and understood, and largely if not exclusively (in raw form at least) numerical. Data from such sources has come to be called "hard" data in the sense of these well-understood properties. For the type of operations involving traditional military engagements, where ISR must be conducted covertly and at a distance, the use of such ISR sensors and hard data has worked reasonably well.

However, subsequent to the end of the Cold War, the nature of defense and military operations has changed dramatically, from so-called "conventional" operations to what today are called "irregular" and "asymmetric" operations. These environments are characterized by a number of complicating features:

- They can be quite complex, involving terrorist, criminal, insurgent, and warfighting mixed operations
- They typically have no clearly defined or identifiable adversaries
- Hostiles/adversaries are mixed in with neutral, friendly persons or forces
- The goals involve not only destructive goals ("kinetic" actions) but establishment of influence and indirect effects

among other factors; such definitions are controversial and it is not our intent to be precise here but to give a flavor for these distinctions. Some of the other subtleties of these environments are that there are now improved data sources to better understand an "enemy" (a first principle of warfare), not only from a military point of view but from a socio-cultural point of view. Further, as noted above, friendly forces are often embedded in certain of these environments, which permits direct and close observation by these forces (humans, not sensors). Experiences in Iraq and Afghanistan and other places in the world in dealing with intelligence and security problems are typical of these new problems, and have required the (ongoing) formulation of new paradigms of intelligence analysis and dynamic decision-making. Broadly, these problems fall into the categories of counter-terrorism and counter-insurgency (COIN) as well as stability operations. Depending on the phases of counter-insurgency or other operations, the nature of decision-making ranges from conventional military-like to socio-political (sometimes also characterized as "hard" and "soft" decisions). Because of this wide spectrum of action, the nature of information support required for analysis has an equally wide range. Since automated DIF processes provide some of the support to such decision-making, requirements for DIF process design must address these varying requirements, resulting in considerable challenges in DIF process design. One important driving factor for DIF process design is the new heterogeneity of the information supportive of DIF process design; these factors are discussed in the Sect. 2.

2 Heterogeneity of Supporting Information

2.1 Observational Data

As remarked above, the experiences in Iraq and Afghanistan, and in other similar involvements have also shown that some of the key observational and intelligence data in such operations comes not only from traditional sensor systems but from dismounted soldiers or other human observers reporting on their patrol activities. These data are naturally communicated in language in the form of various military and intelligence reports and messages. Such data, in textual, linguistic form, are entirely different than hard sensor data, as they are much more ambiguous, yet they can also be much more semantically rich; they are "soft" data in the sense that they are both largely uncalibrated and their content is much harder to fully understand (deep understanding begs the age-old challenge of forming automated methods for natural language understanding). Such "Soft" data finds its way into DIF processes as both structured and unstructured digitized text, and this input modality creates new challenges to DIF process designs, contrasted with more traditional DIF applications involving the use of highly-calibrated, numerically precise observational data from sensors. Combined with the data from the usual repertoire of "hard" or sensor data from various radio frequency (RF) sensors, video and other imaging systems, as well as SIGINT and satellite imagery, the observational data stream is a composite of data of highly different quality, sampling rates, content, and structure.

One main deficiency and critical path is on the soft data/human observation side, since it is generally agreed that DIF for observational data provided by hard, physical-science type sensors is much more mature; some have in fact argued that capabilities for Level 1 Fusion with hard data input is rather mature and that limited research investments should be made in this area. Although additional hard/Level 1 fusion research remains to be done, we generally concur with these judgments and believe that the first requirement is to define and prototype a viable processing paradigm for soft data fusion both for single and multiple input streams, so that the critical pre-estimation functions of Common Referencing and DA can be constructed. If we examine a notional processing diagram for multiple streams of human observational data expressed in linguistic terms (this is just one category of soft data), we envision something like the process in Fig. 2 (this is similar to a prototype of this process we have developed at our research center [2]):

In this depiction, each human observer processes the energy received from their sensing capability into a Perception-Cognition cycle, and a mental process judges how to express the observation in language (Linguistic Framing), resulting in a linguistic utterance, and the chosen instance of language. This utterance may be audio and need to be converted to digital text, and then formed into a message (that

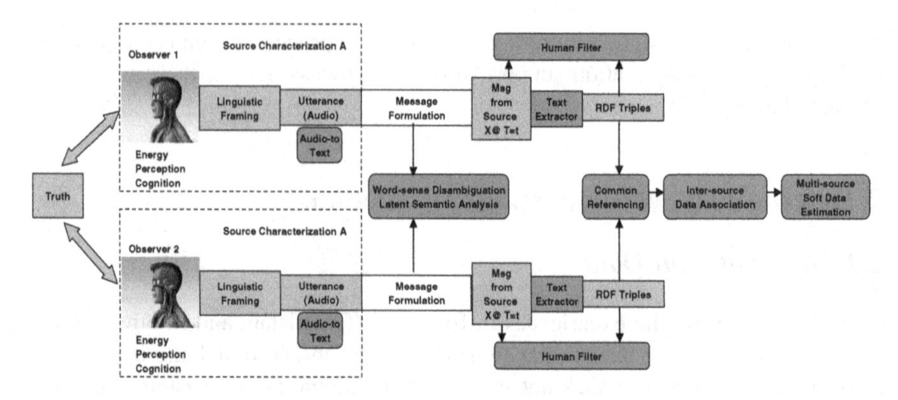

Fig. 2 Notional multi-message-stream soft data fusion process

may be sent over network communications channels, not shown). Today, the received message is typically parsed by a state-of-the-art Text Extractor, yielding for example RDF Triples of Subject-Verb-Object phrases, or some other representation (see [3] for the "propositional graph" approach we are using in our research). In virtually every military application, the message stream and/or the triples would be filtered through a human observer who functions as a first-level Quality Control process. Each filtered triples stream then comprises the raw data input into a downstream Data Fusion process. The meta-data (time-tags, uncertainty, etc) and the semantic content of these triples need to be framed in a normalized way by processing through the CR function, and then associated to determine if they relate to the same Entity in the true, unknown world, so that multisource, fusion-based estimation processes can exploit their informational content.

2.2 *Open Source and Social Media Data*

Soft or Hard data can also find its way into modern DIF processes in the form of monitored Open Source and Social Media feeds such as newswire feeds, Twitter and Blog sources judged to be possibly helpful. Getting such data into a DIF system will require automated web crawlers and related capabilities, and subsequent natural language processing capabilities, as much of these data are also represented in language.

2.3 *Contextual Data*

Modern problems also afford (and demand) the use of additional data and information beyond just observational data. A major category of such data and information is

Contextual Information. Contextual Information is that information that can be said to "surround" a situation of interest in the world (many definitions and characterizations exist but we will not address such issues here). It is information that aids in understanding the (estimated) situation and also aids in reacting to the situation, if a reaction is required. Contextual Information can be relatively or fully static or can be dynamic, possibly changing along the same timeline as the situation (e.g., weather). It is also likely that the full characterization and specification of Contextual Information may not be able to be known at system/algorithm design time, except in very closed worlds. Thus, we envision an "a priori" framework of exploitation of Contextual Information that attempts to account for the effects on situational estimation of that Contextual Information (CI henceforth) that is known at design time. Even if such effects are known at design time, there is a question of the ease or difficulty involved in integrating CI effects into a fusion system design or into any algorithm designs. This issue is influenced in part by the nature of the CI and the manner of its native representation, e.g., as numeric or symbolic, and the nature of the corresponding algorithm; for example, cases can arise that involve integrating symbolic CI into a numeric algorithm. Strategies for a priori exploitation of CI may thus require the invention of new hybrid methods that incorporate whatever information an algorithm normally employs in estimation (usually observational data) with an adjunct CI exploitation process. Note too that CI may, like observational data, have errors and inconsistencies itself, and accommodation of such errors is a consideration for hybrid algorithm design. Similarly, we envision the need for an "a posteriori" CI exploitation process, due to at least two factors: (1) that all relevant CI may not be able to be known at system/algorithm design time, and may have to be searched for and discovered at runtime, as a function of the current situation estimate, and (2) that such CI may not be of a type that was integrated into the system/algorithm designs at design time and so may not be able to be easily integrated into the situation estimation process. In this case then we envision that at least part of the job of posteriori CI exploitation would involve checking the consistency of a current situational hypothesis with the newly-discovered (and situationally-relevant) CI. There are yet other system engineering issues. The first is the question of accessibility; CI must be accessible in order to use it, but accessibility may not be a straightforward matter in all cases. One question is whether the most-current CI is available; another may be that some CI is controlled or secure and may have limited availability. The other question is one of representational form. CI data can be expected to be of a type that has been created by "native" users; for example, weather data, important in many fusion applications as CI, is generated by meteorologists, for meteorologists (not for fusion system designers). Thus, even if these data are available, there is likely to be a need for a "middleware" layer that incorporates some logic and algorithms to both sample these data and shape them into a form suitable for use in fusion processes. In even simpler cases, this middleware may be required to reformat the data from some native form to a useable form. In spite of some a priori mapping of how CI influences or constrains the way in which situational inferences or estimates can be developed, which may serve certain environments, the defense and security type applications, with their various dynamic and uncertain types of CI, demand a more adaptive approach. Given a nominated situational hypothesis Hf from a fusion process or "engine", the first question is: what CI type information is relevant to this hypothesis? Relevant CI is only that information that influences our interpretation or understanding of Hf. Presuming a "relevancy filter" can be crafted, a search function would explore the available CI and make this CI available to an "posteriori" reasoning engine. That reasoning engine would then use: (1) a CI-guided subset of Domain Knowledge, and (2) the retrieved CI to reason over Hf to first determine consistency of Hf with the relevant CI. If it is inconsistent, then some type of adjudication logic will need to be applied to reconcile this inconsistency between: (1) the fusion process that produced Hf and (2) the posteriori reasoning process that judges it as inconsistent. If however Hf is judged as consistent with the additional CI, an expanded interpretation of Hf could be developed, providing a deeper situational understanding. This overall process, which can be considered a "Process Refinement" operation, would be a so-called "Level 4" process in the context of the JDL Data Fusion Process Model (see [1]), that is, as an adaptive operation for fusion process enhancement. The overall ideas discussed here are elaborated in [4].

2.4 Ontological Data

DIF processes and algorithms have historically been developed in a framework that has assumed the a priori availability of a reliable body of procedural and dynamic knowledge about the problem domain; that is, knowledge that supports a more direct approach to temporal reasoning about the unfolding patterns of interest in the problem domain. In COIN and other complex problems, such a priori and reliable knowledge is most often not available—the Tactics, Techniques and Procedures ("TTP's") of modern-day adversaries are highly adaptive and extremely hard to model with confidence. The US DARPA COMPOEX Program [5] attempted to develop such models but only achieved partial success, experiencing gaps in the overall modeling space of such desired behavioral models. We label these types of problems as "weak knowledge" problems, implying that only fragmentary a priori behavioral model type knowledge is available to aid in DIF based reasoning, inferencing, and estimation.

Ontological information however, that does not attempt to overtly form such comprehensive behavioral and temporal models but does include temporal primitives along with structural/syntactic relations among entities, can be specified a priori with reasonably good confidence, and thus provides a declarative knowledge base to support DIF reasoning and estimation. Note that such knowledge is also represented in language and is available as digital text, in the same way as data from messages, documents, Twitter, etc. The use of ontological information in DIF systems can be varied; ontological information can augment observed data, can aid in asserting possible relationships, help in directing search and also in sensor management (to acquire expected information based on ontological relations), and yet other ways. Importantly, specified ontologies can also serve as providing consistent and grounded semantic terminology for any given system. In our current research, we employ ontologies primarily for augmenting observational data with asserted ontological data whose relevance is algorithmically determined using "spreading activation" and then integrated to enrich the evidential basis for reasoning [6]. The broader implications of ontologies for intelligence analysis are described in [7], that comes from our university's National Center for Ontological Research (see http://ncorwiki.buffalo.edu/index.php/Main_Page).

2.5 *Learned Information*

Finally, there is the class of information that could be learned (online) from all of the above sources if the DIF process is designed with a Data Mining/Inductive or Abductive Learning functional component. Very little research and prototyping of such dual-inferencing-process type DIF systems has been done although the conceptualization of such DIF schemes and architectures has been put forward some time ago by Waltz (e.g., [8]), as shown in Fig. 2. Any DIF system that incorporates such dual-inferencing schemes will encounter the challenge of knowledge management; whether and how any runtime learned knowledge gets integrated into runtime operations, or gets saved for later operations, or any other scheme for employment of learned knowledge is a challenge for storing, managing, and integrating that knowledge. The runtime integration of learned information raises a number of both algorithmic issues as well as architectural issues. For example, if meaningful patterns of behavior can be learned and can be measured/judged as persistent or enduring, such patterns could be incorporated in a dynamically-modifiable knowledge base to be reused. In Fig. 3, Waltz shows that the management of such knowledge evolving from what he calls Data Mining operations is handled by the "Level 4", process refinement function of the traditional JDL DIF process.

Such learning processes will also not be perfect and have some uncertainty that also needs to be factored into the traditional CR and DA functions of the target fusion process.

The heterogeneity of data and information as just described also creates new challenges and complexities for the traditional functions of DIF as depicted in Fig. 1. In the next section, we address the impacts of these modern defense/security problems and of data heterogeneity on the DIF functions of Data Alignment or Common Referencing, and on DA.

3 Common Referencing and Data Association

As pointed out in Fig. 1, CR is that traditional DIF system function that is sometime called "Alignment" and is the function that normalizes these input sources for any given fusion application or design. CR addresses such things as coordinate system normalization, temporal alignment issues, and uncertainty alignment issues

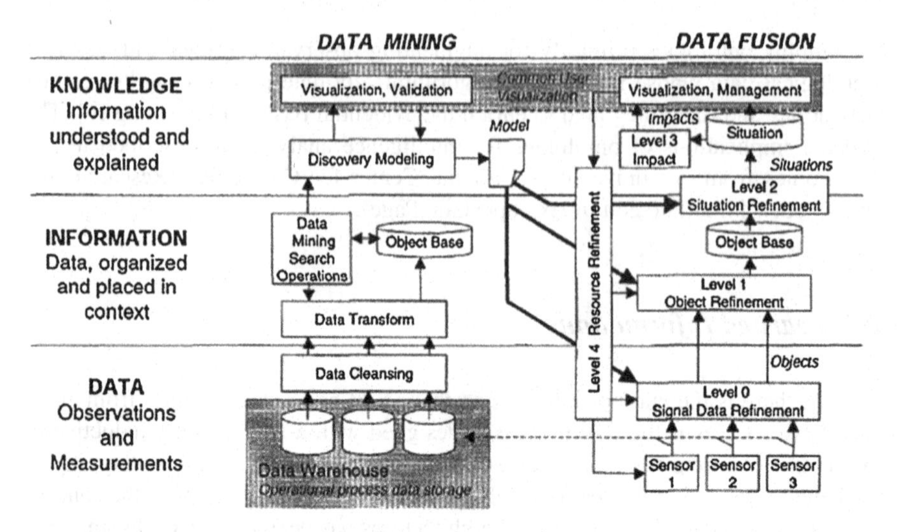

Fig. 3 Notional fusion process architecture combining data mining and data fusion (from [8])

across the input streams, among other issues. With the highly-disparate input streams described above, the design of required CR techniques is a non-trivial challenge. There are at least two major CR issues that this heterogeneous data represent: temporal alignment and uncertainty alignment. Consider a textual input message whose free text, in just a few lines, could have past-present-future tense expressions, e.g., "3 days ago I saw....", "past precedents lead me to believe that tomorrow I should see....", etc. Other sources can also have varied temporal structures regarding their input. Such data lead to the issue of what the DIF community has called "OOSM: out-of- sequence-measurements" for hard/sensor data but the issue carries over to all sources as well. Dealing with these issues requires complex temporal alignment techniques for CR and also raises the issue of retrospective fusion processing operations to correct for delayed inputs (if warranted; this is a design choice). For example, such process designs impute the need to set a threshold for allowable delays (how far back in time will we adjust for), and this also sets a requirement for memory capacity to save all data in that window to allow undoing and redoing the inferences when such time-late or past-referenced data arrives. Temporal alignment methods we have used for Soft data are described in [9].

The uncertainty alignment requirement evolves due to the high likelihood that any uncertainty in the widely disparate sources described above will be represented in inconsistent forms. Consider the basic differences between the uncertainty in sensor (hard) data and textual (soft) data; sensor data uncertainty is sensibly always expressed in probabilistic form whereas, due to the problem of imprecise adjectives and adverbs in language, linguistic uncertainty is often expressed in possibilistic (fuzzy) terms. It can be expected that uncontrolled Open Source or Social Media data may use yet other uncertainty formalisms to express or tag inputs (e.g., beliefs and subjective confidence measures). Transformation and normalization of disparate forms of uncertainty is a specialized topic in the uncertainty/statistical literature (e.g., [10]), and is among the high-priority issues in the DIF community [11]. It should be noted that such transformations can only be developed by invoking some statistical type qualities that are preserved across the transform, such as some form of total uncertainty; that is, the transform of some probability value does not create an "equivalent" value of a probability in, say, a possibilistic space; instead the transformed value is one that satisfies some statistical constraint about which the transform is structured. For the interested reader, seminal papers on the probability-possibility transformation issue are in [12–14]. In our research, we have addressed the probabilistic- possibilistic transformation issue in an approach that satisfies the consistency and preference preservation principles [15], resulting in the most specific distribution for a specified portion of a probabilistic representation; this yields a truncated triangular transformation in our case [16].

Regarding the DA function, that some consider the heart of a fusion process, these highly-varied data raise the level of DA complexity in significant ways. The soft data category, that inherently is reporting about Entities and (judged) Relationships, and is inherently in semantic format (language/words), raises the important issue of how to measure semantic similarity of such elements as reported in these various input streams. Such scores are needed in the "Hypothesis Evaluation" step of the DA process (see [17] on these DA subfunctions). But there are further DA complications that arise due to the soft data: linguistic phrases have verbs that reflect inter-Entity (noun) relationships; also of note is that the Natural Language Processing (NLP) community has employed graphical methods for the representation of linguistic structures. As a result, the DA process now involves inter-association of both Entities (nouns) and Relations (verbs), and of graphical structures. This requirement extends to the hard data as well since that data needs to be cast in a semantic framework in order to enable the overall DA process for the combined Hard and Soft data. Developing DA methods for graphical structures represents an entirely new challenge for the DA function. In such approaches for these applications, a scoring approach also needs to be developed to assess Relational similarity as well as Entity similarity, and a composite association scheme for these graphical substructures needs to be evolved. Historical approaches to DA have often employed solution methods drawn from assignment problems in Operations Research. When association is required between many non-graphical data sources (i.e., among entities and attributes, as in the multisensor-multitarget tracking DA problem), this can be handled by such methods as the multidimensional assignment problem [18, 19]. The main difference between the multidimensional assignment problem and graph-based association is how topological information from the graphs is used. Our research center has attacked this problem and has developed research prototype algorithms, as described in [20] where the graph association problem is formulated as a binary linear program and a heuristic for solving the multiple graph association is developed using a Lagrangian relaxation approach to address issues involving a between-graph transitivity requirement.

In virtually all computer applications involving the estimation or inferencing about some state of affairs such as a "situation", there is the issue of constructing computer-based processes (software) that is able to work with notions of "meaning". Dealing with notions of meaning becomes more difficult in DIF processes as one attempts to build methods for so-called "high-level" fusion, involving more abstract hypotheses such as situations and threats, etc. In modern problems and with hard and soft data sources, these problems are aggravated; some aspects of these issues are discussed in the next section.

4 Semantics

The introduction of linguistic information, as well as the transformation of sensor+algorithm estimation process outputs (hard fusion outputs) into a semantic frame, also adds to the complexity of DIF process design and development. Semantic complexity is also added by the very nature of modern intelligence and security problems wherein the situations of interest relate to both military operations and also socio-political behaviors and entities. Clear meanings of such notions of interest in modern intelligence or ISR problems such as "patterns of life", "rhythm of the city", "radicalization" as patterns or situations of interest—to be estimated by DIF systems—have proven difficult to specify in clear semantic terms, that is, to specify their meaning with adequate specificity for computer-based processes. While the use of ontologies helps in this regard, standardization issues remain when considering networked and distributed systems, which are typical in the modern era. For example, in distributed intelligence or military systems there is typically no single point of architectural authority that can mandate a single ontological framework for the network. For large-scale real systems there is also the problem of large legacy systems that were never designed with ontological formalisms in mind; this creates a "retrofit" problem of adjusting the semantic framework of that system to some new ontological standard, which can be a costly and complex operation.

It must also be noted that the way in which all textual/linguistic information gets into a DIF system is through processing in some type of NLP or text extraction system. Such systems serve as a front-end filter for the admission of fundamental entity and relationship data, the raw Soft data of the system, and so any imperfections in such extractions bound the capture of semantically-grounded evidential information for the subsequent reasoning and estimation processes; that is, the meaning of the text can be lost. While errors in hard sensor data are typically known with reasonable accuracy due to sensor calibrations, the errors in text extraction and NLP systems are either weakly known or unknown, sometimes as a result of proprietary constraints. Other strategies to deal with the complexities of semantics involve the use of controlled languages, to bound the grammatical structures and also the extent of the vocabulary that has to be dealt with. A good example for military/intelligence applications is the "Battle Management Language" or BML [21] that has been under development since about 2003 for both Command and Control simulation studies but also for DIF applications (e.g., [22, 23]).

There is a corresponding need to better understand the nature of semantic (and syntactic) complexity in language, and also to develop measures and metrics that aid in developing better NLP processes and controlled languages. There is a reasonably rich literature on these topics (e.g., [24]) that should be exploited in regard to the integrated design of DIF systems that today have to deal with a wide range of semantic difficulties.

As hinted at in our discussion regarding DA, many of these current problems involve graphical data representations and therefore impose the use of graphically-based algorithmic techniques. Some of these issues are addressed in the next section.

5 Graphical Representations and Methods

There are a number of reasons that, for COIN and asymmetric warfare-type problems, graphs are becoming a dominant representational form for the information in and the processes involved in DIF systems. In the information domain, many of the components discussed in Sect. 2 are textual/linguistic and to capture this information in digital form, graphs are the representational form of choice. The problem domain is also described in the ontologies that are also typically couched in graphical forms. Note that ontologies describe inter-entity relations of various types. Note too that the inferences and estimates of interest in these problems are of the "higher-level" type in the sense of the JDL Model of Information Fusion, that is, estimates of situations and threat states. These higher-level states—the conditions of interest for intelligence and security applications—are also best described as graphs, since situations can in the most abstract sense be considered as a graph of entities and relations.

As a result, it is not unexpected to see that the core functions of DIF such as DA as previously described, are employing graphical methods in these fusion function operations. The U.S. Army's primary intelligence support system, the Distributed Common Ground Station-Army (DCGS-A) employs a "global graph" approach to capture all of the evidentiary information that supports DIF and other intelligence analysis operations; see [25] and Fig. 4 that shows the top-level structure of this graphical concept.

Developing a comprehensive understanding of these problems thus involves a logical synthesis of the many situational substructures or subgraphs in these problem domains; the fusion-process-generated subgraphs can be thought of situational components or hypotheses. The subgraphs are somewhat thematic and can be thought of as revolving about the "PMESII" notion of the heterogeneity of the classes of information of interest in such problems (PMESII stands for Political, Military, Economic, Social, Infrastructure, and Information categories). Thus, it is also not surprising to see Social Network Analysis tools—that are by the way graph-theoretic and graph-centric—employed in support of intelligence analysis, here with the focus on the Social and Infrastructure patterns and subgraphs of the problem space.

In our own work for such problems, we considered that it would be broadly helpful in analysis to enable a subgraph-querying capability as a generalized analysis tool.

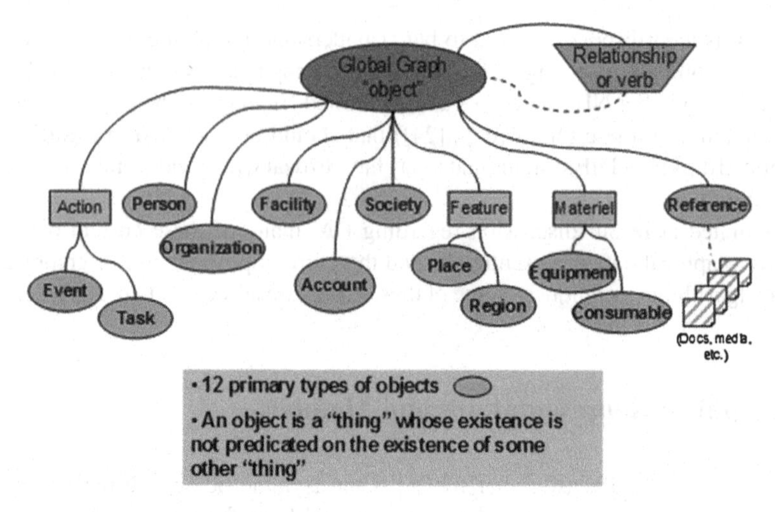

Fig. 4 U.S. Army's "Global Graph" concept for DCGS-A (from [25])

In such an approach, the analyst forms a query in text that can be transformed to a graph (we call these "template" graphs in that they are subgraph structures of interest—a textual/graphical question in effect) that is then searched for in the associated-evidence graph that is formed by the DA process. This search operation is a stochastic, inexact graph-matching problem, since the nodes and arcs of the evidential data set have uncertainty values associated with them (or perhaps the template graph as well, if the query has stochastic/uncertain aspects), and also because what is sought is the best match to the query, not an exact match, since there may be no exact match in such unpredictable problem situations. Other complexities arise in trying to realize such capability, such as executing such operations incrementally for streaming data, and also doing them in a computationally-efficient way since the graphs can get quite large. As a consequence of several PhD efforts, we have realized today a rather mature graph-matching capability for intelligence analysis that is implemented in a cloud-based process; see [26–28], among other of our works.

6 Analytics, Sensemaking, and Decision-Making

We have noted previously that for the problems of interest here, those so-called "irregular" and "asymmetric" problems, the amount and reliability of a priori knowledge about the problem spaces is typically very limited. By and large, this means that analyzing the associated, multisource evidential data involves a mixture of strategies as has been suggested in Fig. 2. System designers and analysts must understand that there will be no singular tool or analytical technique that provides the "answer" at the level of abstraction desired. Such analysis environments are not entirely new to intelligence and military ISR analyses but these modern problems impose new and additional difficulties in analysis methods and strategies. Commanders and analysts do not approach these environments totally absent of knowledge, and they usually have some type of focal topics and issues of interest. For commanders and analysts both, it is usual to have a set of Priority Intelligence Requirements (PIR's) that are ideally interrelated to anticipated Course of Action (COA) decision options. However, the action space for these problems involves the range of political, economic, military, paramilitary, psychological, and civic actions, i.e., not only "kinetic" actions involving the use of weaponry. As remarked previously, the decision space can thus also be labeled as "soft" in that the decision-space includes such decisions as those resulting in the realization of desired levels of influence (e.g., onto tribal leaders etc). It can be seen immediately from this definition that both the understanding of a current situation and its various elements, and the space of possible decisions and actions both have a much larger dimensionality than traditional military decision-making in force-on-force operations. Collectively, the broad elements of this action space can be broken into "direct" and "indirect" classes of actions, where direct actions are those focused on adversarial structure in the traditional military sense, and indirect actions those focused on undermining support to the adversaries while simultaneously attacking them militarily. It can also be argued that the End States of any decision sequence are "Effects" created by the sequence of actions (the COA). The concepts of Effects Based Operations (EBO), not a new term but actively revisited for these modern problems (e.g. [29]), shows that many references suggest that EBO is a viable concept for irregular/asymmetric problems, in part because effects are soft-type results, and subsume behavioral end-states, reflecting a human focus. One simple taxonomy of Effects is shown in Fig. 5 (from [30]), a main distinction being "Physical" versus "Behavioral", which could be equated to "Kinetic" versus "Non-kinetic".

The development of an interlinked COA to create these behavioral, non-kinetic Effects as end-states is very difficult and involves a web of interdependencies that make EBO a process involving notions of Complex Adaptive Systems (CAS). Smith [31] elaborates on this in various ways, and this CAS notion is also discussed in [32] that emphasizes the non-deterministic aspect of any Course of Action producing an intended Effect. Smith [31] has an extended development of the Effects-Based approach for asymmetric operations, and in consideration of what Smith calls an action-reaction cycle model (sensibly equivalent to Situation Management) puts forward a linked process that specifically shows the influences of understanding the Social Domain as part of the "Sensemaking" process that ultimately drives the COA development.

A very important notion (see [33]) is that the COA development process starts with a projected "Plausible Future" state so that actions are taken not necessarily on the basis of the current situation but one that *is* expected to exist at the time actions are taken on it, i.e., so that the situational state and actions onto it are as synchronous as possible. Note that, ideally, the DIF system should be supportive of some type of situational projection of such plausible future states, as part of an analysis suite. Supported under Air Force Research Laboratory funding, we have explored the ideas involved with, and the prototyping of automated DIF techniques for such estimation

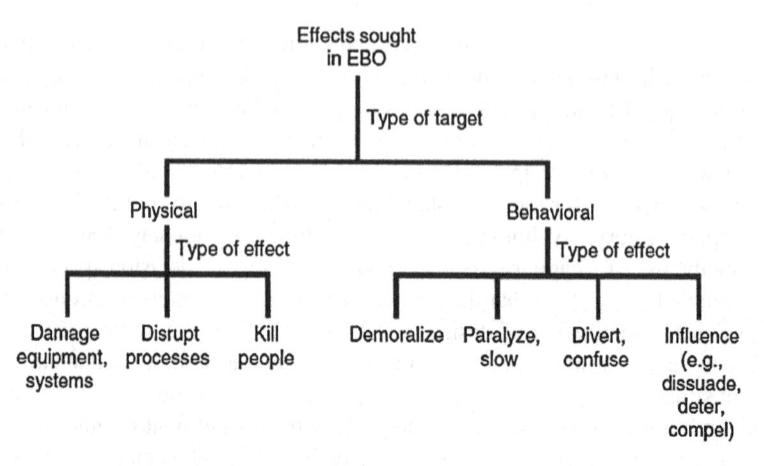

Fig. 5 Sample taxonomy of effects (from [30])

of plausible futures; see [34]. Additional remarks on the issues surrounding DIF, decision-making, COA development in the counterinsurgency environment can be seen in Llinas [35].

We see, as shown later, what today are called Sensemaking processes, as lying between DIF and DM processes, in a stage wherein "final" situation assessments and understandings (in the human mind) are developed. Thus, our view of this meta-process is as a three-stage operation: DIF as an automated process that nominates algorithmically-formed situational hypotheses (including nominations of "plausible future situations"), Sensemaking that dynamically interacts with DIF and human judgment in a kind of mixed-initiative operation to produce a final situational hypotheses upon which then DM operations are triggered. While there is also a substantive literature on Sensemaking, we address here three models: those of Pirolli and Card [36], of Klein et al. [37], and of Kurtz and Snowden [38]. The first two have many similarities and so we will show a figure of just one. These models depict Sensemaking as an iterative operation involving a hopefully-converging dynamic between a supporting information-space and an evolving situation hypothesis space. Here, the former is considered to be an automated DIF process and the latter is seen as occurring in the human mind, possibly aided by automated utilities. In [36], the overall Sensemaking process is "organized into two major loops of activities: (1) a "foraging" loop that involves processes "aimed at seeking information, searching and filtering it, and reading and extracting information, possibly into some schema, and (2) a sensemaking loop that involves iterative development of a mental model (a conceptualization) from the schema that best fits the evidence." The Klein et al. Sensemaking model [37], called a Data-Frame Model, has many similarities to the process characteristics just described. The Kurtz and Snowden model, organized around their framework called Cynefin [38] is really based on the idea that categorizing the nature of the problem at hand, thereby partitioning it (in a "divide and conquer" strategy) and applying appropriate solution methods, is a part of the Sense-

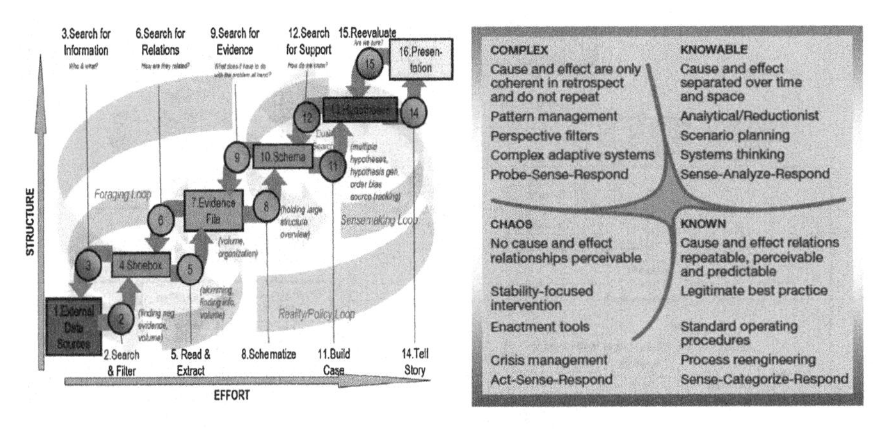

Fig. 6 Pirolli and card (*left*) and Kurtz and Snowden (*right*) models of Sensemaking (from [36, 38])

making process. Cynefin partitions problems into four categories called: Known (soluble by known methods), Knowable (soluble by analytical/reductionist techniques), Complex (soluble by what [38] calls Probe-Sense-Respond iterative discovery type processes), and Chaos (soluble by actions to reduce disorder, sensing the results, and responding or acting again). Cynefin takes a broader view of the states of complexity as ranging from order-to complexity-to chaos than do the models of Pirolli, Card or Klein; we include them because our concerns are for modern irregular, asymmetric warfare applications that often can have such properties. Diagrams showing the Pirolli/Card and Kurtz/Snowden models are provided in Fig. 6.

7 Connected Processes

So how do these processes interact, as we are asserting here? Fig. 7 shows a functional characterization of how:

- DIF, a largely-automated inferencing/estimation providing process that offers:
  - Algorithmically-developed situational estimates
  - Organized raw observational data—note these are hard (sensor) and soft (linguistic)
  - Controllable collection management of observational data
  - An Analytical Suite of useful but typically disparate tools
- Sensemaking, a semi-automated, human-on-the-loop process that:
  - Considers the DIF-provided estimates
  - Forages over these hypotheses as well as the data (e.g. drill-down etc)
  - Assesses the "Cynefin-category" nature of the problem at hand

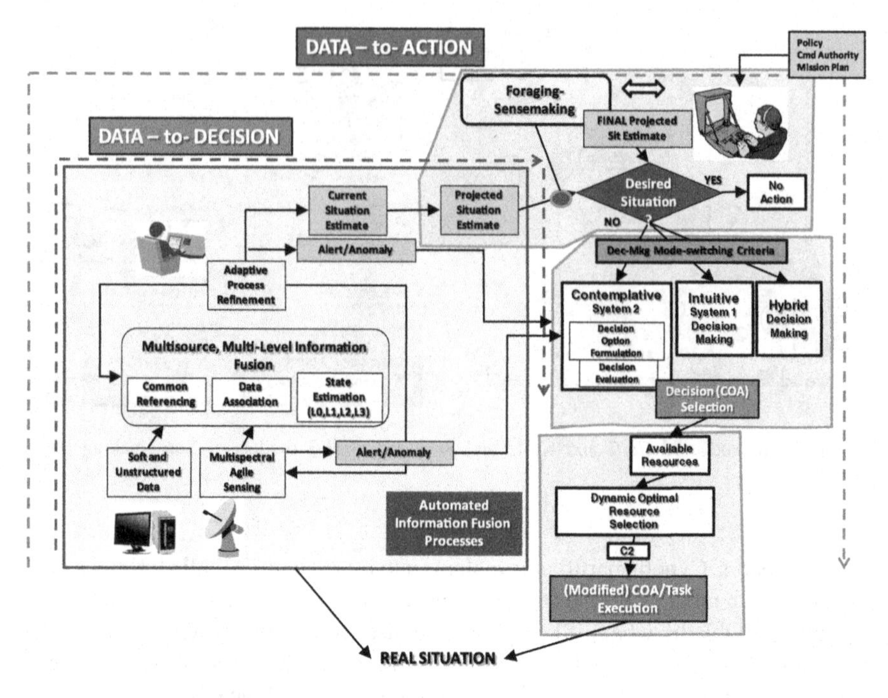

Fig. 7 Interconnected/dependent DIF-Sensemaking-DM-Resource Mgmt processes

  - Considers possible Policy, Authority, and Mission factors
  - Culminates in a "Final Adjudicated Situation Hypothesis" that is also judged as to acceptability; if not, this hypothesis is the starting point for decision-making and action-taking to "manage the situation"

- Decision-making, also a semi-automated, human-on-the-loop process that:
  - Operates in a System 1 (intuitive), 2 (contemplative, analytic) or "hybrid/mixed" DM mode
  - Yields a selected Course of Action
  - That triggers a Resource Optimization process to define specific resources that physically enable the selected COA onto the real-world situation

In [33], some of the issues regarding inter-process interdependencies were discussed (such as temporal dependencies), although that paper's focus was on the various metrics involved across these processes. Another point we will make in this chapter is that yet another consideration related to decision-making is that most models of DM depict it as an analytical, contemplative process (analytical DM or ADM). It is important we think to realize that the DM community also discusses intuitive DM (IDM) that has considerably different properties than ADM. If we examine the disparate features of ADM and IDM, shown below in Table 1, we see that DIF process designs will need to be quite different to service the distinct functionalities

Table 1 Comparative features of IDM/System 1 and ADM/System 2 DM modalities (from [43])

Intuitive	Analytical
Experiential-inductive	Hypothetico-deductive
Bounded rationality	Unbounded rationality
Heuristic	Normative reasoning
Gestalt effect/pattern recognition	Robust decision making
Modular (hard-wired) responsivity	Acquired, critical, logical thoght
Recognition-primed/thin slicing	Multiple branching, arborization
Unconscious thinking theory	Deliberate, purposeful thinking

for each DM mode. Thus, it can be argued that the DIF process should, ideally, be informed of the DM modality that users are in at any moment, so that, assuming the DIF design can be made DM-mode-sensitive, switch its operating mode to best service any DM mode at the moment. The DIF research community has conducted very minimal research on designing DM-mode-sensitive DIF processes; we see only two papers in the recent literature addressing this topic (see [39, 40]). In regard to IDM in particular, it could be argued that Case-Based Reasoning (CBR) techniques (similar to Klein's RPD process, that enable intuitive, experientially-based inferencing and DM) might be a preferred inferencing mode in DIF for the IDM modality. While there are similarities between IDM and CBR/RPD, there is an important distinction for (probably most) modern operational domains about the notion of novelty in situations, and the true underlying capability of a human to deal with situations that are "seriously different" from their experience base. Naturalistic decision making using the RPD model fails in theory if there is a lack of experience or when encountering a completely novel scenario [41]. A review of most IDM models suggests that the inherent limits of IDM are the decision-makers personal range of situational experience combined with what has been "implicitly learned". Any presented situation that is not adequately similar to this body of experience requires adaptation and learning. Boin et al. [42], state that "if the situation is radically different from those stored in memory, a somewhat different kind of sense-making process will be necessary."

Another dependency area is between the DIF and Sensemaking processes. Clearly the Foraging function within Sensemaking implies that the DIF process will have to be open to, and enable, a range of queries that will be in regard to: raw or processed observational data, DIF functional operations (e.g., DA[1]), and nominated situational

[1] The Army's DCGS-A future system requirements for example include user-modifiable DA capabilities.

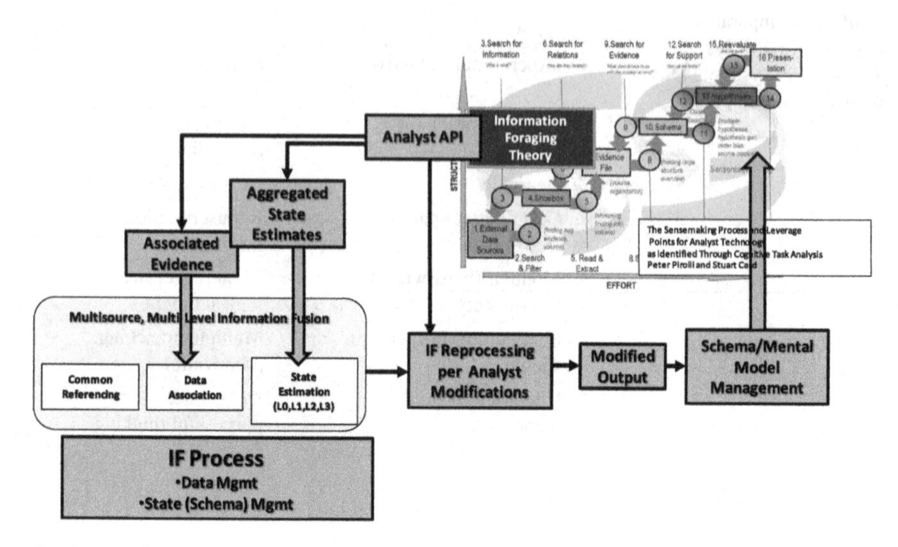

Fig. 8 Notional functional operations of the DIF-Sensemaking interactions

hypotheses, among possibly other runtime interactive operations. The notion of this interaction is depicted in Fig. 8, showing an analyst API that allows runtime modification of either or both of the Association and Estimation functions, then followed by IF reprocessing to generate new results that then get absorbed (possibly with automated support, not shown) into the analyst's schema and mental models.

8 The Human Role in DIF, Sensemaking, and DM

This general process model can be seen to have at least two human points of involvement (assuming that the analyst is not the decision-maker). In our prototype hard-soft DIF system, we also have a possible role for a human in regard to editing the automated text extraction process for the soft/message data stream due to the considerable difficulty in achieving high quality extraction with automated methods. It can be appreciated that the complexity of natural language understanding, the complexities of the problems domain and the hard-soft fusion process all impute a serious consideration for placement of human intelligence in system design. The human role in DIF processes has been discussed for some time in the DIF community, and there are some works addressing the issues [44, 45]. In our judgment however, this general issue has been inadequately addressed at the community level, probably as a result of the DIF community having a quantitative bias, as can be seen in any review of community publications. The assertions and discussion here expands the challenge of addressing not only the human role in DIF but in Sensemaking and Decision-Making as well. The larger issue is a meta-system design question across the DIF-Sensemaking-DM meta-process, as regards the placement of human intelligence and judgment for interpretation, control, and decision-making. The usual issues of quality of interpretation, quality of decision-making, quality of control versus timeliness need to be dealt with in developing approaches to designing this meta-system.

9 Summary

The world is dynamic in many ways. Looking at world politics and technology, no one should be surprised that there have been dramatic changes in the nature of security aspects driven by world politics; over the span of a decade or so, there should similarly not be any surprise that technology has advanced considerably. It is in this setting that this chapter was written, to offer perspectives on what those meta-changes have implied for the design and development of DIF systems as they sit in the interdependent environment with sensemaking and either analysis or decision-support systems. DIF system designers need to both take a larger view of their system's design but also reach out to and collaborate with those designing the related major functional capabilities for sensemaking and analysis and decision-making. DIF has always been a multidisciplinary area of study; this larger view further complicates that aspect but it is the opinion taken here that those interdependencies are inescapable, and that effective and efficient DIF designs can only be realized in the context discussed herein.

References

1. Bowman, C.L.: The dual node network (DNN) data fusion & resource management (DF&RM) architecture AIAA intelligent systems conference, Chicago, 20–22 Sept 2004
2. Prentice, M., Kandefer, M., Shapiro, S.C.: Tractor: a framework for soft information fusion. In: Proceedings of the 13th international conference on information fusion (fusion 2010), Edinburgh, Scotland, UK, July 2010
3. Prentice, M., Shapiro, S.C.: Using propositional graphs for soft information fusion. In: Proceedings of the 14th international conference on information fusion (fusion 2011), Chicago IL, USA, July 2011
4. Gómez-Romero, J., García, J., Kandefer, M., Llinas, J., Molina, J.M., Patricio, M.A., Prentice, M., Shapiro, S.C.: Strategies and techniques for use and exploitation of contextual information in high-level fusion architectures. In: 13th international conference on information fusion (fusion 2010), Edinburgh, Scotland, UK, July 2010
5. A. Kott, P.S. Corpac, COMPOEX Technology To Assist Leaders in Planning And Executing Campaigns In Complex Operational Environments, 12th International Command and Control Research and Technology Symposium: Newport. Rhode Island, USA (2007)
6. Kandefer, M., Shapiro, S.C.: Evaluating spreading activation for soft information fusion. In: Proceedings of the 14th international conference on information fusion (fusion 2011), Chicago, IL, USA, 5–8 July 2011
7. Smith, B., et al. Ontology for the intelligence analyst, in crosstalk. J. Def. Softw. Eng. 18–25, (2012)

8. Waltz, E.: Information understanding: integrating data fusion and data mining processes. In: Proceedings of the IEEE international symposium on circuits and systems, ISCAS '98, Ann Arbor, Michigan, USA, 1998
9. McMaster, D.: Temporal alignment in soft information processing. In: Proceedings of the 14th international conference on information fusion (fusion 2011), Chicago, Illinois, USA, July 2011
10. Klir, G.J.: A principle of uncertainty and information invariance. Int. J. Gen. Syst. **17**, 249–275 (1990)
11. Blasch, E., Llinas, J., Lambert, D., Valin, P., Das, S., Hong, C., Kokar, M., Shahbazian, E.: High level information fusion developments, issues, and grand challenges—fusion10 panel discussion. In: International conference on info fusion (fusion10), Edinburgh. Scotland, UK, July 2010
12. Oussalah, M.: On the probability/possibility transformations: a comparative analysis. Int. J. Gen. Syst. **29**(5), 671–718 (2000)
13. Geer, J.F., Klir, G.J.: A mathematical analysis of information- preserving transformations between probabilistic and possibilistic formulations of uncertainty. Int. J. Gen. Syst. **20**, 143–176 (1992)
14. Klir, G., Parviz, B.: Probability-possibility transformations: a comparison. Int. J. Gen. Syst. **21**(1), 291–310 (1992)
15. Dubois, D., Prade, H.: Unfair coins and necessity measures: a possibilistic interpretation of histograms. Fuzzy Sets Syst. **10**, 15–20 (1983)
16. Gross, G., Nagi, R., Sambhoos, K.: A fuzzy graph matching approach in intelligence analysis and maintenance of continuous situational awareness. J. Inf. Fusion **18**, 43-61 (2013)
17. Hall, D.L., Llinas, J.: Handbook of multisensor data fusion. CRC Press, Boca Raton (2001)
18. Poore, A., Lu, S., Suchomel, B.: Data association using multiple-frame assignments. In: Liggins M., Hall D., Llinas J. (eds.) Handbook of multisensor datafusion, chapter 13,2nd ed., pp. 299–318. CRC Press, Boca Raton, 2009
19. Poore, A., Rijavec, N.: A lagrangian relaxation algorithm for multidimensional assignment problems arising from multitarget tracking. SIAM J Optim **3**, 544–563 (1993)
20. Tauer, G., Nagi, R., Sudit, M.: The graph association problem: mathematical models and a lagrangian heuristic. Nav. Res. Logist. **60**(3), 251–268 (2013)
21. Mark Pullen, J., et al.: Joint battle management language (JBML)—US contribution to the C-BML PDG and NATO MSG-048 TA. IEEE European simulation interoperability workshop, Genoa, Italy, June 2007
22. Schade, U., Biermann, J., Frey, M., Kruger, K.: From battle management language (BML) to automatic information fusion. In: Proceedings of the IF and GIS, pp. 84–95, (2007)
23. Lee, H., Zeigler, B.P.: SES-based ontological process for high level information fusion. J. Def. Model. Simul. Appl. Methodol. Technol. **7**(4), (2010)
24. Pollard, S., Biermann, A.W.: A measure of semantic complexity for natural language systems. In: Proceedings of NLP complexity workshop: syntactic and semantic complexity in natural language processing systems, Stroudsburg, Pennsylvania, USA, 2000
25. Walsh, D.: Relooking the JDL model for fusion on a global graph. In: National symposium on sensor and data fusion, Las Vegas, Nevada, USA, July 2010
26. Stotz, A., et al.: Incremental graph matching for situation awareness. In: 12th international conference on information fusion (fusion 2012), Seattle, WA, USA, 6–9 July 2009
27. Sambhoos, K. et al.: Enhancements to high level data fusion using graph matching and state space search. Inf. Fusion **11**(4), 351–364 (2010)
28. Gross, G.: Continuous preservation of situational awareness through incremental/stochastic graphical methods. In: Proceedings of the 14th international conference on information fusion (fusion 2011), Chicago, IL, USA, 5–8 July 2011
29. Davis, P.K.: Effects based operations: a grand challenge for the analytical community. RAND Report, Santa Monica, CA (2001)
30. Smith, E.A.: Complexity, networking, and effects-based approaches to operations, US DoD Command and control research program publication. www.dodccrp.org (2006)

31. Smith, E.A.: Effects-based operations: applying network centric warfare in peace, crisis and war. CCRP, Washington, DC (2002)
32. Senglaub, M.: Course of action analysis within an effects-based operational context, SANDIA REPORT SAND2001-3497 Nov 2001
33. Llinas, J.: Quantitative aspects of situation management: measuring and testing situation management concepts, intelligent sensing, situation management, impact assessment, and cybersensing. Proc. SPIE **7352** (2009)
34. Holsopple, J., Yang, S.: Fusia: future situation and impact awareness. In: Proceedings of fusion 2008: the international conference on information fusion, Cologne, Germany, July 2008
35. Llinas, J.: Situation management in counterinsurgency operations: an overview of operational art and relevant technologies. In: Proceedings of the 14th international conference on information fusion (fusion 2011), Chicago, IL, USA, July 2011
36. Pirolli, P., Card, S.K.: The sensemaking process and leverage points for analyst technology. In: Proceedings of the 2005 international conference on intelligence analysis, McLean, Virginia, USA, 2005
37. Klein, G., Moon, B., Hoffman, R.F.: Making sense of sensemaking I: alternative perspectives. IEEE Intell. Syst. **21**(4), 70–73 (2006)
38. Kurtz, C.F, Snowden, D.J.: The new dynamics of strategy: sense-making in a complex and complicated world. IBM Syst. J. **42**(3), 462–483. (2003)
39. Louvieris, P., et al. Smart decision support system using parsimonious information fusion. In: Proceedings of the 7th international conference on information fusion(fusion 2005), Philadelphia, PA, USA, July 2005
40. Solano, M.A., et al.: High-Level fusion for intelligence applications using recombinant cognition synthesis. Inf. Fusion Jl **13**(1), (2012)
41. Diaz, S.K.: Where do I start? Decision making in complex novel environments, naval postgraduate school thesis, Sept 2010
42. Boin, A., et al.: The politics of crisis management: public leadership under pressure. Cambridge University Press, Cambridge (2005)
43. Croskerry, P.A.: Universal model of diagnostic reasoning. Acad. Med. **84**(8), 1022–1028, (2009)
44. Blasch, E.P., Plano, S.: JDL level 5 fusion model: user refinement issues and applications in group tracking. In: Proceedings SPIE 4729, signal processing, sensor fusion, and target recognition XI, p. 270, 31 July 2002
45. Hall, D.L., Jordan, J.M.: Human-centered information fusion. Artech House, Norwood, MA (2010)

Intrusion Detection with Type-2 Fuzzy Ontologies and Similarity Measures

Robin Wikström and József Mezei

Abstract Intrusions carry a serious security risk for financial institutions. As new intrusion types appear continuously, detection systems have to be designed to be able to identify attacks that have never been experienced before. Insights provided by knowledgeable experts can contribute to a high extent to the identification of these anomalies. Based on a critical review of the relevant literature in intrusion detection and similarity measures of interval-valued fuzzy sets, we propose a framework based on fuzzy ontology and similarity measures to incorporate expert knowledge and represent and make use of imprecise information in the intrusion detection process. As an example we developed a fuzzy ontology based on the intrusion detection needs of a financial institution.

1 Introduction

Intrusion detection systems are becoming more and more important in controlling network security as the number of intrusion events is increasing rapidly due to the widespread use of internet. A general intrusion detection system operates as decision support system by making use of the (real-time) information and event reports describing previous intrusion cases to identify potential dangerous activities. There are two main approaches to intrusion detection: misuse detection (known patterns of intrusion are compared to present activities) and anomaly detection (activities that deviate from normal system behaviour but cannot be matched to any previous cases) [2, 56].

R. Wikström (✉) · J. Mezei
Department of Information Technologies, IAMSR, ÅBo Akademi University, Joukahaisenkatu 3-5B, 20520 Turku, Finland
e-mail: rowikstr@abo.fi

J. Mezei
e-mail: jmezei@abo.fi

R.R. Yager et al. (eds.), *Intelligent Methods for Cyber Warfare*,
Studies in Computational Intelligence 563, DOI 10.1007/978-3-319-08624-8_7

As sensors and other data collection methods are evolving continuously, intrusion detection systems are forced to constantly process increasing amounts of information. A fair part of this information consists of imprecise and vague knowledge [41]. A fuzzy ontology is basically an ontology that employs fuzzy logic for dealing with imprecise knowledge [6]. Using fuzzy ontologies for analysing this knowledge is important for identifying possible intrusions [20]. This is especially true regarding anomalies, as it becomes possible to identify cases in a database that are similar in a fuzzy sense. As an extension of traditional fuzzy ontologies, type-2 fuzzy ontologies are not limited to a crisp value for defining the membership function of different concepts therefore offering more possibilities to model uncertainty compared to type-1 fuzzy ontologies [36].

Similarity measures have successfully been used for several anomaly detection implementations [14]. Detecting anomalies is an important method for finding unwanted behaviour, not only in intrusion detection but also in e.g. fraud detection and military surveillance. Kernel based similarity measures (cosine and binary [40]) together with text processing techniques were applied for example for detecting host-based intrusion detection [51].

Ning and Xu [47] noticed that Intrusion Detection Systems (IDSs) are producing an increasing amount of alerts, containing a fair share of false alerts, regardless, one needs to process this data. By applying similarity measures on the collected intrusion alerts, they generated the similarity between different attacks strategies. Their basic assumption is that knowing the attack strategy, one can predict the coming moves of the attacker. The use of expert knowledge can play a crucial role in identifying anomalies and assessing the potential loss that can be caused by an intrusion. One could even state that it is necessary to include experts in intrusion detection systems, as fully automated systems seem to be impossible to achieve [13]. Some systems are able to detect malware and intrusions based on behavioural patterns, however, few come even close to automatically deciding whether the spotted abnormality is a malware or not, and therefore depend on experts for making the final decision [29].

In this chapter, we will provide an extensive literature review concerning intrusion detection systems in the financial context and similarity measures for interval-valued fuzzy sets. Based on the analysis of the literature, we will address three important issues that are not widely considered in intrusion detection systesms: (i) making use of expert knowledge in identifying anomalies; (ii) systematic representation of (imprecise) information concerning previous intrusion cases; (iii) identifying the potential causes of an intrusion in the presence of these imprecise descriptions. Our proposal is to use fuzzy ontologies to represent the available information in terms of interval-valued fuzzy sets: the combination of similarity analysis and expert opinions provides a promising tool to identify and measure the related risks of misuses and also anomalies. We use financial institutions and their operating environment as the example case to describe the model in details. The system can provide information to the users concerning two types of decisions: (i), identifying suspicious activities that can indicate intrusion, and (ii), recommendation on the countermeasures to be undertaken in a given case (which requires the estimation of the possible losses caused by the intrusion).

The chapter is structured as follows: Sect. 2 presents relevant concepts and definitions of intrusion detection systems focusing on financial institutions. Section 3 discusses the role of expert (linguistic) knowledge in information systems and a possible representation in the form of a fuzzy ontology. A discussion on the most important similarity measures for interval-valued fuzzy sets is provided in Sect. 4. In Sect. 5, an *OWA* operator-based distance is presented for interval-valued fuzzy numbers that can be used to calculate similarities. Section 6 presents a fuzzy ontology based on a financial institution taxonomy. Finally, some conclusions are given in Sect. 7.

2 Intrusion Detection Systems and Financial Institutions

The following section introduces relevant concepts and definitions of Intrusion Detection Systems in financial institutions.

2.1 Intrusion Detection

Today, as roughly 1/3 of the earth's population[1] have access to the internet and the penetration rate is rapidly increasing, naturally not only the private users are active online, but also an increasing amount of businesses. With increased amounts of users and business connected to the internet, the risk of intrusions and other complications is amplified. As a consequence, in this context, intrusion detection systems are constantly becoming more important.

Applications and software that help users to protect their devices from viruses and malware constitute an important research topic. Ontologies have turned out to be an useful method to be used for intrusion detection tasks, as they offer possibilities to analyse, for instance, patterns generated by intruders; this way also previously unknown attack methods can be detected [37].

Dai et al. [18] observe that it is a well known fact that hackers tend to be one step ahead of systems created for protection. This results in an endless circle of data losses and a constant demand for new software to fix previous errors. As hackers and their methods are adaptive, behaviour-based approaches have gained an increased interest in developing systems to protect data, as they are effective when dealing with previously unknown attacks [29].

Malware is the common name used for the software's that perform the attacks on computers. They often employ anti-reverse engineering techniques to avoid being detected by analytically-based software. Wagener et al. [59] propose a possible solution for this problem, by applying similarity and distance measures on malware behaviour to implement a better classification of malware types. Different comparisons of similarity and distance measures in the context of malware have also been

[1] http://www.internetworldstats.com/stats.htm

realized, e.g. by [3]. Due to the complexity of malwares, ontologies and especially fuzzy ontologies have attracted significant interest, regarding their possibilities to aid in these kinds of tasks, notably when dealing with imprecise knowledge.

2.2 *Financial Institutions*

A financial institution can be defined as an institution which offers financial services, working as an intermediary by providing, for example: loans, deposits, currency exchanges and investments. Banks and insurance companies are examples of financial institutions. We chose to use financial institutions as the example case as they tend to be an important object for cyber-attacks. These institutions own sensitive data and also significant amount of monetary funds. It has to be noted that in this context not every attack is conducted for personal monetary gain, e.g. stealing funds, but more as a challenge for achieving credibility in online communities or getting noted by the global media. Financial institutions are attractive targets also for this purpose, as people tend to react when their savings are "in danger". As a result, security systems protecting the institutions are designed in a way that is difficult to break, i.e. the one managing to break it deserves some credit. Recently, there has been a global increase in attacks directed towards financial institutions. As these institutions can be considered to be one of the prime targets even on an nationwide scale, the treat of cyber-terrorism can not be overlooked [26, 48].

The risk of intrusions taking place in the financial sector is consequently increasing steadily. Reports constantly indicate that, for instance, bank website outage hours are increasing every month and that more online banking frauds occur. An old but still relevant financial malware is called Zeus. It was recognized already in 2006, but since then it has been remodelled and re-customized several times, each version requiring more preventive work by the institutions. This malware originates from Russian cybercrime organisations. However, the leakage of the source code in 2011 opened the door for basically anyone to modify Zeus and use it for intrusion purposes. Before the leakage of the code, the software was available only for those willing to pay for it, somewhat limiting the usage. Nowadays, there are communities devoted to sharing and trading "plug-ins" for the Zeus malware. As Zeus is not the only malware available, the risk of intrusions happening to a financial institution is far from unlikely [50]. Recently, there have been several publications about how one could prevent different types of attacks specifically aimed towards the financial sector [35, 49].

3 Expert Based Knowledge in IDS

The role of the human experts in information systems has taken a slightly ambiguous role. Wang et al. [61] and Huang et al. [29], amongst others, state that current systems are unable to completely provide full protection against attacks and intrusions. One of the critical issues is the inclusion or exclusion of the human experts. The goal seems to be to exclude the expert as much as possible, however, currently all systems require human input at some stages of the process [13].

Experts usually express themselves using linguistic terms, i.e. "the activity is quite low" and "one should block some of the intrusions". These imprecise linguistic terms, fully understandable for other experts, are hard to interpret for a computer as it is designed for computing with precise data. Linguistic modelling [34], fuzzy logic [66] and other methodologies have been proposed as possible solutions for making human-computer communications feasible.

3.1 Fuzzy Ontologies For Intrusion/Malware Detection

Ontologies, in the context of the Semantic Web, provide a structure and blueprint of the tacit data inherent in different domains. An ontology reveals the relations and connections between the different instances, facilitating the reasoning and decision making. Furthermore, ontologies can be combined and reused by/with different internet-based techniques, encouraging interoperability [5, 24]. This, in turn, gives computers the possibility to reason in a more human-like way, as they can grasp some of our tacit knowledge of the world [33]. There has been an increase in using ontologies for the purpose of intrusion and malware detection.

Undercoffer et al. [58] constructed one of the first ontologies for intrusion detection in the context of computer attacks. They used the DAML+OIL ontology modelling language (a precursor to OWL). Before this approach, mainly taxonomies were used for this purpose. Introducing ontologies enabled better meta-data modelling and ontologies can naturally subsume taxonomies. Simmonds et al. [52] developed an ontology for defending against attacks aimed at networks, emphasising that one should also prepare for what happens if the attack is successful and how the designed system reacts in that scenario.

The rapid development of mobile devices created a completely new field vulnerable to intrusions and malwares. Chiang and Tsaur [17] therefore took the first steps towards extending ontologies also towards the protection of mobile devices. They modelled an ontology based on the behaviours of known mobile malware. Hung et al. [30] created an extensive ID ontology, which also included a feature allowing users to model the ontology application from a conceptual level. This broadens the possible range of users, meaning that even non-expert users could contribute to intrusion detection processes.

However, it has several times been stated that traditional, non-fuzzy, ontologies are not suitable to deal with imprecise and vague knowledge [27, 42]. Avoiding imprecise data in the online world is close to impossible, hence, the combination of fuzzy logic and ontologies has recently gained an increased interest from the research community.

In recent years, the combination of fuzzy logic and ontologies has been an emerging topic in intrusion detection [9, 20, 56]. Huang et al. [27–29] developed a

IT2FS-based ontology, as a novel approach for malware behavior analysis (MiT). Using the Fuzzy Markup Language (FML) [1] and the Web Ontology Language (OWL), they managed to create a fully operational system, able to analyse collected data and extract behavioural information.

Tafazzoli et al. [55] created a fuzzy malware ontology for the Semantic Web. The ontology represents relevant concepts inherent in the malware field. The relationships between the different malwares are modelled with the help of fuzzy linguistic terms, such as: **weak relation** and **very good relation**. Considering that it was created with the Semantic Web in mind, it can also be used for sharing information online.

As it can be noticed, there is a fair amount of positive results that have emerged with the fuzzy ontology approach. We believe that it is reasonable to state that more research in how fuzzy ontologies can benefit the task of intrusion detection is needed.

The Web Ontology Language (OWL) is the main language used for creating ontologies on the Semantic Web [25]. By settling on one standard and using it for the Semantic Web and its ontology modelling needs, it makes the co-operation between different domain ontologies (for instance) more straightforward and smoother, facilitating the expansion of OWL to the non-expert users. The fuzzy ontology created in this chapter (and presented in more detail later on) is modelled in OWL.

4 Similarity Measures for Interval-Valued Fuzzy Sets

Similarity measures have become an important technique for handling imprecise information in the context of information systems [60]. The easily embraceable notion behind these measures, comparing how similar two instances are, has made them widely used in various topics. Numerous applications and implementations based on similarity measures exist also for intrusion detection issues.

In this section we provide an extensive overview of the existing similarity measures for interval-valued fuzzy sets. We also included the similarity measures proposed for intuitionistic fuzzy sets by reformulating the definitions using the traditional transformation between interval-valued and intuitionistic fuzzy sets. We will use the following definitions for interval-valued fuzzy sets:

Definition 1 ([22]) An interval-valued fuzzy set A defined on X is given by

$$A = \left\{(x, [\mu_A^L(x), \mu_A^U(x)])\right\}, x \in X,$$

where $\mu_A^L(x), \mu_A^U(x) : X \to [0, 1]; \forall x \in X, \mu_A^L(x) \leq \mu_A^U(x)$, and the ordinary fuzzy sets $\mu_A^L(x)$ and $\mu_A^U(x)$ are called lower fuzzy set and upper fuzzy set of A, respectively.

A starting point for evaluating different similarity measures can be a set of predefined properties that are expected to be satisfied by a measure to be called as similarity. As we will see, in many cases these properties only ensure a basic reliability of the measures: there are many examples for similarity measures that satisfy all the properties but provide non-intuitive values for specific fuzzy sets. In this article, we will adopt the four properties specified in [62]:

1. Reflexivity: $s(A, B) = 1 \Longrightarrow A = B$
2. Symmetry: $s(A, B) = s(B, A)$
3. Transitivity: If $A \leq B \leq C$ than $s(A, B) \geq s(A, C)$
4. Overlapping: If $A \cap B \neq \emptyset$ than $s(A, B) > 0$

Additionally, we can require the similarity measure to be normalized: $0 \leq s(A, B) \leq 1$.

The first group of similarity measures is based on the extension of traditional distance measures (Hamming, Euclidean) to fuzzy sets. This family of similarities has been developed for type-1 fuzzy sets before the 1990s (see [57]), and later on to interval-valued and intuitionistic fuzzy sets. The traditional way to obtain the similarity from a normalized distance measure d is to calculate

$$s(A, B) = 1 - d(A, B).$$

As Zeng and Guo described in a systematic analysis, there are numerous other ways to generate similarity measures from distances.

Theorem 1 ([67]) *Given a real function $f : [0, 1] \to [0, 1]$, if f is a strictly monotone decreasing function, and d is a normalized distance on interval-valued fuzzy sets, then*

$$s(A, B) = \frac{f(d(A, B)) - f(1)}{f(0) - f(1)}$$

is a similarity measure.

For example if $f(x) = \dfrac{1}{1+x}$, then $s(A, B) = \dfrac{1 - d(A, B)}{1 + d(A, B)}$ and for $f(x) = 1 - x^2$ one can obtain the similarity $s(A, B) = 1 - d^2(A, B)$.

Burillo and Bustince [10] were the first ones to extend the Hamming and Euclidean distances to (discrete) interval-valued (and intuitionistic) fuzzy sets in the following way (the normalized distances are presented):

- Hamming distance:

$$d_H(A, B) = \frac{1}{n} \sum_{i=1}^{n} \frac{| A^L(x_i) - B^L(x_i) | + | A^U(x_i) - B^U(x_i) |}{2};$$

- Euclidean distance

$$d_2(A, B) = \left[\frac{1}{2n} \sum_{i=1}^{n} (A^L(x_i) - B^L(x_i))^2 + (A^U(x_i) - B^U(x_i))^2 \right]^{0.5}.$$

In [4], Atannasov presented similar definitions for intuitionistic fuzzy sets. Szmidt and Kacprzyk [53, 54] further improved the definitions for intuitionistic fuzzy sets by incorporating the intuitionistic fuzzy index (and their proposal was extended by using geometric distance and weights in [64]). As Grzegorzewski [23] pointed out, these modifications are not properly motivated and result only in marginal differences and improvements; for this reason, he used the Hausdorff metric to modify the definitions in a natural way that is easy to use for applications. The proposed (normalized) distances can be formulated as:

- Hamming distance:

$$d_H(A,B) = \frac{1}{n}\sum_{i=1}^{n} \max(\mid A^L(x_i) - B^L(x_i) \mid, \mid A^U(x_i) - B^U(x_i) \mid);$$

- Euclidean distance

$$d_2(A,B) = \left[\frac{1}{n}\sum_{i=1}^{n} \max(A^L(x_i) - B^L(x_i))^2, (A^U(x_i) - B^U(x_i))^2)\right]^{0.5}.$$

Zeng and Li [68] proposed the same definition of the Hamming-distance independently as a result of a transformation procedure to connect entropy and similarity measures for interval-valued fuzzy sets. They additionally defined the continuous version of the definition as

$$s(A,B) = \frac{1}{2(b-a)}\int_a^b \mid A^L(x_i) - B^L(x_i) \mid + \mid A^U(x_i) - B^U(x_i) \mid \mathrm{d}x,$$

with the support of the interval-valued fuzzy sets is in the $[a, b]$ interval. Zhang and Fu [69] introduced weight values (functions) in the two cases of the Hamming distance and showed that it is a similarity measure in a more general sense as it can be used for any L-fuzzy sets.

After 2002, as a different direction to extend distance measures, numerous articles were published in the journal Patter Recognition Letters mainly originating from the approach introduced in the article of Dengfeng and Chuntian [19]. They used the middle points of the membership interval of the interval-valued fuzzy sets to obtain a type-1 fuzzy set and then they calculate the distance of the resulting type-1 fuzzy sets. Using the notation

$$f_A(x) = \frac{A^L(x) + A^U(x)}{2},$$

the following formulas can be obtained:

- in the discrete case

$$s_d^p(A, B) = 1 - \frac{1}{n^{1/p}} \left[\sum_{i=1}^{n} (f_A(x_i) - f_B(x_i))^p \right]^{1/p};$$

- in the continuous case

$$s_c^p(A, B) = 1 - \frac{1}{(b-a)^{1/p}} \left[\int_a^b (f_A(x) - f_B(x))^p \mathrm{d}x \right]^{1/p}.$$

Both definitions can be used with weight functions, for example in the continuous case

$$s_{cw}^p(A, B) = 1 - \left[\int_a^b w(x)(f_A(x) - f_B(x))^p \mathrm{d}x \right]^{1/p}.$$

One important disadvantage of this method is that equality of the interval-valued fuzzy sets is only a sufficient but not necessary condition for the similarity measure to take its maximal value. After illustrating this with a simple example, Mitchell [45] proposed an improvement to the method: instead of calculating the similarity of type-1 fuzzy sets obtained as the average of the lower and upper membership functions, the similarity of the two upper fuzzy sets and the two lower fuzzy sets are computed and the overall similarity is the average of these two values.

Noticing the same problem with the Li-Cheng approach, Liang and Shi [39] proposed new families of similarity measures by using a more complex combination of the upper and lower membership functions:

$$f_{l_{AB}}(x) = \frac{\mid A^L(x) + B^L(x) \mid}{2}; \quad f_{u_{AB}}(x) = \frac{\mid A^U(x) + B^U(x) \mid}{2}.$$

Interestingly, this approach in a special case provides a formula similar to the Euclidean-distance based approach originally proposed in [10]. Liang and Shi proposed further modifications by incorporating the median of the interval membership into the f values and showed that the new definitions provide more intuitive results. As a summary of the proposals utilizing distance measures to calculate similarity, Li et al. [38] provided a detailed description of the different proposals published before 2007 (focusing on intuitionistic and vague sets) and created a selection process to identify the best method: they concluded that the measure proposed in [39] is the only one that does not result in counter intuitive values in any case.

The second main group of similarity measures consists of definitions based on set-theoretic measures and arithmetic operations on fuzzy sets. As discussed in [57], for type-1 fuzzy sets, similarity measures belonging to this family are as popular as the distance based similarities, but we can find significantly less measures when investigating interval-valued fuzzy sets. Probably the most general formula for similarity measures was given by Bustince [11]. He proved that the combination of an inclusion grade indicator and any t-norms will result in an interval-valued similarity measure. Using a similar approach, Zhang et al. [70] defined a new inclusion measure and combined it with a t-norm to obtain a similarity value.

An important measure of similarity in set theory is the Jaccard index. In fuzzy set theory, one can find many different generalizations also for interval-valued fuzzy sets. The first approach is to calculate the Jaccard index of the upper membership values and the lower membership values separately and combine them to obtain an overall similarity: Zheng et al. [71] calculated the average as

$$s(A,B)=\frac{1}{2}\left(\frac{\int_X \min(A^U(x),B^U(x))\mathrm{d}x}{\int_X \max(A^U(x),B^U(x))\mathrm{d}x}+\frac{\int_X \min(A^L(x),B^L(x))\mathrm{d}x}{\int_X \max(A^L(x),B^L(x))\mathrm{d}x}\right),$$

while Hwang and Yang [31] used the minimum of the Jaccard index of the lower membership function and the Jaccard index of the complement of the upper memberships as the similarity:

$$s(A,B)=\min\left(\frac{\sum_{x\in X}\min(A^{Uc}(x),B^{Uc}(x))}{\sum_{x\in X}\max(A^{Uc}(x),B^{Uc}(x))},\frac{\sum_{x\in X}\min(A^L(x),B^L(x))}{\sum_{x\in X}\max(A^L(x),B^L(x))}\right).$$

A different approach making use of the Jaccard index is of calculating the similarity directly from the interval-valued memberships and not as a combination of upper and lower values. This method results in the following formula defined by Wu and Mendel [62] (this is an improved version of the previously defined vector similarity measure [63] as the authors noticed that it does not satisfy the overlapping property:)

$$s(A,B)=\frac{\int_X \min(A^U(x),B^U(x))+\min(A^L(x),B^L(x))\mathrm{d}x}{\int_X \max(A^U(x),B^U(x))+\max(A^L(x),B^L(x))\mathrm{d}x}.$$

The theory of similarity for general type-2 fuzzy sets is still in the early stages, we only mention two approaches that can naturally be applied to interval-valued fuzzy sets (as special cases of general type-2 fuzzy sets): Zheng et al. [72] defined a similarity measure employing the footprint of uncertainty and secondary membership function of type-2 fuzzy sets motivated by a clustering application. McCulloh et al. [43] created a framework to extend any similarities of interval-valued fuzzy sets to general type-2 fuzzy sets.

Another commonly used approach is to calculate the similarity of specific type-1 fuzzy sets and aggregate them into the overall similarity of interval-valued fuzzy sets. Mitchell [46] proposes to choose N embedded fuzzy sets randomly and calculate the average similarity value (any type-1 similarity measure can be used). Motivated by risk analysis problems, Chen et al. [15, 16] proposed several similarity measures based on arithmetic operations by comparing the upper and lower similarities. Feng and Liu [21] used the similarity for the upper and lower membership functions and the kernel function of interval-valued fuzzy sets to obtain a new similarity measure.

5 Similarity of Interval-Valued Fuzzy Sets Based on the OWAD Operator

In this section we will define a similarity measure for interval-valued fuzzy numbers (IVFN's) based on the concept of the ordered weighted averaging distance operator (*OWAD*) introduced by Xu and Chen [65]:

Definition 2 An OWAD operator of dimension n is a mapping $OWAD : \mathbb{R}^n \times \mathbb{R}^n \to [0, 1]$ that has an associated weighting vector W with $\sum_{j=1}^{n} W_j = 1$ and $W_j \in [0, 1]$ such that:

$$OWAD\left(\langle\mu_1^{(1)}, \mu_1^{(2)}\rangle, \ldots, \langle\mu_n^{(1)}, \mu_n^{(2)}\rangle\right) = \sum_{j=1}^{n} w_j D_j,$$

where D_j represents the jth largest of the $|\mu_i^{(1)} - \mu_i^{(2)}|$.

In order to extend the definition to the family of IVFN's, we use the mean value of an interval-valued fuzzy number to measure the distance of IVFN's.

Definition 3 ([12]) The mean (or expected) value of $A \in$ IVFN is defined as

$$E(A) = \int_0^1 \alpha(M(U_\alpha) + M(L_\alpha))\mathrm{d}\alpha, \tag{1}$$

where U_α and L_α are uniform probability distributions defined on $[A^U]^\alpha$ and $[A^L]^\alpha$, respectively, and M stands for the probabilistic mean operator.

The distance of two IVFN's, d : IVFN $\times$ IVFN $\to \mathbb{R}$, is defined as

$$d(A, B) = |E(A) - E(B)|. \tag{2}$$

The distance (2) satisfies the four properties of a distance measure:

1. Non-negativity: $|E(A) - E(B)| \geq 0$
2. Commutativity: $|E(A) - E(B)| = |E(B) - E(A)|$
3. Reflexivity: $|E(A) - E(A)| = 0$
4. Triangle inequality: $|E(A) - E(B)| + |E(B) - E(C)| \geq |E(A) - E(C)|$.

Definition 4 ([44]) A Quasi IVFN-IOWAD operator of dimension n is a mapping f : IVFN^n $\times$ IVFN^n $\times$ IVFN^n $\to \mathbb{R}$ that has an associated weighting vector W of dimension n with $w_j \in [0, 1]$ and $\sum_{j=1}^{n} w_j = 1$, such that:

$$f(\langle U_1, A_1, B_1\rangle, \langle U_2, A_2, B_2\rangle, \ldots, \langle U_n, A_n, B_n\rangle) = g^{-1}\left(\sum_{j=1}^{n} w_j g(D_j)\right), \quad (3)$$

where D_j is the $d(A_i, B_i)$ value of the triplet $\langle U_i, A_i, B_i\rangle$ having the jth largest U_i and $g : \mathbb{R} \to \mathbb{R}$ is a continuous, strictly monotone function.

Theorem 1 *If f is an Quasi IVFN-IOWAD operator, then the following properties are satisfied:*

1. *f is commutative:*

$$f(\langle U_1, A_1, B_1\rangle, \langle U_2, A_2, B_2\rangle, \ldots, \langle U_n, A_n, B_n\rangle) = f(\langle U_1^{'}, A_1^{'}, B_1^{'}\rangle, \langle U_2^{'}, A_2^{'}, B_2^{'}\rangle, \ldots, \langle U_n^{'}, A_n^{'}, B_n^{'}\rangle),$$

where $(\langle U_1^{'}, A_1^{'}\rangle, \langle U_2^{'}, A_2^{'}\rangle, \ldots, \langle U_n^{'}, A_n^{'}\rangle)$ is any permutation of the arguments.

2. *f is monotone: if $d(A_i^1, B_i^1) \geq d(A_i^2, B_i^2)$ for all i, then*

$$f(\langle U_1, A_1^1, B_1^1\rangle, \langle U_2, A_2^1, B_2^1\rangle, \ldots, \langle U_n, A_n^1, B_n^1\rangle) = f(\langle U_1, A_1^2, B_1^2\rangle, \langle U_2, A_2^2, B_2^2\rangle, \ldots, \langle U_n, A_n^2, B_n^2\rangle).$$

3. *f is idempotent: if $d(A_i, B_i) = d(A_j, B_j) = d, \forall i, j$, then*

$$f(\langle U_1, A_1, B_1\rangle, \langle U_2, A_2, B_2\rangle, \ldots, \langle U_n, A_n, B_n\rangle) = d.$$

4. *f is bounded:*

$$\min_i \{d(A_i, B_i)\} \leq f(\langle U_1, A_1, B_1\rangle, \langle U_2, A_2, B_2\rangle, \ldots, \langle U_n, A_n, B_n\rangle) \leq \max_i \{d(A_i, B_i)\}.$$

Proof The proofs are straightforward consequences of the definition and the arithmetic operations on interval-valued fuzzy sets, we only prove the boundedness. It can be proven by comparing the aggregated value to the minimum and maximum as follows:

$$\min_i \{d(A_i, B_i)\} = g^{-1}\left(g(\min_i \{d(A_i, B_i)\})\right)$$
$$= g^{-1}\left(\sum_{j=1}^{n} w_j g(\min_i \{d(A_i, B_i)\})\right) \leq g^{-1}\left(\sum_{j=1}^{n} w_j g(D_j)\right)$$
$$= f(\langle U_1, A_1, B_1\rangle, \langle U_2, A_2, B_2\rangle, \ldots, \langle U_n, A_n, B_n\rangle)$$

and

$$\max_i \{d(A_i, B_i)\} = g^{-1}\left(g(\max_i \{d(A_i, B_i)\})\right)$$
$$= g^{-1}\left(\sum_{j=1}^{n} w_j g(\max_i \{d(A_i, B_i)\})\right) \geq g^{-1}\left(\sum_{j=1}^{n} w_j g(D_j)\right)$$
$$= f(\langle U_1, A_1, B_1\rangle, \langle U_2, A_2, B_2\rangle, \ldots, \langle U_n, A_n, B_n\rangle).$$

Note 1 One special case of this definition is the generalized IVFN-IOWAD operator, where $g(x) = x^{\alpha}, \alpha \in \mathbb{R}$, and it takes the following form:

$$\left(\sum_{j=1}^{n} w_j D_j^{\alpha}\right)^{\frac{1}{\alpha}}.$$

Definition 5 ([44]) An IVFN-IOWAD operator of dimension n is a mapping $f : \mathbb{R}^n \times \text{IVFN}^n \times \text{IVFN}^n \rightarrow \mathbb{R}$ that has an associated weighting vector W of dimension n with $w_j \in [0, 1]$ and $\sum_{j=1}^{n} w_j = 1$, such that:

$$f(\langle u_1, A_1, B_1\rangle, \langle u_2, A_2, B_2\rangle, \ldots, \langle u_n, A_n, B_n\rangle) = \sum_{j=1}^{n} w_j D_j, \quad (4)$$

where D_j is the $d(A_i, B_i)$ value of the triplet $\langle u_i, A_i, B_i\rangle$ having the jth largest u_i, where u_i is the order inducing variable and A_i, B_i are the argument variable represented in the form of IVFN's.

Example 1 To illustrate the definition, we will calculate the OWA- distance of trapezoidal-shaped IVFN's (the upper and lower fuzzy numbers are trapezoidal fuzzy numbers) choosing $g(x) = x$, which is a special case of the definition, an IVFN-IOWAD operator. The upper and lower fuzzy numbers can be represented as $A^L = (a, b, \alpha, \beta)$ and $A^U = (c, d, \theta, \tau)$ respectively, where $[a, b]$ and $[c, d]$ stand for the central intervals, (α, β) and (θ, τ) denotes the left and right width of the fuzzy numbers. The mean value of a triangular IVFN can be expressed as

$$E(A) = \frac{a+b}{4} + \frac{c+d}{4} + \frac{\beta-\alpha}{12} + \frac{\tau-\theta}{12}.$$

In the example, we suppose that new data is available (concerning a potential intrusion) and it is compared to two previous intrusion cases stored in the database. One expert evaluates the cases based on three criteria, and this evaluation will be used to calculate the distance between the new observation and the two stored cases. The expert's evaluation is described in Table 1.

Table 1 The evaluation of the expert for the different cases

Criteria	New case	Case 1	Case 2
C_1	$A_1^U = (4, 6, 2, 2)$	$B_1^{1,U} = (3, 6, 1, 1)$	$B_1^{2,U} = (5, 5, 3, 3)$
C_1	$A_1^L = (4, 5, 1, 1)$	$B_1^{1,L} = (4, 5, 1, 1)$	$B_1^{2,L} = (5, 5, 2, 2)$
C_2	$A_2^U = (8, 10, 3, 3)$	$B_2^{1,U} = (7, 8, 3, 4)$	$B_2^{2,U} = (9, 11, 2, 2)$
C_2	$A_2^L = (9, 9, 2, 1)$	$B_2^{1,L} = (7, 7, 2, 4)$	$B_2^{2,L} = (10, 11, 2, 2)$
C_3	$A_3^U = (2, 4, 1, 1)$	$B_3^{1,U} = (5, 7, 2, 2)$	$B_3^{2,U} = (4, 6, 1, 1)$
C_3	$A_3^L = (3, 4, 1, 1)$	$B_3^{1,L} = (6, 7, 2, 1)$	$B_3^{2,L} = (5, 6, 2, 1)$

The corresponding order inducing variables for the criteria (we use crisp values in this example) are $u_1 = 5, u_2 = 7, u_3 = 2$. The weights are defined as $W = (0.4, 0.2, 0.4)$. The aggregation can be calculated as

$$f_i(\langle 5, A_1, B_i^1\rangle, \langle 7, A_2, B_i^2\rangle, \langle 2, A_3, B_i^3\rangle) = 0.4|E(A_2 - B_i^2)| \\ + 0.2|E(A_1 - B_i^1)| + 0.4|E(A_3 - B_i^3)|.$$

for $i = 1, 2$. The obtained values are $f_1 = 1.78$ and $f_2 = 1.35$, which indicate that the new case is more similar to Case 2 from the database, as the distance between these two instances is smaller than the distance between the new case and Case 1.

To use the distance measure for obtaining similarities, we need to normalize the aggregated values by dividing by the factor

$$\sup\{x \in \cup_i (supp(A_i) \cup supp(B_i))\} - \inf\{x \in \cup_i (supp(A_i) \cup supp(B_i))\}$$

and compute $s = 1 - d$, where d stands for the Quasi IVFN-IOWAD operator.

6 The Financial Institution Ontology

With the ambition to illustrate how fuzzy ontologies and similarity measures could be used for intrusion detection purposes, we created a simple fuzzy ontology. Although intrusion methods are seldom limited to a specific context, our ontology was adapted to fit relevant risks associated with financial institutions. This ontology was then used as the base for creating a simple application, showing a practical example on how fuzzy ontologies can aid in intrusion detection by generating the risk for certain intrusions to occur. In Sect. 6.1 we demonstrate this by presenting a couple of scenarios, pointing out where the fuzzy ontology could contribute, Sect. 6.2 presents the structure of the ontology whereas the technical parts of the application are presented more in Sect. 6.3.

6.1 Intrusion Scenarios

As to clarify the context and functions of the application proposed, we describe a couple of scenarios, showing how the fuzzy ontology could aid in detecting possible intrusions.

Scenario 1.

The first scenario addresses a malware attack, presumably from the widely used Zeus malware. In this scenario, the advice generated by the system is assessed and combined by human experts, whereas a final result is produced, based on both the ontology result and the expert assessments.

As the surveillance system notices an abnormal behaviour, the recorded values are processed by our proposed intrusion detection system. This generates a result showing how likely the detected abnormality is an intrusion attempt. In other words, this example would display the following result:

Value 1 is 95 % similar to Zeus_Intrusion_nr45
Value 2 is 59 % similar to Zeus_Intrusion_nr32
Value 3 is 67 % similar to Gauss_Intrusion_nr2
Value 4 is 85 % similar to Zeus_Intrusion_nr45
Value 5 is 75 % similar to Zeus_Intrusion_nr5

It is ***Highly*** *likely that the detected intrusion is a* ***Zeus-based malware****.*

The values represent different measures relevant to the behaviour of the intrusion. Regarding the Zeus Malware, they could represent: amount of hazardous .php files detected; amount of hazardous .exe files detected; amount of functions reporting a malfunction. The detected files are compared with different lists containing hazardous files frequent in different types of intrusions.

A human expert would then asses the results generated by the ontology. The expert does also have the option to see not only the most similar case, but also the whole list of generated similarities. In this case, the expert could notice that Zeus_Intrusion_nr45 and Zeus_Intrusion_nr47 had a 65 and 64 % similarity to value 4, respectively. This aids in the experts' decision making, making it possible to embrace the whole picture and decide in favour of the proposed analysis. The defence systems would then take the appropriate measures, being more efficient, as the intrusion method is likely to be known.

Scenario 2.

The second scenario is assumed to be a denial-of-service attack (Dos), conducted with the purpose of overloading the institutions online system, creating chaos that would consume both time and money to be sorted out. The number of Dos attacks has increased lately, with the main goal of punishing the target by making their online system crash. This scenario excludes the experts, as Dos require immediate action and therefore can not wait for human input.

Online systems can easily define what the normal range of data traffic is, using historical data, and also defining when the crucial limits are reached. One could use fuzzy interval values to model when the values are closing in on the critical limits. In other words, we can use linguistic terms, such as: low risk, medium risk and high risk for indicating how close to the critical limit the amount of data traffic is. This means that one can observe even small risks, where several slightly suspicious factors (which would not have been noticed in a non-fuzzy system) together can detect possible intrusions. Risks or changes that otherwise would have been unnoticed.

For example, a bank usually registers 1000 logins per hour in their online banking system, the record is 1500 and the minimum 500. In other words, it usually moves between 500 and 1500. By using type-2 fuzzy sets, one can define that if the value goes over 1500, it is considered to be "Highly trafficked" and as it reaches closer to 2000, it becomes more and more "Critical". However, the system does not need to shut down if the number of logins exceeds a critical limit; if several similar measures are starting to reach a critical level, the system can conclude that a possible attack is occurring. The fuzzy ontology then can be used to define what kind of Dos attack is most likely taking place and adjust the counter measures according to that knowledge, for instance, by quickly shutting down the system before it crashes and wait for human maintenance persons to make the final decision. In this way one could avoid costly maintenance conducted after a real crash has happened.

6.2 The Fuzzy Ontology

The ontology was created with Protégé [32], the main modelling software for OWL ontologies. Fuzzy datatypes where added to the ontology using the Fuzzy OWL plug-in [7]. The plug-in is an important step towards including fuzzy logic in OWL and making fuzzy logic available for general users. The intrusion risks included in the ontology were collected from different computer security companies and reports, e.g. from The Kaspersky Lab[2] and S2sec.[3]

Figure 1 shows a overview of the ontology. The ontology is structured by fuzzy classes according to the intrusion type, e.g. **Social_Engineering** and **Malware_and _Viruses**. Each of these general classes have more specified subclasses, such as: **Phisihing** and **Win32.** . The subclasses are populated with individuals, representing specific intrusions, such as the famous Zeus malware and previous recorded intrusion attempts. All the individual instances have a set of recorded values or behaviours showing how the intrusion was conducted. Using similarity measures these values are compared with the new intrusions.

[2] http://www.kaspersky.com/

[3] http://www.s21sec.com/

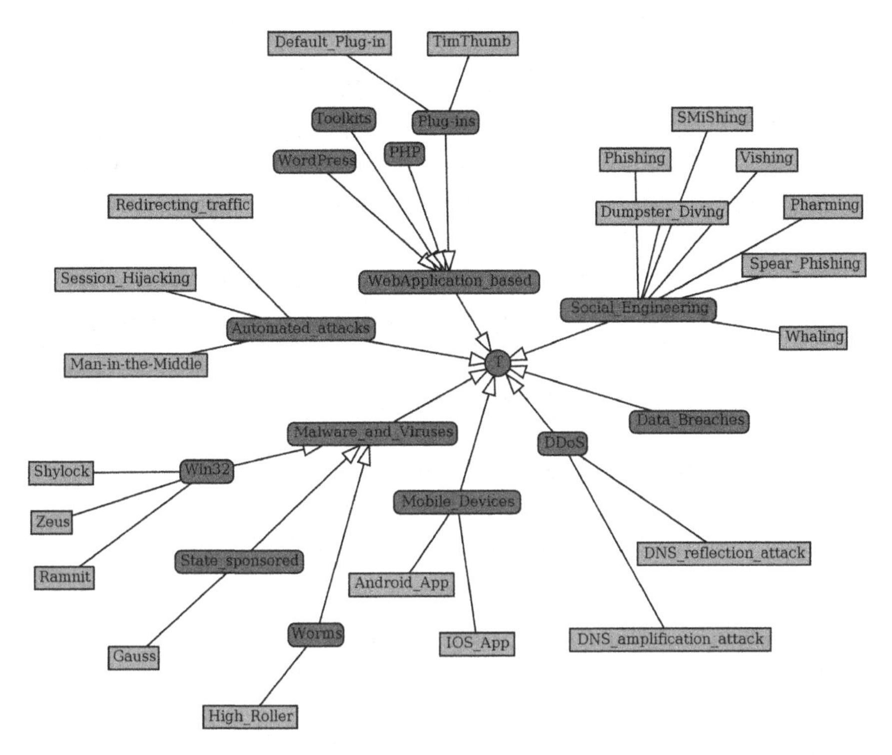

Fig. 1 The structure of the fuzzy ontology

6.3 F.I. Application

Using the programming language Java, an application, retrieving information from the OWL ontology, was constructed [8]. Java makes it possible to connect the structure with other techniques, for instance HTML, meaning that one can run the application online. The application functions in the following way:

- The user decides among a couple of pre-defined example treats that have been "registered". The function of registering and comparing the possible intrusion would be automatic in a real world intrusion detection system (Fig. 2a).
- The chosen intrusion is modelled with interval type-2 fuzzy sets. The previously stored intrusions are retrieved from the ontology and the similarities to the current intrusion are computed.
- The results of the computation, i.e. how likely the detected abnormally is an intrusion and in that case which previous intrusion it resembles, is presented to the user, see Fig. 2b.
- The user has the possibility to view also other similar intrusions, screenshot shown in Fig. 2c, offering, for instance, for the experts that should make the final decision an opportunity to get a more comprehensive picture of the situation.

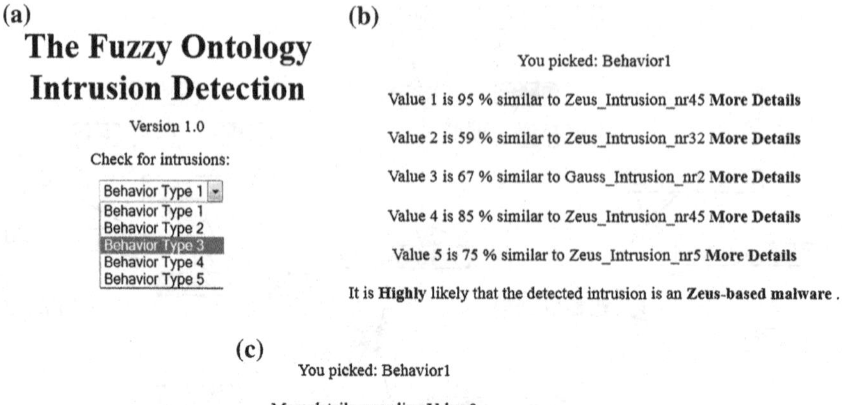

Fig. 2 Example of a fuzzy intrusion detection user interface: **a** The initial choice, **b** First retrieved results, **c** More detailed results

It has to be acknowledged that the functions in the application are only basic, however, the structure of the application and the techniques it builds on, e.g. OWL, fuzzyDL [7], Java and HTML to make it suitable for extending and combining with numerous other applications.

7 Conclusion

Intrusion detection is becoming more and more essential to handle the risks associated with network activities. New intrusion detection systems should be capable of preventing organisation not only from an increasing numbers of attacks but also from more and more sophisticated intrusion strategies. One promising solution would be to exercise expert insights in the detection process. As experts have accumulated an extensive knowledge of their field, in many cases they can point out some irregularities which can indicate anomalies that otherwise could not be identified by automatic intrusion detection systems and would result in significant losses.

In many cases, tacit knowledge of experts can be expressed only using linguistic (imprecise) terms. In our proposal the combination of fuzzy logic and ontologies can transform expert knowledge into a systematic description processable by computational methods. The fusion of type-2 fuzzy ontologies and similarity measures to identify possible causes of intrusions provides benefits to organisations that are not achievable by other methods.

To support the results of the ontology and provide additional information that can be essential to identify anomalies, expert opinions expressed in terms of linguistic information and modelled by interval-valued fuzzy numbers are employed. As numerous intrusions can occur at the same time, the proposed system can provide estimations on the seriousness of different activities in terms of potential losses. Based on this, the decision makers can assign the limited available resources in a way that is optimal: by assigning more resources to the more serious cases, the potential loss can be minimized. Our proposal can be extended by incorporating more detailed database of intrusions for testing purpose and also by using different types of similarity measures.

References

1. Acampora, G., Loia, V.: Using FML and fuzzy technology in adaptive ambient intelligence environments. Int. J. Comput. Intell. Res. **1**(1), 171–182 (2005)
2. Anderson, J.P.: Computer security threat monitoring and surveillance. Technical Report James P. Anderson Company, Fort Washington, Pennsylvania (1980)
3. Apel, M., Bockermann, C., Meier M.: Measuring similarity of malware behavior. In: IEEE 34th Conference on Local Computer Networks (LCN 2009), pp. 891–898 (2009)
4. Atannasov, K.: Intuitionistic Fuzzy Sets: Theory and Applications. Physica-Verlag, New York (1999)
5. Berners-Lee, T.: Semantic web on XML. http://www.w3.org/2000/Talks/1206-xml2k-tbl/ (2000)
6. Bobillo, F.: Managing vagueness in ontologies. PhD Thesis, University of Granada, Spain (2008)
7. Bobillo, F., Straccia, U.: Fuzzy ontology representation using OWL 2. Int. J. Approximate Reasoning **52**(7), 1073–1094 (2011)
8. Bobillo, F., Straccia, U.: Aggregation operators for fuzzy ontologies. Appl. Soft Comput. **13**(9), 3816–3830 (2013)
9. Botha, M., von Solms, R.: Utilising fuzzy logic and trend analysis for effective intrusion detection. Comput. Secur. **22**(5), 423–434 (2003)
10. Burillo, P., Bustince, H.: Entropy on intuitionistic fuzzy sets and on interval-valued fuzzy sets. Fuzzy Sets Syst. **78**(3), 305–316 (1996)
11. Bustince, H.: Indicator of inclusion grade for interval-valued fuzzy sets. application to approximate reasoning based on interval-valued fuzzy sets. Int. J. Approximate Reasoning **23**(3), 137–209 (2000)
12. Carlsson, C., Fullér, R., Mezei J.: Project selection with interval-valued fuzzy numbers. In: IEEE 12th International Symposium on Computational Intelligence and Informatics (CINTI), pp. 23–26 (2011)
13. Catania, C.A., Garino, C.G.: Automatic network intrusion detection: current techniques and open issues. Comput. Electr. Eng. **38**(5), 1062–1072 (2012)
14. Chandola, V., Banerjee, A., Kumar, V.: Anomaly detection: a survey. ACM Comput. Surv. (CSUR) **41**(3), 15 (2009)
15. Chen, S.-J., Chen, S.-M.: Fuzzy risk analysis based on measures of similarity between interval-valued fuzzy numbers. Comput. Math. Appl. **55**(8), 1670–1685 (2008)
16. Chen, S.-M., Chen, J.-H.: Fuzzy risk analysis based on similarity measures between interval-valued fuzzy numbers and interval-valued fuzzy number arithmetic operators. Expert Syst. Appl. **36**(3), 6309–6317 (2009)

17. Chiang, H.-S., Tsaur, W.: Mobile malware behavioral analysis and preventive strategy using ontology. In: IEEE Second International Conference on Social Computing (SocialCom), pp. 1080–1085 (2010)
18. Dai, S.-Y., Fyodor, Y., Kuo, S.-Y., Wu, M.-W., Huang Y.: Malware profiler based on innovative behavior-awareness technique. In: IEEE 17th Pacific Rim International Symposium on Dependable Computing (PRDC), pp. 314–319 (2011)
19. Dengfeng, L., Chuntian, C.: New similarity measures of intuitionistic fuzzy sets and application to pattern recognitions. Pattern Recogn. Lett. **23**(1), 221–225 (2002)
20. Dickerson, J.E., Juslin, J., Koukousoula, O., Dickerson, J.A.: Fuzzy intrusion detection. In: IEEE 9th joint IFSA World Congress and 20th NAFIPS International Conference, vol. 3, pp. 1506–1510 (2001)
21. Feng, Z.-Q., Liua, C.-G.: On similarity-based approximate reasoning in interval-valued fuzzy environments. Informatics **36**, 255–262 (2012)
22. Gorzałczany, M.: A method of inference in approximate reasoning based on interval-valued fuzzy sets. Fuzzy Sets Syst. **21**(1), 1–17 (1987)
23. Grzegorzewski, P.: Distances between intuitionistic fuzzy sets and/or interval-valued fuzzy sets based on the hausdorff metric. Fuzzy Sets Syst. **148**(2), 319–328 (2004)
24. Hendler, J.: Agents and the semantic web. Intell. Syst. **16**(2), 30–37 (2001)
25. Horridge, M., Krötzsch, M., Parsia, B., Patel-Schneider, P., Rudolph, S.: OWL 2 web ontology language, primer. W3C Working Group (2009)
26. Hua, J., Bapna, S.: The economic impact of cyber terrorism. J. Strateg. Inf. Syst. **22**(2), 175–186 (2013)
27. Huang, H.-D., Acampora, G., Loia, V., Lee,C.-S., Hagras, H., Wang, M.-H., Kao, H.-Y., Chang J.-G.: Fuzzy markup language for malware behavioral analysis. In: On the Power of Fuzzy Markup Language, pp. 113–132. Springer (2013)
28. Huang, H.-D., Acampora, G., Loia, V., Lee, C.-S., Kao, H.-Y.: Applying FML and fuzzy ontologies to malware behavioural analysis. In: IEEE International Conference on Fuzzy Systems, pp. 2018–2025 (2011)
29. Huang, H.-D., Lee, C.-S., Wang, M.-H., Kao, H.-Y.: IT2FS-based ontology with soft-computing mechanism for malware behavior analysis. Soft Comput. **18**(2), 267–284 (2014)
30. Hung, S.-S., Liu, D.S.-M.: A user-oriented ontology-based approach for network intrusion detection. Comput. Stan. Interfaces **30**(1–2), 78–88 (2008)
31. Hwang, C.-M., Yang, M.-S.: New similarity measures between interval-valued fuzzy sets. In: Proceedings of the 15th WSEAS International Conference on Systems, pp. 66–70 (2011)
32. Knublauch, H., Fergerson, R., Noy, N., Musen, M.: The Protégé OWL plugin: an open development environment for semantic web applications. The Semantic Web-ISWC **2004**, 229–243 (2004)
33. Lau, A., Tsui, E., Lee, W.: An ontology-based similarity measurement for problem-based case reasoning. Expert Syst. Appl. **36**(3, Part 2):6574–6579 (2009)
34. Lawry, J.: A framework for linguistic modelling. Artif. Intell. **155**(1–2), 1–39 (2004)
35. Leder, F.S., Martini, P.: Ngbpa next generation botnet protocol analysis. In: Emerging Challenges for Security, Privacy and Trust, pp. 307–317. Springer (2009)
36. Lee, C., Wang, M., Hagras, H.: A type-2 fuzzy ontology and its application to personal diabetic-diet recommendation. IEEE Trans. Fuzzy Syst. **18**(2), 374–395 (2010)
37. Li, W., Tian, S.: An ontology-based intrusion alerts correlation system. Expert Syst. Appl. **37**(10), 7138–7146 (2010)
38. Li, Y., Olson, D.L., Qin, Z.: Similarity measures between intuitionistic fuzzy (vague) sets: a comparative analysis. Pattern Recogn. Lett. **28**(2), 278–285 (2007)
39. Liang, Z., Shi, P.: Similarity measures on intuitionistic fuzzy sets. Pattern Recogn. Lett. **24**(15), 2687–2693 (2003)
40. Liao, Y., Vemuri, V.R.: Using text categorization techniques for intrusion detection. In: USENIX Security Symposium, vol. 12 (2002)
41. Liu, W.: Research of data mining in intrusion detection system and the uncertainty of the attack. In: International Symposium on Computer Network and Multimedia Technology, pp. 1–4 (2009)

42. Lukasiewicz, T., Straccia, U.: Managing uncertainty and vagueness in description logics for the semantic web. Web Semantics: Science, Services and Agents on the World Wide Web **6**(4), 291–308 (2008)
43. McCulloch, J., Wagner, C., Aickelin, U.: Extending similarity measures of interval type-2 fuzzy sets to general type-2 fuzzy sets. In: IEEE International Conference on Fuzzy Systems, pp. 1–8 (2013)
44. Mezei J., Wikström, R.: OWAD operators in type-2 fuzzy ontologies. In: Proceedings of the 2013 Joint IFSA World Congress NAFIPS Annual Meeting, number ISBN: 978-1-4799-0347-4, pp. 848-853 (2013)
45. Mitchell, H.: On the dengfeng-chuntian similarity measure and its application to pattern recognition. Pattern Recogn. Lett. **24**(16), 3101–3104 (2003)
46. Mitchell, H.B.: Pattern recognition using type-II fuzzy sets. Inf. Sci. **170**(2), 409–418 (2005)
47. Ning, P., Xu, D.: Learning attack strategies from intrusion alerts. In: Proceedings of the 10th ACM Conference on Computer and Communications Security, pp. 200–209. ACM, (2003)
48. Park, W.H.: Risk analysis and damage assessment of financial institutions in cyber attacks between nations. Math. Comput. Model. **58**(11–12), 18–45 (2012)
49. Riccardi, M., Oro, D., Luna, J., Cremonini, M., Vilanova, M.: A framework for financial botnet analysis. In: eCrime Researchers Summit (eCrime), pp. 1–7 (2010)
50. Riccardi, M., Pietro, R.D., Palanques, M., Vila, J.A.: Titans revenge: detecting Zeus via its own flaws. Comput. Networks **57**(2):422–435 (2013) (Botnet Activity: Analysis, Detection and Shutdown.)
51. Sharma, A., Pujari, A.K., Paliwal, K.K.: Intrusion detection using text processing techniques with a kernel based similarity measure. Comput. Secur. **26**(7–8), 488–495 (2007)
52. Simmonds, A., Sandilands, P., van Ekert, L.: An ontology for network security attacks. In: Applied Computing, pp. 317–323. Springer (2004)
53. Szmidt, E., Kacprzyk, J.: On measuring distances between intuitionistic fuzzy sets. Notes on IFS **3**(4), 1–13 (1997)
54. Szmidt, E., Kacprzyk, J.: Distances between intuitionistic fuzzy sets. Fuzzy Sets Syst. **114**(3), 505–518 (2000)
55. Tafazzoli, T., Sadjadi, S.H.: Malware fuzzy ontology for semantic web. Int. J. Comput. Sci. Network Secur. **8**(7), 153–161 (2008)
56. Tajbakhsh, A., Rahmati, M., Mirzaei, A.: Intrusion detection using fuzzy association rules. Appl. Soft Comput. **9**(2), 462–469 (2009)
57. Turksen, I., Zhong, Z.: An approximate analogical reasoning schema based on similarity measures and interval-valued fuzzy sets. Fuzzy Sets Syst. **34**(3), 323–346 (1990)
58. Undercoffer, J., Joshi, A., Pinkston, J.: Modeling computer attacks: an ontology for intrusion detection. In: Recent Advances in Intrusion Detection, pp. 113–135. Springer, (2003)
59. Wagener, G., Dulaunoy, A., et al.: Malware behaviour analysis. J. Comput. Virol. **4**(4), 279–287 (2008)
60. Wang, C., Entropy, AQu: similarity measure and distance measure of vague soft sets and their relations. Inf. Sci. **244**, 92–106 (2013)
61. Wang, G., Hao, J., Ma, J., Huang, L.: A new approach to intrusion detection using artificial neural networks and fuzzy clustering. Expert Syst. Appl. **37**(9), 6225–6232 (2010)
62. Wu, D., Mendel, J.: A comparative study of ranking methods, similarity measures and uncertainty measures for interval type-2 fuzzy sets. Inf. Sci. **179**(8), 1169–1192 (2009)
63. Wu, D., Mendel, J.M.: A vector similarity measure for linguistic approximation: interval type-2 and type-1 fuzzy sets. Inf. Sci. **178**(2), 381–402 (2008)
64. Xu, Z.: Some similarity measures of intuitionistic fuzzy sets and their applications to multiple attribute decision making. Fuzzy Optim. Decis. Making **6**(2), 109–121 (2007)
65. Xu, Z., Chen, J.: Ordered weighted distance measure. J. Syst. Sci. Syst. Eng. **17**(4), 432–445 (2008)
66. Zadeh, L.A.: Fuzzy logic = computing with words. IEEE Trans. Fuzzy Syst. **4**(2), 103–111 (1996)

67. Zeng, W., Guo, P.: Normalized distance, similarity measure, inclusion measure and entropy of interval-valued fuzzy sets and their relationship. Inf. Sci. **178**(5), 1334–1342 (2008)
68. Zeng, W., Li, H.: Relationship between similarity measure and entropy of interval valued fuzzy sets. Fuzzy Sets Syst. **157**(11), 1477–1484 (2006)
69. Zhang, C., Fu, H.: Similarity measures on three kinds of fuzzy sets. Pattern Recogn. Lett. **27**(12), 1307–1317 (2006)
70. Zhang, H., Zhang, W.: Inclusion measure and similarity measure of intuitionistic and interval-valued fuzzy sets. In: Proceedings of the 2007 International Conference on Intelligent Systems and Knowledge Engineering (ISKE2007) (2007)
71. Zheng, G., Wang, J., Zhou, W., Zhang, Y.: A similarity measure between interval type-2 fuzzy sets. In: International Conference on Mechatronics and Automation (ICMA), pp. 191–195 (2010)
72. Zheng, G., Xiao, J., Wang, J., Wei, Z.: A similarity measure between general type-2 fuzzy sets and its application in clustering. In: IEEE 8th World Congress on Intelligent Control and Automation (WCICA), pp. 6383–6387 (2010)

A Multi-objective Genetic Algorithm Based Approach for Effective Intrusion Detection Using Neural Networks

Gulshan Kumar and Krishan Kumar

Abstract In this paper, a novel multi-objective genetic algorithm (MOGA) based approach is proposed for effective intrusion detection based on benchmark datasets. The proposed approach can generate a pool of non-inferior individual solutions and ensemble solutions thereof. The generated ensembles can be used to detect the intrusions accurately. For intrusion detection problem, the proposed MOGA based approach could consider conflicting objectives simultaneously like detection rate of each attack class, error rate, accuracy, diversity etc. The proposed approach can generate a pool of non-inferior solutions and their ensemble thereof having optimized trade-offs values of multiple conflicting objectives. In this paper, a three phase MOGA based approach is proposed to generate solutions with a simple chromosome design in first phase. In first phase, a Pareto front of non-inferior individual solutions is approximated. In the second phase of the proposed approach, entire solution set is further refined to determine effective ensemble solutions considering solution interaction. In this phase, another improved Pareto front of ensemble solutions over that of individual solutions is approximated. The ensemble solutions in improved Pareto front reported improved detection results based on benchmark datasets for intrusion detection. In third phase, a combination method like majority voting method is used to fuse the predictions of individual solutions for determining prediction of ensemble solution. Benchmark datasets namely KDD cup 1999 and ISCX 2012 dataset are used to demonstrate and validate the performance of the proposed approach for intrusion detection. The proposed approach can discover individual solutions and ensemble solutions thereof with good support and detection rate from benchmark datasets (in comparison with well-known ensemble methods like bagging and boosting). In addition, the proposed approach is a generalized classification approach that is applicable to the problem of any field having multiple conflicting objectives and a dataset can be represented in the form of labeled instances in terms of its features.

G. Kumar (✉) · K. Kumar
Shaheed Bhagat Singh State Technical Campus, Ferozepur, Punjab, India
e-mail: gulshanahuja@gmail.com

K. Kumar
e-mail: k.salujasbs@gmail.com

R.R. Yager et al. (eds.), *Intelligent Methods for Cyber Warfare*,
Studies in Computational Intelligence 563, DOI 10.1007/978-3-319-08624-8_8

1 Introduction

The industry faces the challenges of a fast changing trends of attacking the Internet resources, inability of conventional techniques to protect the Internet resources from a variety of attacks, and biases of individual techniques towards specific attack class(es). Developing effecting techniques is necessary for securing valuable Internet resources from attacks. Nowadays, conventional protection techniques such as firewalls, user authentication, data encryption, avoiding programming errors and other simple boundary devices are used as the first line of defense for security of the systems. Some attacks are prevented by the first line of defense where as some bypass them. Such attacks must be detected as soon as possible so that damage may be minimized and appropriate corrective measures may be taken. Several techniques from different disciplines are being employed for the accurate intrusion detection systems (IDSs). Detection Rate (DR) and False Positive Rate (FPR) are two key indicators to evaluate the capability of an IDS. Many efforts are being done to improve DR and FPR of the IDSs [47]. In beginning, the research focus was on rule based and statistical IDSs. But, with large datasets, the results of these IDSs become un-satisfactory. Thereafter, a lot of Artificial Intelligence (AI) based techniques have been introduced to solve the problem due to their advantages over the other techniques [41, 60]. The AI based techniques have reported certain improvements in the results to detect the intrusions. Many researchers analyzed various AI based techniques empirically and compared their performance for detection of intrusions. Findings of representative empirical comparative analysis are as follows: Most of the existing techniques strive to obtain a single solution that lacks classification trade-offs [22]; Low detection accuracy and high false alarm rate; No single technique is capable enough to detect all classes of attacks to an acceptable level of false alarm rate and detection accuracy [41, 49]; Some of the existing techniques fall into local minima. For global minima, these techniques are computationally expensive; The existing techniques are not capable to model correct hypothesis space of the problem [20]; Some existing techniques are unstable in nature such as neural networks show different results with different initializations due to the randomness inherent in the training procedure; Different techniques trained on the same data may not only differ in their global performances, but they may show strong local differences also. Each technique may have its own region in the feature space where it performs the best [30]; Delay in the detection of intrusions due to the processing of a large size of high dimensional data [9, 60]; and NB, MLP and SVM techniques are found to be most promising in detecting the intrusions effectively [35]. It is also noticed from the literature of AI based techniques that most of the existing intrusion detection techniques report poor results in terms of DR and FPR towards some specific attack class(es). Even, Artificial Neural Networks (ANNs), Naive Bayes (NB) and Decision Trees (DT) have been popularly applied to intrusion detection (ID), but these techniques have provided poor results, particularly towards the minor attack class(es) [10, 31]. The poor results may be due to an imbalance of instances of a specific class(es) or the inability of techniques to represent a correct hypothesis of the problem based on available training data.

In order to improve the low DR and high FPR, focus of the current research community in the field of intrusion detection (ID) is on ensemble based techniques. Because, there is a claim in the literature that ensemble based techniques generally outperform the best individual techniques. Moreover, several theoretical and empirical reasons including statistical, representational and computational reasons exist that also advocate the use ensemble based techniques over the single techniques [19]. This paper aims to develop a multi-objective genetic algorithm (MOGA) based approach for intrusion detection to generate a pool of non-inferior individuals solutions and combine them to generate ensemble solutions for improved detection results. The pool of solutions provides classification trade-offs to the user. Out of pool of solutions, the user can select an ideal solution as per application specific requirements.

Paper Overview: Sect. 2 presents the related work and identifies the research gaps in the field. A novel MOGA based approach for effective intrusion detection is proposed in Sect. 3. This section also explains implementation detail of the proposed approach including brief description of multi layer perceptron (MLP), benchmark datasets, performance metrics followed by experimental setup, results of the proposed approach using MLP as a base classifier. Finally, the concluding remarks along with the scope for future work are listed at the end of this paper in Sect. 4.

2 Related Work

Ensemble techniques/classifiers have been recently applied to overcome the limitations of a single classifier system in different fields [19, 34, 42]. Such attention is encouraged by the theoretical [19] and experimental [21] studies, which illustrate that ensembles can improve the results of traditional single classifiers. In general, an ensemble construction of base classifiers involves generating a diverse pool of base classifiers [6], selecting an accurate and diverse subset of classifiers [57], and then combining their outputs [42]. These activities correspond to ensemble generation, ensemble selection and ensemble integration phases of ensemble learning process [38]. Most of the existing ensemble classifiers aim at maximizing the overall detection accuracy by employing multiple classifiers. The generalizations made concerning ensemble classifiers are predominantly suitable in the field of ID. As Axelsson [4] notes, "In reality there are many different types of intrusions, and different detectors are needed to detect them". Use of multiple classifiers is supported by the statement that if one classifier fails to detect an attack, then another should detect it [43]. However, to create an efficient ensemble, we are still facing numerous difficulties: How can we generate diverse base classifiers? Then, once these base classifiers have been generated, should we use all of them or should we select a sub-group of them? If we decide to select a subgroup, how do we go about it? Then, once the sub-group has been selected, how can we combine the outputs of these classifiers?

Previous studies in the field of intrusion detection have attempted various techniques to generate effective ensembles such as bagging, boosting, and random sub-space etc. The researchers proposed a multi classifier based system of Neural Networks (NNs) [24]. The different neural networks were trained using different features of KDD cup 1999 dataset. They concluded that a multi strategy combination technique like belief function outperforms other representative techniques. Multi classifier system of NNs was also advocated by Sabhnani and Serpen [50]. The authors reported improved results over single techniques. The researchers used weighted voting to compute the output of ensemble of CART and BN and reported improved results for intrusion detection [1, 11]. Perdisci et al. [48] proposed a clustering based fusion method that reduces the volume of alarms produced by the IDS. The reduced alarms provides a concise high level description of attacks to system administrator. The proposed method uses correlation between alarms and meta alarms to reduce the volume of alarms of the IDSs. A hierarchical hybrid system was also proposed in [61]. But, the proposed system leads to high false positive rate. Chen et al. [12] used different features of dataset to generate ensemble solutions based on evolutionary algorithms. Toosi and Kahani [56] proposed a neuro-fuzzy classifier to classify instances of KDD cup 1999 dataset into five classes. But, a great time consuming is a big problem. Hu and Damper [28] proposed a adaBoosting ensemble method that uses different features to generate diverse set of classifiers. No doubt, the proposed method reported improved performance but it suffers from limitation of incremental learning. It requires continuous retraining for changing environment. Zainal et al. [62] proposed a heterogeneous ensemble of different classifiers and used weighted voting method for combining their predictions. Wang et al. [58] proposed an approach based on NN and fuzzy clustering. Fuzzy clustering helps to generate homogeneous training subsets from heterogeneous training dataset which are further used to train NN models. They reported improved performance in terms of detection precision and stability. Clustering based hybrid system was also advocated by Muda et al. [45] for intrusion detection. The system was unable to detect the intrusions of U2R and R2L attack classes. Khreich et al. [33] proposed a iterative boolean combination (IBC) technique for efficient fusion of the responses from any crisp or soft detector trained on fixed-size datasets in the ROC space. However, IBC does not allow to efficiently adapt a fusion function over time when new data becomes available, since it requires a fixed number of classifiers. The IBC technique was further improved as incremental Boolean combination (incrBC) by the authors in [34]. The incrBC is a ROC-based system to efficiently adapt ensemble of HMM (EoHMMs) over time, from new training data, according to a learn-and-combine approach without multiple iterations. Govindarajan and Chandrasekaran [26] suggested a hybrid architecture of NNs for intrusion detection. They used weighted voting method compute the final prediction of system.

However, the models developed based on these techniques attempted to obtain a single solution. They lack in providing classification trade-offs for application specific requirements. Most of the models provided biased results towards specific attack class(es).

In contrast, genetic algorithm (GA) is the most widely used technique in data mining and knowledge discovery [23]. Applying GA is valuable for its robustness in performing global search in search space compared with other representative techniques. Several researchers employed single and multiple objective genetic algorithms for finding a set of non-inferior solutions for the problem of ID. Such initiative was carried by Parrott et al. [46] by suggesting an evaluation function which was later known as Parrot function. They proposed to use accuracy of each target class as a separate objective in their evaluation function for MOGA. Here, accuracy of each class refers to correctly classified instances of that class. The Parrot function was further adopted in [2] and [3] to generate an ensemble of base classifiers. The generation of the ensemble was completed in two stages using modified NSGA-II [18]. In the first stage, a set of base classifiers was generated. Second stage optimized the combination of base classifiers using a fixed combining method. Both of these methods differ in their function evaluation. The former study proposed to optimize the classifiers by minimizing the aggregated error of each class and maximize diversity among them. Since, the error on each class is not treated as separate objectives, this is similar to a general error measure such as MSE (mean square error), which have the same issues as the implementation of Parrot function, being biased towards the major class(es). In the second phase of the approach proposed in [2] and [3], the objectives are to minimize the size of the ensemble and maximize the accuracy. Consequently, the drawback of their approach is to create a single best solution based on general performance metrics. The same concept was further extended by Engen [22] by conducting similar experiments with different evaluation functions for creating ensemble of ANNs as base classifiers in the presence of imbalanced datasets using NSGA-II. He used 3-class classification by using ANNs and MOGA. He proved that MOGA based approach is an effective way to train the ANN which works well for minority attack classes in imbalanced datasets. He proposed two phase process for intrusion detection. In the first phase, he generated a set of base classifiers of ANNs by optimizing their weights assuming a fixed number of hidden layers and the number of neurons per hidden layer in ANN. The second phase generates improved non-dominated front of ensemble solutions based upon base ANN solutions optimized in phase 1. However, the performance of NSGA-II degrades for the real world problems having more than three objectives and large population [55].

3 MOGA Based Approach for Effective Intrusion Detection

A novel MOGA based approach for intrusion detection is proposed. The concept of two tier fitness assignment mechanism consisting of domination rank and diversity measure of solutions (as proposed in [53]) are used to improve the solutions from benchmark datasets. Generally, intrusion detection problem encounters a trade-offs between multiple conflicting criteria such as detection rate of attack classes, accuracy and diversity etc. Considering the multiple criteria of intrusion detection problem, GAs can be used in two ways. The first way to solve a multi-objective problem is to convert multiple objectives into a single objective [13]. The single objective is further optimized by GA to produce a single solution. Generally, prior knowledge about the problem, or some heuristics guide the GA to produce a single solution. By changing the parameters of the algorithm and executing the algorithm repeatedly, more solutions can be produced. This approach has several limitations for multi objective optimization problems. The second way to solve multi objective optimization problems by using GA produces a set of non-inferior solutions. This set of non-inferior solutions represents trade-offs between multiple criteria which is identified as a Pareto optimum front [22, 39]. By incorporating domain knowledge, the user can select a desired solution. Here, GA has produced a set of solutions in Pareto front in a single run without incorporating any domain knowledge or any other heuristic about the problem. Some of the important researches in developing MOGAs are Strength Pareto Evolutionary Algorithm (SPEA2) [63], Pareto-Envelope based Selection Algorithm (PESA-II) [15], Non-dominated Sorting Genetic Algorithm (NSGA-II) [17], Archive based Micro Genetic Algorithm 2 [54] and many more. A comprehensive review of various MOGAs can be further referred in [13, 14, 16].

The proposed approach is developed with particular attention to enhance the detection rate of majority as well as minority attack class(es). A chromosome encoding scheme is proposed to represent the individual classifiers. Further more, the proposed approach is used to find an improved Pareto front consisting of ensemble solutions. The MOGA used in this paper is Archive based Micro Genetic Algorithm 2 (AMGA2) [54], which is an effective algorithm for finding optimal trade-offs for multiple criteria. AMGA2 is a generational algorithm that works with a very small population size and maintains a large external archive of good solutions obtained. Using an external archive that stores a large number of solutions provides useful information about the search space as well as tends to generate a large number of Pareto points at the end of the simulation. At every iteration, a small number of solutions are created using the genetic variation operators. The newly created solutions are then used to update the archive. The strategy used to update the archive relies on the domination level and the diversity of the solutions, and the current size of the archive, and is based on the non-dominated sorting concept borrowed from NSGA-II [18]. This process is repeated until the allowed number of function evaluations is exhausted. We used differential evolution (DE) operator as crossover operator for mating the population. Because, DE has advantage of not requiring a distribution index and it is self-adaptive in that the step size is automatically adjusted depending upon the distribution of the solutions in the search space. After mating the population with crossover operator, it is followed by mutation operator. Modified polynomial mutation operator is used to mutate the offsprings solutions.

3.1 The Proposed Approach

This section describes the proposed approach based on MOGA to create a set of base classifiers and ensembles thereof. The proposed approach follows an overproduce and choose approach that focus on generation of a large number of base classifiers and later on choose the subset of the most diverse base classifiers to generate ensembles. The proposed approach is a three phase approach as described in subsequent paragraphs.

Phase 1 and phase 2 are multi-objective in nature and use MOGA to generate a set of base classifiers and ensembles thereof respectively. These phases of the proposed approach evolve a set of solutions to formulate diverse base classifiers and ensembles thereof using MOGA. The set of base classifiers and their ensembles exhibit classification trade-offs for the user. The diversity among individual solutions and their ensembles is maintained implicitly. The detection rate for each class is treated as a separate objective in both the phase. Here, the MOGA is real-coded, uses cross-over and mutation operators and an elitist replacement strategy.

Phase 1 of the proposed approach is capable to find the optimal Pareto front of non-dominated solutions (depicted in Fig. 1). These solutions formulate the base classifiers as candidate solutions for the ensemble generation in Phase 2. In phase 1, the values in chromosome and its size depends upon the type of base classifier and corresponding encoding scheme. The output of phase 1 is a set of optimized real values for classifiers that formulate the base classifiers of ensembles. The population size is equal to the number of desired solutions input by the user.

Phase 2 generates another improved approximation of optimal Pareto front consisting of a set of non-dominated ensembles based on a set of non-dominated solutions as base classifiers (output of phase 1) which also exhibit classification trade-offs (depicted in Fig. 2). It takes input in the form of archive of non-dominated solutions produced by phase 1 that formulates the base classifiers of the ensembles. The

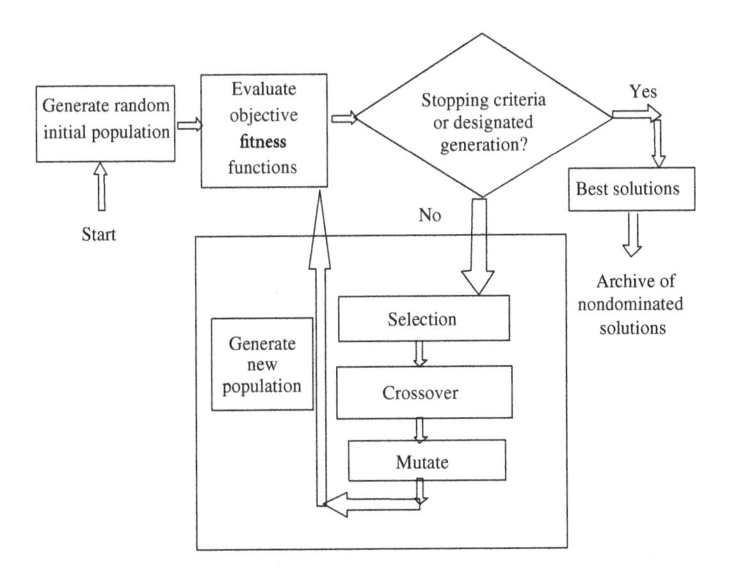

Fig. 1 Phase 1 of the proposed approach

phase evolves ensembles by combining the Pareto front of non-dominated solutions instead of the entire population like other studies [29]. Here, we are interested in those solutions which are non-inferior and exhibit classification trade-offs. The predictions of the base classifiers are combined using the majority voting method. In case of a tie, the winner is randomly chosen. The MOGA method discussed in phase 1 is again applied in phase 2. Here, MOGA is real coded having values from 0 to 1. Value $\geq$0.5 signifies the participation of base classifier in the ensemble and <0.5 signifies non-participation concerned base classifiers in creating the ensembles. The output of phase 2 is an archive of the ensembles of the base classifiers in terms of chromosomes in the range of 0 and 1 (depicted in Fig. 2). Here, value $\geq$0.5 signifies the participation of base classifier in ensemble and <0.5 signifies its non-participation. The set of ensembles provides the classification trade-offs for the user for different objective functions.

Phase 3 of the proposed approach integrates the predictions of base classifiers to get prediction of the final ensemble. As depicted in Fig. 3, the phase takes two inputs (1) archive of non-dominated base solutions (output of phase 1); and (2) one chromosome from the archive of ensembles as chosen by the user depending on requirements (output of Phase 2). The user may adopt static or a dynamic strategy to choose an appropriate ensemble from a pool of ensembles (evolved in Phase 2). Here in this work, we selected the ensemble classifier using a static strategy based on its performance on the training data in terms of pre-defined performance metrics. Based on the values of the chromosome, corresponding predictions of base classifiers

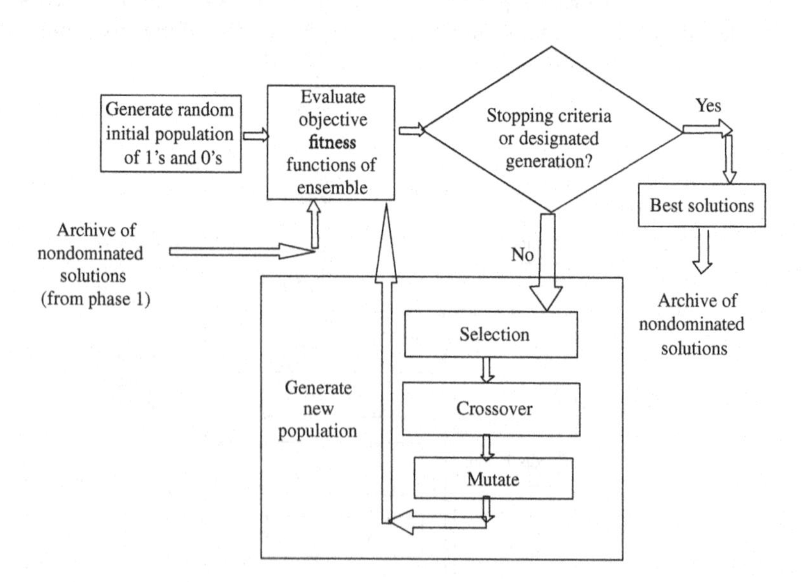

Fig. 2 Phase 2 of the proposed approach

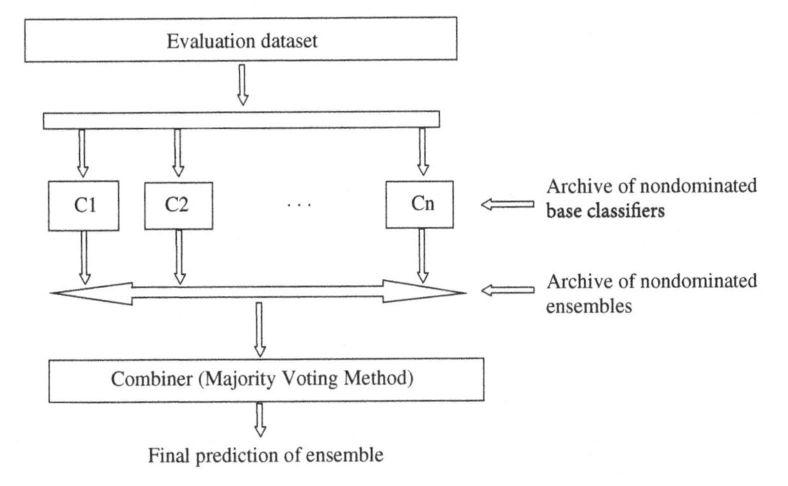

Fig. 3 Phase 3 of the proposed approach

are integrated to get a final prediction of the ensemble. In order to test the proposed approach, test dataset is directly fed to different base classifiers. Their predictions are combined in this phase to give the final output of the ensemble. In this work, we computed the final prediction of ensemble by using the majority voting method because of its popularity as depicted in Fig. 3.

The phases of the proposed approach address key issues of the current research in the field of ensembles. The issues addressed are (1) generation of a set of non-inferior solutions that exhibit classification trade-offs to formulate base classifiers of the ensemble; (2) generation of a set of non-inferior ensemble solutions that exhibit classification trade-offs; and (3) integration of predictions of the base classifiers to get final prediction of the ensemble.

3.2 Implementation

To evaluate the proposed approach, it is implemented in VC++. MLP is used as a base classifier as per finding of state of art literature in the field of ID. The performance of the proposed technique is evaluated based on benchmark datasets for ID namely KDD cup 1999 and ISCX 2012 dataset. During the optimization of multiple criteria by AMGA2, detection rate of each attack class in the dataset is used as a separate objective. Majority voting method is used to integrate the predictions of base classifiers to get prediction of the final ensemble. The results of experiments are computed on a Windows PC with Core i3-2330M 2.20 GHz CPU and 2 GB RAM. Following sub-sections describes the details of GA, MLP, benchmark dataset, and performance metrics used in the experiments.

3.2.1 Genetic Algorithm

GA are population based search techniques that have been identified to perform better that than the classical heuristics or gradient approaches [25]. GAs provides better solutions particularly for multi models, non-differentiable, or discontinuous functions. Generally, GA experiences following steps:

1. Generate a random population of individuals that represents solution to the underlying problem.
2. Evaluate the population by computing their fitness function of each individual.
3. Elevate high quality individuals by selecting them from entire population.
4. Generate new population containing individuals created by applying variation operators of cross-over and mutation.
5. Repeat the above steps till termination criteria is satisfied.

A large number of methods have been developed to implement steps for GAs. However, major issues consist of representation of individuals, fitness evaluation mechanism, variation operators of cross-over and mutation, and deciding the termination criteria.

3.2.2 Multi Layer Perceptron

An MLP is a network of simple neurons called perceptrons [5]. The perceptron computes a single output from multiple real-valued inputs by forming a linear combination according to its input weights and then possibly putting the output through some non-linear activation function. In other words, MLPs are feed forward Artificial Neural Networks (ANNs) that may be trained with the standard back propagation algorithm [5] or by using other alternative techniques. They are supervised networks, so they require a desired response to be trained. They learn how to transform input data into a desired response, so they are widely used for pattern classification. With one or two hidden layers, they can approximate virtually any input-output map. They have been shown to approximate the performance of optimal statistical classifier in difficult problems.

The MLP used in this paper is composed of three neuron layers, namely, the input layer, the output layer and the hidden layer as shown in Fig. 4. Although the MLP can have more than one hidden layer, having more than one hidden layer is rarely beneficial and can lead to gross over parametrization [22]. For a particular instance i of training/test dataset, the input layer of the MLP used for intrusion detection receives the input vector T from training dataset. The input vector T has general format

$$T_i = (t_{i,1}, t_{i,2} \ldots\ldots\ldots\ldots\ldots\ldots\ldots t_{i,n}) \tag{1}$$

Here, is the jth feature of ith instance of training/test dataset. Total number of input neurons in input layer is equal to total features of training/test dataset for intrusion

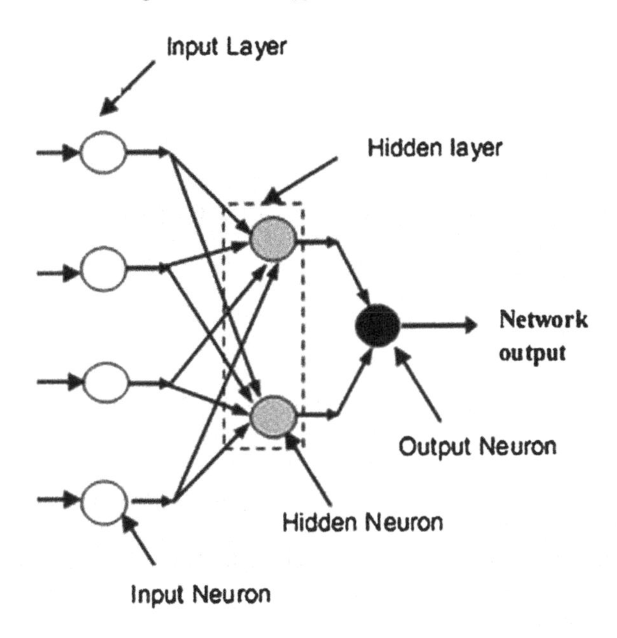

Fig. 4 Structure of MLP

detection. The output layer contains the output neurons. The output neurons are equals to number of classes in dataset. A hidden layer is a middle layer. This layer adds a degree of flexibility to the performance of the ANN that enables it to deal efficiently with complex nonlinear problems. Each neuron in the single hidden layer receives the same input vector of N elements from the neurons of the input layer, as defined by Eq. (1), and produces the output. The input-output transformation in each hidden neuron is achieved by a mathematical non-linear transfer (or activation) function. The general form of activation function is

$$Y_{i,k} = f(\sum_{j=1}^{N} W_{j,k} * T_{i,j} + b_k) \tag{2}$$

where $Y_{i,k}$ is the output of kth neuron in hidden layer for ith instance of dataset, f() is an activation function, is the connection weight assigned to kth hidden neuron and jth neuron in input layer and is the bias of kth hidden neuron. In literature, many activation functions are proposed [22]. The most widely used activation functions is the sigmoid function which can be expressed as

$$Y_{i,k} = \frac{1}{1 + \exp(-\sum_{j=1}^{N} W_{j,k} * T_{i,j} - b_k)} \tag{3}$$

The neurons in output layer produce the final network output. These output neurons receives an input array in form of Eq. 4.

$$Z_i = (Y_{i,1}, Y_{i,2}.....................Y_{i,n}) \tag{4}$$

The input-output transformation for this output neuron is similar to that of the hidden neurons

3.2.3 Benchmark Datasets

The performance of the proposed approach is measured based on benchmark datasets. In the literature, various benchmark datasets are proposed for validation of the IDSs. As per statistics of a survey of 276 papers published between 2000 and 2008 conducted by Tavallaee [52], most of the researchers used publicly available benchmark datasets for evaluating their network based approaches. It is observed that KDD cup 1999 [32] data set is the most widely data set used for validation of an IDS [41, 52] in spite of many criticisms [7, 22, 44]. The raw training dataset contains about 4 GB of TCP connection data in the form of 5 million connection records. Similarly, test data set contains about 2 million records. KDD cup 1999 dataset utilizes TCP/IP level information and embedded with domain-specific heuristics, to detect intrusions at the network level. KDD dataset contains four major classes of attacks: Probe, Denial of Service (DoS), User-to-Root (U2R) and Remote-to-Local (R2L) attacks The labeled connection records consist of 41 features and 01 attack type. The labeled connection records consist of 22 different attack types categorized into 04 classes whereas unlabeled dataset consist of 20 known and 17 unknown attack types. The 41 features can be divided into three categories viz: Basic features of individual TCP connections, Content features within a connection suggested by domain knowledge and Traffic features computed using a two-second time window.

In a thorough study of KDD cup 1999 dataset, Tavallaee [52] observed that there are some inherent problems. He refined the KDD cup 1999 dataset and named it as NSL-KDD dataset. As the number of connection records in training and test NSL-KDD data set is very large, so it's practically very difficult to use the whole data set. Thus, in order to conduct unbiased learning and testing of the proposed approach, we used subsets of the dataset containing different proportions of normal and attack instances. The statistics of selected subsets of NSL-KDD datasets used in our experiments is as depicted in Table 1. Here, we selected 10 most prominent features in ITFS data subset by applying feature selection technique described in [36, 37].

In order to overcome the limitations of KDD cup 1999 dataset, [51] presented a new dataset for validation of an IDS at Information Security Center of eXcellence (ISCX). The dataset is available in the packet capture form. Features are extracted from the packet format by using tcptrace utility (downloaded from www.tcptrace.org) and applying the following command.

tcptrace csv –l filename1.7z > filename1.csv

Table 1 Statistics of subsets of KDD cup 1999 dataset as Training and Test data subsets

Dataset	Mode	Number of features	Class	Number of instances	Total instances
KDD 1	Training	41	Normal	1000	
			Probe	100	
			DoS	100	
			U2R	11	
			R2L	100	1311
	Test	41	Normal	500	
			Probe	75	
			DoS	75	
			U2R	50	
			R2L	50	750
KDD 2	Training	41	Normal	13449	
			Probe	2289	
			DoS	9234	
			U2R	11	
			R2L	209	25192
	Test	41	Normal	2152	
			Probe	2402	
			DoS	4342	
			U2R	200	
			R2L	2754	11850
ITFS	Training	41, 10	Normal	10000	
KDD			Probe	32316	
			DoS	23467	
			U2R	52	
			R2L	1126	66961
	Test	41, 10	Normal	5000	
			Probe	4166	
			DoS	17761	
			U2R	228	
			R2L	13448	40603

where filename is the name of the 7z (packet capture) file. From resulting csv files, we selected features which are most widely used features in the literature as proposed by Brugger [8]. The data instances including normal as well as attack instances are randomly selected to create a subset of the benchmark dataset for our experiments. The selected dataset is further preprocessed by converting discrete feature values to numeric ones as described in [40]. The statistics of selected ISCX 2012 data subset are depicted in Table 2.

Table 2 Statistics of subset of ISCX 2012 dataset as Training and Test data subset

Dataset	Mode	Number of features	Class	Number of instances	Total instances
ISCX	Training	9	Normal	4125	
2012			Attack	578	4703
	Test	9	Normal	64127	
			Attack	577	4704

3.2.4 Performance Metrics

In order to evaluate the effectiveness of the IDS, we measure its ability to correctly classify events as normal or intrusive along with other performance objectives, such as economy in resource usage, resilience to stress and ability to resist attacks directed at the IDS [27]. Measuring this ability of the IDS is important to both industry as well as research community. It helps us to tune the IDS in a better way as well as compare different IDSs. There exist many metrics that measure different aspects of the IDS, but no single metric seems sufficient to objectively measure the capability of the IDS. Most widely used metrics by intrusion detection research community are True Positive Rate (TPR) and False Positive Rate (FPR). Or False Negative rate $FNR = 1 - TPR$ and True Negative Rate $TNR = 1 - FPR$ can also be used alternatively. Based upon values of these two metrics only, it is very difficult to determine better IDS among different IDSs. For example, one IDS reporting, $TPR = 0.8$; $FPR = 0.1$, while at another IDS, $TPR = 0.9$; $FPR = 0.2$. If only values TPR and FPR are given, then it is very difficult to determine the better IDS. To solve this problem, Gu et al. [27] proposed a new objective metric called Intrusion Detection Capability (CID) considering base rate, TPR and FPR collectively. CID possesses many important features. For example, (1) it naturally takes into account all the important aspects of detection capability, i.e., FPR, FNR, Positive Predictive Value (PPV) [4], Negative Predictive Value (NPV), and base rate (the probability of intrusions); (2) it objectively provides an essential measure of intrusion detection capability; and (3) it is very sensitive to IDS operation parameters such as base rate, FPR and FNR. Detail of CID can be further studied in [27]. Keeping these points in view, we computed TPR, FPR and CID to evaluate the performance of the proposed technique and compare it with other representative techniques in the field.

3.2.5 Design of Experiments

In this investigation, we used AMGA2 as a multi objective genetic algorithm because of its benefits over other representative algorithms [54]. The implementation of AMGA2 algorithm takes following input parameters.

- Number of function evaluations
- Number of desired solutions

- Random seed
- Output file

Rest of parameter like mutation rate, crossover rate, etc. is automatically tuned by the AMGA2 algorithm.

The proposed approach involves three phases to create the ensemble as described in Sect. 3.1. In phase 1 (ensemble generation phase), AMGA2 optimizes an archive of diverse base classifiers that exhibit classification trade-offs. The values in chromosome represent the weights of MLP. The size of chromosome is equal to the number of weights of MLP which is further dependent structure of the MLP (i.e. input nodes, hidden layers, number of hidden nodes per layer and output nodes). Each chromosome represents a MLP classifier in terms of its weights. The output of phase 1 is a set of optimized real values of the weights of MLPs that formulate the base classifiers for the ensembles. In phase 2 (ensemble selection phase), AMGA2 is again used to create an archive of the ensembles that also exhibits classification trade-offs. In phase 3 (ensemble integration phase), the predictions of selected base classifiers are combined to compute the final prediction of the ensemble using the majority voting method. The parameters used as input by the user to AMGA2 are depicted in Table 3. Other simulation parameters tuned automatically by AMGA2 for KDD cup 1999 dataset and the ISCX 2012 dataset are presented in Tables 4 and 5 respectively. For investigation of MLP as a base classifier, the structure of MLP used is as depicted in Table 6.

Table 3 Parameters of AMGA2 input by the user

Number of function evaluations	25000
Number of desired solutions	100
Random seed	0.1

Table 4 Simulation parameters tuned by AMGA2 for KDD cup 1999 dataset

Parameter	Value
Maximum allowed size of archive	Number of desired solutions input by the user
Size of initial population	Number of desired solutions input by the user
Size of working population	20
Maximum number of function evaluations	Number of function evaluations input by the user
Probability of crossover	0.1
Probability of mutation	0.01
Index for crossover	0.5
Index for mutation	15

Table 5 Simulation parameters tuned by AMGA2 for ISCX 2012 dataset

Parameter	Value
Maximum allowed size of archive	Number of desired solutions input by the user
Size of initial population	Number of desired solutionsin-put by the user
Size of working population	8
Maximum number of function evaluations	Number of function evaluations input by the user
Probability of crossover	0.1
Probability of mutation	0.111111
Index for crossover	0.5
Index for mutation	15

Table 6 Configuration of MLP

Input nodes	Number of features of dataset
Hidden layer	1
Number of hidden nodes	30
Output nodes	5

3.3 Results and Discussion

Here, for investigation of MLP as a base classifier, ensemble generation is done by using random initial values of the weights of MLPs. As an output of this phase, we obtained an archive of MLP having optimized values of their weights. In the ensemble selection phase, we selected the MLP classifiers for the final ensemble based upon their performance during the training process (overproduce-and-choose strategy). Finally, the ensemble integration phase involves fusion strategies to combine the predictions of the selected classifiers. We used majority voting method to solve the purpose for its popularity.

In our experiments, we selected the solution for comparison with the other classifiers having a better value of the CID. Alternate solutions from the pool may provide different values of performance metrics. The results of the proposed intrusion detection approach using MLP as a base classifier and the other representative techniques are computed based upon benchmark datasets in terms of confusion matrices and other defined performance metrics. We computed average DR, Average FPR, CID and DR of each target class from the confusion matrices. The representative techniques used in this investigation are MLP trained with back propagation method, their ensembles using bagging and boosting. We utilized WEKA software package [59] to compute the results of MLP trained with back propagation, its ensembles (bagging and boosting). We used default parameters of WEKA for computing the results using MLP and its ensembles.

3.3.1 Results of KDD Cup 1999 Dataset

The proposed approach is applied to various data subsets of KDD cup 1999 dataset that produces a set of non-inferior MLP based ensemble solutions. The performance of ensemble solutions for training and test data of KDD 1 dataset is depicted in Fig. 5.

The performance of ensemble solutions for training and test data of KDD 2 dataset is shown in Fig. 6. The performance of ensemble solutions for training and test data of ITFS-KDD (41 features and 10 features) data subsets is portrayed in Figs. 7 and 8 respectively.

The overview of the classification results of KDD subsets obtained with MLP and its ensembles (using bagging and boosted methods) and our proposed approach (AMGA2-MLP) with respect to different evaluation criteria is as shown in Table 7.

The results indicate that MLP and its ensembles using bagging and boosting demonstrate comparable performance. But, these techniques are more biased towards majority classes and reported poor performance for the minority classes like U2R and

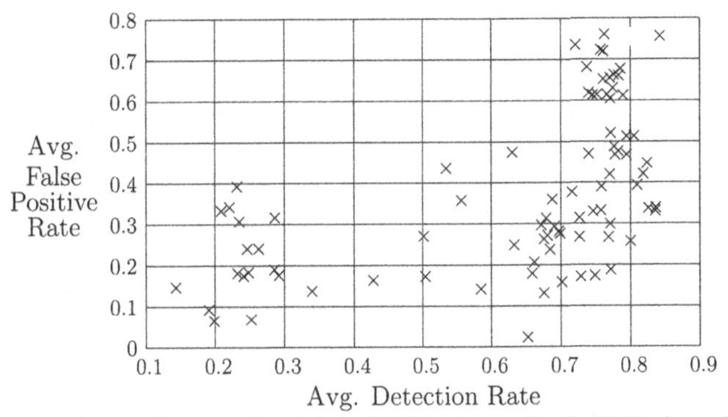

Training performance of non-inferior MLP based ensembles for KDD1 data subset

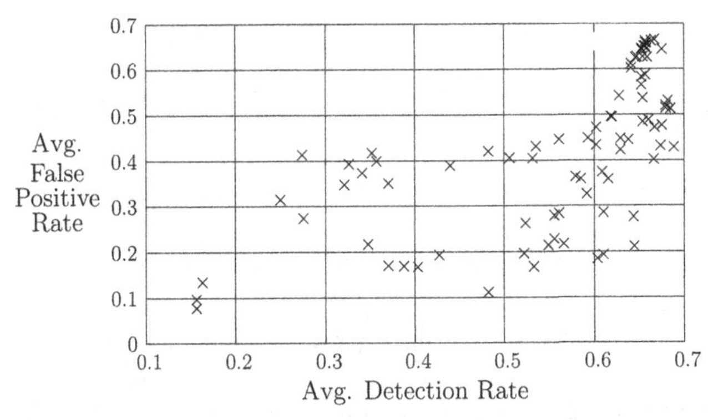

Test performance of non-inferior MLP based ensembles for KDD1 data subset

Fig. 5 Training and Test performance of non-inferior MLP based ensembles for KDD 1 data subset

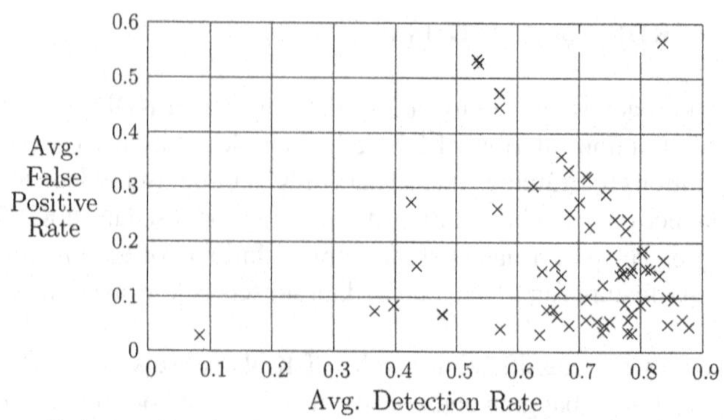

Training performance of non-inferior MLP based ensembles for KDD2 data subset

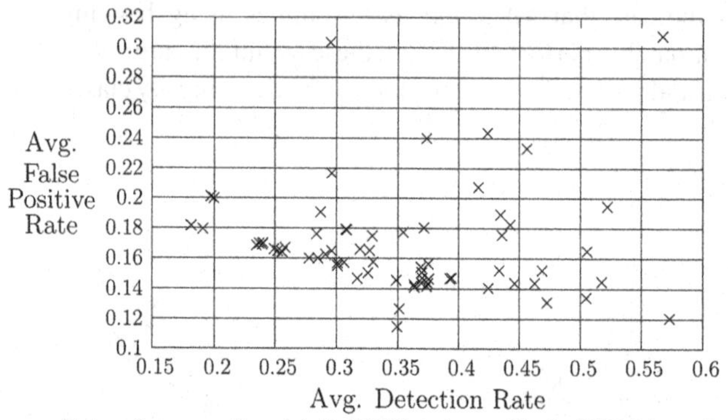

Test performance of non-inferior MLP based ensembles for KDD 2data subset

Fig. 6 Training and Test performance of non-inferior MLP based ensembles for KDD 2 data subset

R2L. MLP trained with our proposed approach is less biased and reported improved results than others for minority as well as majority classes. In case of KDD1 data subset, AMGA2-NB improved the detection of R2L attack class up to 52 % which was detected up to 2 % by the MLP and boosted MLP and 6 % by bagging based ensemble of MLP. Similarly, detection of U2R attack class is also enhanced by 66 % than MLP and its conventional ensemble techniques. In case of KDD2 data subset, MLP and its ensembles based upon bagging and boosting fails to detect U2R and R2L attack classes whereas AMGA2-MLP reported the detection of U2R and R2L attack classes up to 16.5 and 68.5 % respectively. Whereas, detection of the other classes is comparable with the other conventional ensemble techniques. In case of other data sets, the performance of the proposed technique is also comparable to the other representative techniques. Higher values of CID of our proposed technique revealed in Table 7 (in most of the cases) indicate that it outperformed the other techniques considered in this investigation.

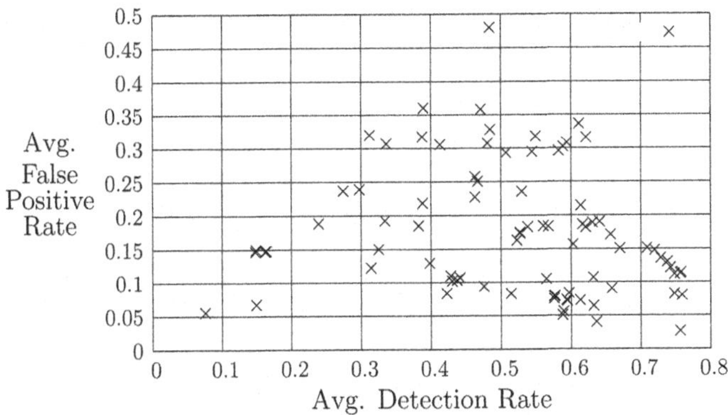

Training performance of non-inferior MLP based ensembles for ITFS-KDD (41 features) data subset

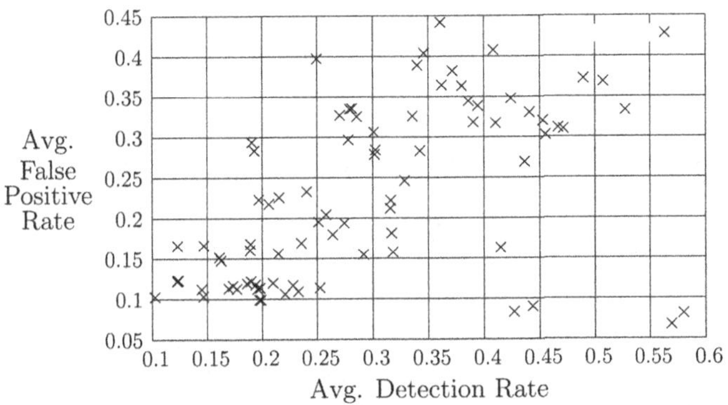

Test performance of non-inferior MLP based ensembles for ITFS-KDD (41 features) data subset

Fig. 7 Training and Test performance of non-inferior MLP based ensembles for ITFS (41 features) data subset

3.3.2 Results of ISCX 2012 Dataset

The performance of ensemble solutions for training and test data of ISCX 2012 dataset is depicted in Fig. 9. The detection results of the techniques are presented for the subset of ISCX 2012 dataset in Table 8. It can also be observed from the reporting results that AMGA2-MLP (The MLP trained with the proposed approach) reported superior performance than MLP and its bagging based ensemble and comparable performance that of boosting based ensemble of MLP. AMGA2-MLP reported the detection of normal and attack classes upto 96.9 and 97.7 % respectively. Higher value of CID indicates that our proposed approach outperformed the other techniques for the ISCX 2012 dataset considered in this investigation.

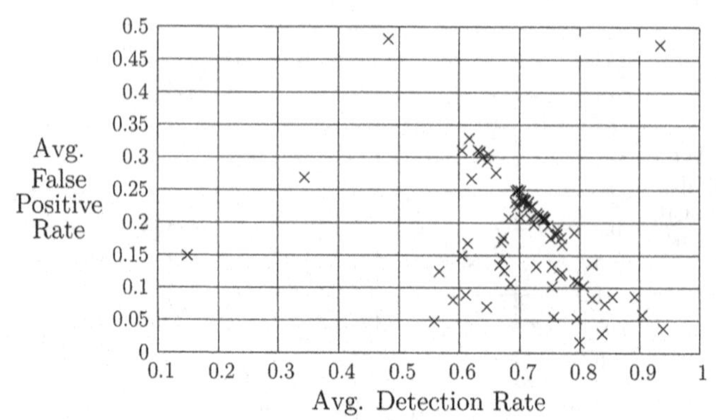

Training performance of non-inferior MLP based ensembles for ITFS-KDD (10 features) data subset

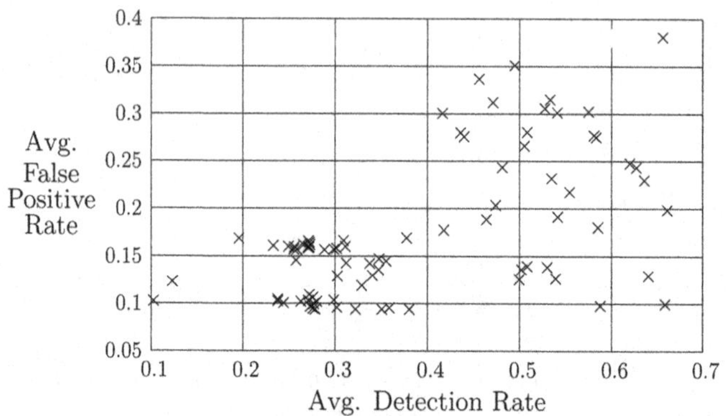

Test performance of non-inferior MLP based ensembles for ITFS-KDD (10 features) data subset

Fig. 8 Training and Test performance of non-inferior MLP based ensembles for ITFS (10 features) data subset

3.3.3 Discussion

The results obtained in this paper highlight clearly the benefits of training the MLP and its ensembles by using the proposed multi-objective genetic algorithm based approach. The proposed technique helps to improve the detection results especially for minority attack classes than that of other conventional ensemble approaches. The percentage improvement of the results of the proposed approach over other approaches is depicted in Table 9. The reporting results indicate that the proposed approach helps to enhance the average detection rate, reduce average false positive rate and overall increase in CID values over the other approaches.

In case of KDD cup 1999 dataset, MLP trained with the proposed approach helps to enhance the detection of minority attack classes like U2R and R2L attack classes which was very poorly detected by MLP trained using back propagation method.

Table 7 Overview of classification results of KDD cup 1999 subsets using MLP as a base classifier

Dataset	Technique	Avg. DR	Avg. FPR	CID	Normal	Probe	DoS	U2R	R2L
KDD1	MLP	0.720	0.374	0.086	0.898	0.627	0.507	0.100	0.020
	Bagged-MLP	0.736	0.334	0.116	0.894	0.827	0.520	0.020	0.060
	Boosted-MLP	0.720	0.374	0.086	0.898	0.627	0.507	0.100	0.020
	AMGA2-MLP	0.645	0.212	**0.142**	0.726	0.760	0.000	0.760	0.520
KDD2	MLP	0.447	0.135	0.073	0.897	0.453	0.526	0.000	0.000
	Bagged-MLP	0.435	0.138	0.066	0.908	0.414	0.507	0.000	0.000
	Boosted-MLP	0.447	0.135	0.073	0.897	0.453	0.526	0.000	0.000
	AMGA2-MLP	0.573	0.120	**0.143**	0.566	0.697	0.456	0.165	0.685
ITFS-KDD	MLP	0.510	0.071	0.134	0.971	0.958	0.360	0.136	0.405
(41 features)	Bagged-MLP	0.437	0.437	0.000	0.000	0.000	1.000	0.0000	0.000
	Boosted-MLP	0.510	0.071	0.134	0.971	0.958	0.360	0.136	0.405
	AMGA2-MLP	0.570	0.068	**0.170**	0.835	0.911	0.299	0.031	0.733
ITFS-KDD	MLP	0.580	0.202	0.087	0.859	0.818	0.652	0.101	0.314
(10 features)	Bagged-MLP	0.591	0.192	0.097	0.863	0.876	0.679	0.013	0.294
	Boosted-MLP	0.580	0.202	0.087	0.859	0.818	0.652	0.101	0.314
	AMGA2-MLP	0.659	0.099	**0.199**	0.744	0.836	0.884	0.118	0.284

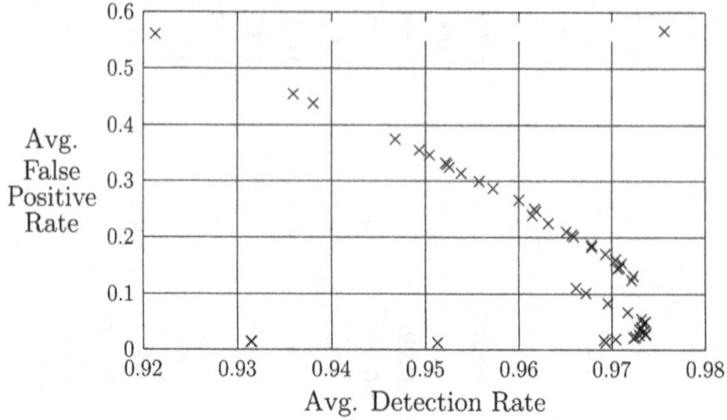

Training performance of non-inferior MLP based ensembles for ISCX 2012 data subset

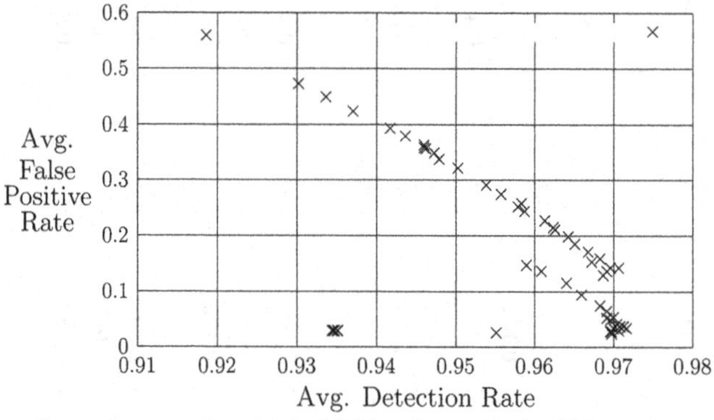

Test performance of non-inferior MLP based ensembles for ISCX 2012 data subset

Fig. 9 Training and Test performance of non-inferior MLP based ensembles for ISCX 2012 data subset

Table 8 Overview of classification results of ISCX 2012 subset using MLP as a base classifier

Dataset	Technique	Avg. DR	Avg. FPR	CID	Normal	Attack
ISCX 2012	MLP	0.906	0.660	0.049	0.998	0.248
	Bagged MLP	0.906	0.660	0.049	0.999	0.246
	Boosted MLP	0.947	0.083	0.560	0.952	0.912
	AMGA2-MLP	0.970	0.024	**0.778**	0.969	0.977

MLP and its ensemble based on the conventional techniques like bagging and boosting are biased towards majority classes, so reported poor results for minority attack classes. Whereas, the findings of the proposed approach are that they are less biased towards majority attack classes. Thus, the proposed approach is applicable where there is class imbalance and detection of all classes especially minority attack classes is equally important, as expected in many application domains including intrusion

Table 9 Percentage improvement of the results of the proposed approach using MLP as a base classifier

Classifier	MLP			Bagged	MLP		Boosted	MLP	
Dataset	DR	FPR	CID	DR	FPR	CID	DR	FPR	CID
KDD1	−10.42	−43.32	65.12	−12.36	−36.53	22.41	−10.42	−43.32	65.12
KDD2	28.19	−11.11	95.89	31.72	−13.04	116.67	28.19	−11.11	95.89
ITFS 41	11.76	−4.23	26.87	30.43	−84.44	–	11.76	−4.23	26.87
ITFS 10	13.62	−50.99	128.74	0.12	−0.48	1.05	13.62	−50.99	128.74
ISCX 2012	7.06	−96.36	1487.76	7.06	−96.36	1487.76	2.43	−71.08	38.93

detection. It is observed from the literature that MLPs trained with back propagation methods are often used for classification tasks as they are universal approximation algorithms. But, the results of this investigation indicate that back propagation method and other similar methods for training are not appropriate in all scenarios especially where detection of majority as well as minority attack classes is equally important. In case of ISCX 2012 dataset, results similar to KDD cup 1999 dataset are also obtained. MLP trained with back propagation method and its bagging based ensemble demonstrated poor results for detection of attack class. Whereas, AMGA2-MLP enhanced average DR to 0.97 (0.906 in case of MLP) and reduced average FPR to 0.024 (0.66 in case of MLP) approximately. It is also observed that most of the conventional techniques provide a single solution and lacks in providing classification trade-offs. Whereas, the proposed approach provides a pool of solutions to the problem. Out of this pool, the user can select any one solution based on its better value for CID and his/her application specific requirements. Other solutions with different values of CID may offer different detection results for the same problem that helps to exhibit the different classification trade-offs. Hence, the results depicted above sections proved the superiority of the proposed multi-objective genetic algorithm based approach and validated its applicability for proper training of the MLP for intrusion detection.

In a nut shell, the empirical investigation and comparison of the results indicate the following:

- The proposed approach outperforms the individual representative techniques in terms of identified performance metrics.
- There are indications in the literature that bagging and boosting learn better from imbalanced data. However, the experiments here have demonstrated that these algorithms remain biased towards the majority class(es).
- Using MLP as a base classifier, the proposed approach is able to enhance DR by 28 % , reduce FPR by 51 % approximately over the results of MLP trained using back propagation method and its ensemble using boosting technique based on KDD cup 1999 dataset. However, an improvement of results is noticed upto 30 % in DR and 84 % in FPR approximately over bagging based ensemble of MLP for KDD cup 1999 dataset. For ISCX 2012 dataset, the results of the proposed

technique are improved upto 7 % in DR and 96 % in FPR approximately over MLP and its ensemble using bagging technique.

- The ensembles evolved with the proposed technique provides better solutions, and also achieves a higher detection accuracy.
- Higher values of CID for the proposed technique proved the superiority over the existing individual techniques and their ensembles using bagging and boosting.
- The proposed approach is capable to produce a pool of solutions that address the limitations of the existing techniques, striving to obtain a single solution in which there is no control on classification trade-offs (for application specific requirements).
- The proposed approach is a generalized classification approach that is applicable to the problem of any field having multiple conflicting objectives and a dataset can be represented in the form of labeled instances in terms of its features.

4 Concluding Remarks

In this paper, a novel multi objective genetic algorithm based approach is proposed for effective intrusion detection. The proposed approach is capable of producing a pool of non inferior individual solutions and ensemble solutions thereof which exhibit classification trade-offs for the user. By using certain heuristics or prior domain knowledge, a user can select an ideal solution as per application specific requirements. The proposed approach attempts to tackle the issues of low DR, high FPR and lack of classification trade-offs in the field of ID. The proposed approach consists of encoding of chromosomes that provides optimized values of weights of MLPs. AMGA2 is employed to build multi objective optimization model that generates individual solutions and ensemble solutions thereof with simultaneous consideration of detection rate of each attack class in the dataset. A three phased multi-objective genetic algorithm based approach can rapidly generate numerous individual solutions and ensemble solutions thereof with simple chromosome design in first phase of the proposed approach. The entire solutions are further refined to obtain ensemble solutions in second phase of the approach. The predictions of individual solutions are fused together to compute final prediction of the ensemble using majority voting method in phase 3 of proposed approach.

Benchmark datasets namely KDD cup 1999 and ISCX 2012 dataset for intrusion detection are used to demonstrate and validate the performance of the proposed approach based on MLP as a base classifier. The proposed approach can discover an optimized set of individual MLPs and ensemble of MLPs thereof with good support and detection rate from benchmark datasets (in comparison with well-known ensemble methods like bagging and boosting). The optimized set of MLPs and ensemble of MLPs exhibit the classification tradeoffs for the users. The user may select an ideal solution as per application specific requirements. Using MLP as a base classifier, the proposed approach is able to enhance DR by 28 % , reduce FPR by 51 % approximately over the results of MLP trained using back propagation method and its ensemble using boosting technique based on KDD cup 1999 dataset. However, an improvement of results is noticed upto 30 % in DR and 84 % in FPR approximately over bagging based ensemble of MLP for KDD cup 1999 dataset. For ISCX 2012 dataset, the results of the proposed technique are improved upto 7 % in DR and 96 % in FPR approximately over MLP and its ensemble using bagging technique. Higher values of CID for the proposed approach proved the superiority over the existing individual techniques and their ensembles using bagging and boosting.

The major issue in the proposed approach is that it takes long time to compute fitness functions in various generations. It may be overcome by computing the function values in parallel. Here, we computed the results by limiting the population size and number of generations of MOGA. More experiments may be conducted by using different values of these parameters. The proposed approach is validated using small subsets of benchmark datasets only, whereas its applicability can be tested by conducting more experiments with real network traffic in the field of ID. The proposed approach utilized static method for selecting an appropriate ensemble solution whereas dynamic selection method may lead to more fruitful results.

References

1. Abraham, A., Thomas, J.: Distributed intrusion detection systems: a computational intelligence approach. Applications of Information Systems to Homeland Security and Defense, pp. 105–135. Idea Group Inc., Publishers, USA (2005)
2. Ahmadian, K., Golestani, A., Analoui, M., Jahed, M.: Evolving ensemble of classifiers in low-dimensional spaces using multi-objective evolutionary approach. In: Proceedings of 6th IEEE/ACIS International Conference on Computer and Information Science (ICIS), pp. 217–222. IEEE (2007)
3. Ahmadian, K., Golestani, A., Mozayani, N., Kabiri, P.: A new multi-objective evolutionary approach for creating ensemble of classifiers. In: Proceedings of IEEE International Conference on Systems, Man and Cybernetics (ISIC), pp. 1031–1036. IEEE (2007)
4. Axelsson, S.: Intrusion detection systems: a survey and taxonomy. Technical report (2000)
5. Bishop, C.: Pattern Recognition and Machine Learning, vol. 4. Springer, New York (2006)
6. Breiman, L.: Bias, variance, and arcing classifiers (technical report 460). Department of statistics. University of California at Berkeley (1996)
7. Brown, C., Cowperthwaite, A., Hijazi, A., Somayaji, A.: Analysis of the 1999 darpa/lincoln laboratory ids evaluation data with netadhict. In: Proceedings of IEEE Symposium on Computational Intelligence for Security and Defense Applications (CISDA), pp. 1–7. IEEE (2009)
8. Brugger, S.: Data mining methods for network intrusion detection. University of California at Davis (2004). www.citeseerx.ist.psu.edu/viewdoc/download?doi=10.1.1.88.3127&rep=rep1&type=pdf
9. Chandola, V., Banerjee, A., Kumar, V.: Anomaly detection: a survey. ACM Comput. Surv. (CSUR) **41**(3), 15 (2009)
10. Chawla, N.: C4. 5 and imbalanced data sets: investigating the effect of sampling method, probabilistic estimate, and decision tree structure. In: Proceedings of the ICML Workshop on Learning from Imbalanced Datasets II, vol. 3 (2003)
11. Chebrolu, S., Abraham, A., Thomas, J.: Feature deduction and ensemble design of intrusion detection systems. Comput. Secur. **24**(4), 295–307 (2005)
12. Chen, Y., Abraham, A., Yang, B.: Hybrid flexible neural-tree-based intrusion detection systems. Int. J. Intell. Syst. **22**(4), 337–352 (2007)

13. Coello, C.: An updated survey of ga-based multiobjective optimization techniques. ACM Comput. Surv. (CSUR) **32**(2), 109–143 (2000)
14. Coello, C., et al.: A comprehensive survey of evolutionary-based multiobjective optimization techniques. Knowl. Inf. syst. **1**(3), 129–156 (1999)
15. Corne, D., Jerram, N., Knowles, J., Oates, M., et al.: Pesa-ii: Region-based selection in evolutionary multiobjective optimization. In: Proceedings of the Genetic and Evolutionary Computation Conference (GECCO'2001). Citeseer (2001)
16. Deb, K.: Multi-objective optimization. Multi-objective Optimization using Evolutionary Algorithms, pp. 13–46. Wiley, New York (2001)
17. Deb, K., Agrawal, S., Pratap, A., Meyarivan, T.: A fast elitist non-dominated sorting genetic algorithm for multi-objective optimization: Nsga-ii. Lect. Notes Comput. Sci. **1917**, 849–858 (2000)
18. Deb, K., Anand, A., Joshi, D.: A computationally efficient evolutionary algorithm for real-parameter optimization. Evol. Comput. **10**(4), 371–395 (2002)
19. Dietterich, T.: Ensemble methods in machine learning. Multiple Classifier Systems, pp. 1–15. Springer, Heidelberg (2000)
20. Dietterich, T., Bakiri, G.: Error-correcting output codes: a general method for improving multiclass inductive learning programs. In: Proceedings of Santa fe Institute Studies in the Sciences of Complexity, vol. 20, pp. 395–395. Citeseer (1994)
21. Dos Santos, E.M.: Static and dynamic overproduction and selection of classifier ensembles with genetic algorithms. Ph.D. thesis, Montreal (2008)
22. Engen, V.: Machine learning for network based intrusion detection: an investigation into discrepancies in findings with the kdd cup'99 data set and multi-objective evolution of neural network classifier ensembles from imbalanced data. Ph.D. thesis, Bournemouth University (2010)
23. Fung, K., Kwong, C., Siu, K., Yu, K.: A multi-objective genetic algorithm approach to rule mining for affective product design. Expert Syst. Appl. **39**(8), 7411–7419 (2012)
24. Giacinto, G., Roli, F.: An approach to the automatic design of multiple classifier systems. Pattern Recogn. Lett. **22**(1), 25–33 (2001)
25. Giannopoulos, N., Moulianitis, V., Nearchou, A.: Multi-objective optimization with fuzzy measures and its application to flow-shop scheduling. Eng. Appl. Artif. Intell. **25**, 1381–1394 (2012)
26. Govindarajan, M., Chandrasekaran, R.: Intrusion detection using neural based hybrid classification methods. Comput. Netw. **55**(8), 1662–1671 (2011)
27. Gu, G., Fogla, P., Dagon, D., Lee, W., Skorić, B.: Measuring intrusion detection capability: An information-theoretic approach. In: Proceedings of the 2006 ACM Symposium on Information, Computer and Communications Security, pp. 90–101. ACM (2006)
28. Hu, R., Damper, R.: A no panacea theorem for classifier combination. Pattern Recogn. **41**(8), 2665–2673 (2008)
29. Ishibuchi, H., Nojima, Y.: Evolutionary multiobjective optimization for the design of fuzzy rule-based ensemble classifiers. Int. J. Hybrid Intell. Syst. **3**(3), 129–145 (2006)
30. Jain, A., Duin, R., Mao, J.: Statistical pattern recognition: a review. IEEE Trans. Pattern Anal. Mach. Intell. **22**(1), 4–37 (2000). doi:10.1109/34.824819
31. Jo, T., Japkowicz, N.: Class imbalances versus small disjuncts. ACM SIGKDD Explor. Newsl. **6**(1), 40–49 (2004)
32. KDD: Kdd cup 1999 dataset (1999). http://kdd.ics.uci.edu/databases/kddcup99/kddcup99.html
33. Khreich, W., Granger, E., Miri, A., Sabourin, R.: Iterative boolean combination of classifiers in the roc space: an application to anomaly detection with hmms. Pattern Recogn. **43**(8), 2732–2752 (2010)
34. Khreich, W., Granger, E., Miri, A., Sabourin, R.: Adaptive roc-based ensembles of hmms applied to anomaly detection. Pattern Recogn. **45**(1), 208–230 (2012)
35. Kumar, G., Kumar, K.: Ai based supervised classifiers: an analysis for intrusion detection. In: Proceedings of International Conference on Advances in Computing and Artificial Intelligence, pp. 170–174. ACM (2011)

36. Kumar, G., Kumar, K.: A novel evaluation function for feature selection based upon information theory. In: Proceedings of 24th Canadian Conference on Electrical and Computer Engineering (CCECE), pp. 000,395–000,399. IEEE (2011)
37. Kumar, G., Kumar, K.: An information theoretic approach for feature selection. Secur. Commun. Networks **5**(2), 178–185 (2012). doi:10.1002/sec.303
38. Kumar, G., Kumar, K.: The use of artificial-intelligence-based ensembles for intrusion detection: a review. Appl. Comput. Intell. Soft Comput. **2012**, 1–20 (2012). doi:10.1155/2012/850160
39. Kumar, G., Kumar, K.: The use of multi-objective genetic algorithm based approach to create ensemble of ann for intrusion detection. Int. J. Intell. Sci. **2**(24), 115–127 (2012). doi:10.4236/ijis.2012.224016
40. Kumar, G., Kumar, K., Sachdeva, M.: An empirical comparative analysis of feature reduction methods for intrusion detection. Int. J. Inf. Telecommun. Technol. **1**(1), 44–51 (2010)
41. Kumar, G., Kumar, K., Sachdeva, M.: The use of artificial intelligence based techniques for intrusion detection: a review. Artif. Intell. Rev. **34**(4), 369–387 (2010)
42. Kuncheva, L.I.: Combining pattern classifiers: methods and algorithms (kuncheva, li; 2004)[bibbookreview]. IEEE Trans. Neural Netw. **18**(3), 964–964 (2007)
43. Lee, W., Stolfo, S., Mok, K.: Adaptive intrusion detection: a data mining approach. Artif. Intell. Rev. **14**(6), 533–567 (2000)
44. McHugh, J.: Testing intrusion detection systems: a critique of the 1998 and 1999 darpa intrusion detection system evaluations as performed by lincoln laboratory. ACM Trans. Inf. Syst. Secur. **3**(4), 262–294 (2000)
45. Muda, Z., Yassin, W., Sulaiman, M., Udzir, N., et al.: A k-means and naive bayes learning approach for better intrusion detection. Inf. Technol. J. **10**(3), 648–655 (2011)
46. Parrott, D., Li, X., Ciesielski, V.: Multi-objective techniques in genetic programming for evolving classifiers. In: Proceedings of IEEE Congress on Evolutionary Computation, vol. 2, pp. 1141–1148. IEEE (2005)
47. Patcha, A., Park, J.M.: An overview of anomaly detection techniques: existing solutions and latest technological trends. Comput. Netw. **51**(12), 3448–3470 (2007). doi:10.1016/j.comnet.2007.02.001. http://www.sciencedirect.com/science/article/pii/S138912860700062X
48. Perdisci, R., Giacinto, G., Roli, F.: Alarm clustering for intrusion detection systems in computer networks. Eng. Appl. Artif. Intell. **19**(4), 429–438 (2006)
49. Re, M., Valentini, G.: Integration of heterogeneous data sources for gene function prediction using decision templates and ensembles of learning machines. Neurocomputing **73**(7–9), 1533–1537 (2010)
50. Sabhnani, M., Serpen, G.: Application of machine learning algorithms to kdd intrusion detection dataset within misuse detection context. In: Proceedings of International Conference on Machine Learning: Models, Technologies, and Applications, vol. 1, pp. 2009–215 (2003)
51. Shiravi, A., Shiravi, H., Tavallaee, M., Ghorbani, A.A.: Toward developing a systematic approach to generate benchmark datasets for intrusion detection. Comput. Secur. **31**(3), 357–374 (2012)
52. Tavallaee, M.: An adaptive hybrid intrusion detection system. Ph.D. thesis, University of new brunswick (2011)
53. Tiwari, S.: Development and integration of geometric and optimization algorithms for packing and layout design. Ph.D. thesis, Clemson University (2009)
54. Tiwari, S., Fadel, G., Deb, K.: Amga2: improving the performance of the archive-based micro-genetic algorithm for multi-objective optimization. Eng. Optim. **43**(4), 377–401 (2011)
55. Tiwari, S., Koch, P., Fadel, G., Deb, K.: Amga: an archive-based micro genetic algorithm for multi-objective optimization. In: Proceedings of Genetic and Evolutionary Computation conference (GECCO-2008), Atlanta, USA, pp. 729–736 (2008)
56. Toosi, A.N., Kahani, M.: A new approach to intrusion detection based on an evolutionary soft computing model using neuro-fuzzy classifiers. Comput. Commun. **30**(10), 2201–2212 (2007). doi:10.1016/j.comcom.2007.05.002. http://www.sciencedirect.com/science/article/pii/S0140366407001855

57. Tsoumakas, G., Angelis, L., Vlahavas, I.: Selective fusion of heterogeneous classifiers. Intell. Data Anal. **9**(6), 511–525 (2005)
58. Wang, G., Hao, J., Ma, J., Huang, L.: A new approach to intrusion detection using artificial neural networks and fuzzy clustering. Expert Syst. Appl. **37**(9), 6225–6232 (2010)
59. Witten, I., Frank, E., Hall, M.: Data Mining: Practical Machine Learning Ttools and Techniques. Morgan Kaufmann, San Francisco (2011)
60. Wu, S., Banzhaf, W.: The use of computational intelligence in intrusion detection systems: a review. Appl. Soft Comput. **10**(1), 1–35 (2010)
61. Xiang, C., Yong, P., Meng, L.: Design of multiple-level hybrid classifier for intrusion detection system using bayesian clustering and decision trees. Pattern Recogn. Lett. **29**(7), 918–924 (2008)
62. Zainal, A., Maarof, M., Shamsuddin, S., et al.: Ensemble classifiers for network intrusion detection system. J. Inf. Assur. Secur. **4**, 217–225 (2009)
63. Zitzler, E., Deb, K., Thiele, L.: Comparison of multiobjective evolutionary algorithms: empirical results. Evol. Comput. **8**(2), 173–195 (2000)

Cyber Insider Mission Detection for Situation Awareness

Haitao Du, Changzhou Wang, Tao Zhang, Shanchieh Jay Yang, Jai Choi and Peng Liu

Abstract Cyber insider detection is challenging due to the difficulty in differentiating legitimate activities from malicious ones. This chapter will begin by providing a brief review of exiting works in the machine learning community that offer treatments to cyber insider detection. The review will lead to our recent research advance that focuses on early detection of ongoing insider mission instead of trying to determine whether individual events are malicious or not. Multiple automated software agents are assumed to possess different account privileges on different hosts, to perform different dimensions of a complex insider mission. This work develops an integrated approach that utilizes Hidden Markov Models to estimate the suspicious level of insider activities, and then fuses these suspiciousness values across insider activity dimensions to estimate the progression of an insider mission. The fusion across cyber insider dimensions is accomplished using a combination of Fuzzy rules and Ordered Weighted Average functions. Experimental results based on simulated data show that the integrated approach detects the insider mission with high accuracy and in a timely manner, even in the presence of obfuscation techniques.

Research supported by DARPA Cyber Insider (CINDER, FA8750-11-C-0038) program. The views expressed are those of the authors and do not reflect the official policy or position of the Department of Defense or the U.S. Government. Distribution Statement A—Approved for Public Release, Distribution Unlimited.

H. Du · S. J. Yang (✉)
Department of Computer Engineering, Rochester Institute of Technology, Rochester, NY, USA
e-mail: jay.yang@rit.edu

C. Wang · J. Choi
The Boeing Company, Seattle, WA, USA

T. Zhang · P. Liu
College of Information Sciences and Technology, Pennsylvania State University, University Park, PA, USA

R.R. Yager et al. (eds.), *Intelligent Methods for Cyber Warfare*,
Studies in Computational Intelligence 563, DOI 10.1007/978-3-319-08624-8_9

1 Introduction

Cyber insider threats have attracted much attention within the past decade [1, 3, 5, 8, 10, 11, 13], and raise concerns in various research communities including psychology, criminal justice, computer science and engineering. The key challenge to detect insider threats from the computing perspective lies in the difficulty to differentiate observables that are individually legitimate but together cause threats to critical information loss or operation degradation. This becomes even more challenging when multiple software agents are used in a collusive manner to execute insider activities in different dimensions of an insider mission.

The research undertaken in the past decade on cyber insider detection, for the most part, focuses on determining whether individual actions are malicious or not. This focus has shown to be not successful due to the inherent limitation that insider activities are mostly legitimate and can easily fits, or mimicked to fit, normal behavior profiles. Recognizing this limitation, this chapter discusses an approach that focuses on detection of the progress of an overall insider mission, instead of struggling with finding malicious event observables. Expanding from the multi-perspective notion discussed in Raissi-Dehkordi and Carr [11] and the colluding user roles in Kohli et al. [5], this work assumes that an insider mission is consisted of several dimensions of insider activities. These multi-dimensional insider activities require privileges likely to span across multiple account types and thus a number of software agents are needed to complete the mission. This is not an unreasonable assumption for complex insider missions that are critical and hard to analyze. Note that the objective is *not* to determine whether individual observables are caused by these insider activities; rather, it is to elevate a threat level as early as possible when an insider mission is likely being executed.

To accomplish the above research goal, one needs to go beyond the existing intrusion or misuse detection techniques that either assume malicious behaviors exhibits localized (e.g., per-process, per-user account) deviations from normal behavior or rely on pattern matching against known attack signatures. This chapter will describe an integrated approach that utilizes Hidden Markov Models (HMM) to estimate the suspicious level of insider activities, then fuses these *suspiciousness* values across insider activity dimensions using a combination of Fuzzy systems and Ordered Weighted Average functions to project the progression of an insider mission. The approach combines the benefits of data-driven learning and knowledge-based fusion techniques, to provide a robust system that exhibits early warning capabilities even in the presence of obfuscation techniques used by colluding software agents. The timely detection of cyber insider mission is essential to enhance situation awareness of the overall operation environment. Experimental results based on simulated data show that the integrated approach detects the insider mission with high accuracy and in a timely manner, even in the presence of obfuscation techniques.

2 Related Work

Salem et al. [12] provided a comprehensive survey on cyber insider attack detection in the computer security literature. They categorized the existing works into host-based user profiling and network-based sensing approaches. Host-based user profiling draws similarity to the techniques used for more general human insider behavior profiling works [3, 10, 13]. This set of work is limited, particularly in the cyber space, in that software agents can easily mimic legitimate usage. Relying on differentiating malicious insider cyber observables from legitimate ones is simply inconceivable and impractical.

Early work on cyber insider detection overlapped significantly with the general anomaly-based intrusion detection systems that built upon data mining and machine learning techniques. Singh and Silakari [14] reviewed 18 cyber attack detection systems and identified techniques such as associative rules, Hidden Markov Model (HMM), classification, clustering, Bayes network, Support Vector Machine (SVM), Principle Component Analysis (PCA), neural network, decision tree, and self organizing map. Unlike traditional knowledge-extensive signature based detection techniques, these data mining and machine learning techniques explored large data and machine intelligence to expedite the speed or expand the capability of attack detections.

Exsiting work often focuses on a single aspect of cyber attacks. For example, Liu et al. [6] proposed a multilevel framework as a high-speed transparent network bridge at the edge of the protected network to identify network applications, generate and detect content signatures and detect covert communication. It classified network traffics using statistical and signal processing techniques for signature generation and feature extraction.

Bertino and Ghinita [2] proposed a pattern matching based mechanism to create profiles of nominal user behavior and detect anomalous behavior with respect to database SQL queries. They identified a number of activities that are indicative of data exfiltration by insiders: data identification, retrieval, movement, and exfiltration. Mathew et al. [7] argued that query syntax alone is a poor discriminator of user intent, which is much better rendered by what is accessed. They proposed to model database access patterns profiling the data points that users access, in contrast to analyzing the query expressions. Statistical learning algorithms are trained and tested using a feature-extraction method to model users' access patterns.

Hu and Panda [4] presented a model for detecting insider malicious activities targeted at tampering the contents of files for various purposes. It employs two-dimensional traceability link rule mining to identify intrinsic file dependencies and model file access patterns. Activities that modify data without complying with various file traceability link rules will be identified as suspicious activities.

Raissi-Dehkordi and Carr [11] proposed to extend the notion of profiling by aggregating statistical analysis in multiple system perspectives and performing classification using SVM. Specifically, they analyzed metrics such as user usage behaviors, file server access statistics, and database server access statistics, and established tens of SVMs to perform classification. One of their objectives was to use these multiple SVMs to tackle the colluding insider problem. Their experiments showed a slight improvement by missing around 25 % instead of 30 %. In terms of colluding cyber insider attacks, Kohli et al. [5] discussed a risk assessment framework that shown how multiple insider and even outsider roles can collude to perform attack and cause serious risks.

Cyber insider attackers, in comparison to outsiders, are stealthier to avoid being caught. Yang et al. [17] proposed an enhanced packet matching algorithm to detect stepping-stone insider attacks through comparing outgoing and incoming connections. In such attacks, the insiders use compromised outside computers as stepping-stones to launch their attacks against inside targets. This and similar techniques can be used to detect activities in covering the trace, a dimension often overlooked by existing works.

The notion of evaluating multiple dimensions of a cyber attacks is appealing, as it presents an opportunity to provide a robust solution that does not rely on detecting anomaly in a single aspect of cyber attacks, which can be error prone. Furthermore, modern cyber defense system often implements separation of user and system privileges, and, thus, an insider attack will require multi-dimensional penetrations into, e.g., file system, database, and web application. The approach to be described in the next section employs such a multi-dimensional approach, where HMM is used to generate the suspicious level, defined by a log-likelihood function, for each dimension. The suspicious level detection can be potentially further improved with other techniques. For example, Parveen et al. [9] proposed an ensemble-based data stream mining techniques to classify rare anomalies from dynamic data streams of unbound length. It demonstrated substantially increased classification accuracy over traditional supervised learning methods for real insider threat streams due to automatic adaptation of the models for evolving data. The suspicious levels across dimensions will then be fused by a combined used of Fuzzy rules and Ordered Weighted Average to produce an insider mission score over time. This combination of data-driven anomaly detection and knowledge-driven fusion will be shown to exhibit superior performance.

3 Approaches and Components

The insider mission scenario investigated in this work is described as follows: the ultimate goal of the intrusion is altering sensitive data stored in database. The targeted victim system has an web interface to allow user to query and potentially change the data with approval. In addition, the target system has strict security policy, every change on the data should have a report, which is a file saved in file system. To accomplish the intrusion task, the insider should take actions in different *dimensions*, from reconnaissance (Dim A), tamper data in database (Dim B), tamper data in file system (Dim C), tamper data in web UI (Dim E), watch for sensitive data updates (Dim E), cover the trace (Dim F).

The insider mission identification system has two major components, the event to activity (E2A) module and the cross dimension mission identification (CDMI) module. When an automated software agent, suspicious or not, performs various activities to achieve (both suspicious insider and normal business) mission objectives, it leaves traces, i.e., a sequence of traces staging an attack on the victim system or network. These traces can be tracked by host-based or network-based sensors and reported as *event instances*. The purpose of the E2A module is to map the events into activity space to estimate the degree of suspiciousness by calculating the deviation (log-likelihood) from the internal state machine that models normal behavior. The suspicious activities are categorized into different insider dimensions based on expert knowledge. The CDMI module fuses the output of the E2A module, i.e., possible activities in different insider dimensions and their suspiciousness values to determine a mission score indicating the likelihood of existence of an insider mission.

3.1 Event to Activity Module

The main purpose of the E2A module is to simplify and compress the problem space from the event domain to the activity domain. When insider activities are performed for achieving a mission, each activity will leave traces in the network traffic logs or file systems. These traces will be inspected by network- or host-based security sensors to generate events. In our target insider mission, there are hundreds or thousands of event types, since different sensors often generates different types of events and each sensor may generate multiple types of events. In addition, normal business operations also leave traces and lead to observable events, especially when sensors are tuned to capture events from insider activities that are very similar to normal business activities. Here, the events are observable and available for our mission identification task, but the exact underlying activities are hidden and unknown and need to be inferred from the events.

In general, the same activity may cause multiple observed events, and different activities may cause the same type of observed events. Hence a probabilistic model may be used to infer activities from events. Note that an individual event (instance) by itself usually does not provide sufficient indication of whether it is observed from an insider activity or the normal business operation. Instead, the preceding and succeeding events may provide additional context to help determine how likely a given event is observed when a given type of activity is performed. As a result, the temporal order of events is important in the event-to-activity inference. In the E2A module, Hidden Markov Model (HMM) is used to perform inference for corresponding activities from observed events and to calculate the suspiciousness of inferred activities. Each event type is considered an observable symbol, and each activity type is considered a hidden state in the HMM.

The HMM in the E2A module is initially specified by a group of three security experts with experience of enterprise penetration tests and cyber analytics, and then improved through training using historic data. Specifically, the types of events, the types of activities, and the emission relationship from activities to events (i.e., whether an event can be caused by an activity), are specified based on the target insider mission and business operation environment. Security experts are often knowledgeable and skillful enough to provide such structure knowledge, but may have difficult to specify the exact probabilities in the HMM. The HMM training only requires historic sequences of observed events (without manual labeling of activities) to tune the probabilities. Once created and trained, the HMM can be used to calculate the forward probability for a particular event using only its preceding events. This enables us to support online mission identification as events are observed as a data stream. In HMM inference, one can also calculate the posterior probability for a particular event using the full sequence of events (including both its preceding events and its succeeding events). This enables us to find the optimal probabilities in offline or batch mode mission identification as a comparison baseline to measure the online mission identification method.

In addition, the E2A module estimates the *suspiciousness* of each inferred activity. This value is important for the CDMI module in determining whether the insider mission exists. We use the log-likelihood to estimate the suspiciousness of a given activity. In particular, Let e_i be the ith observed event and the probability associating e_i to each activity a_j is p_{ij}. The best activity match is the activity with the maximum probability $p_i^* = \max_j(p_{ij})$. The suspiciousness of e_i to a_j is defined as

$$L_i \triangleq -(\log(\prod_{k=1}^{i} p_k^*))/i = -(\sum_{k=1}^{i} \log(p_k^*))/i,$$

where p_1^* is set to 1. The suspiciousness value of an inferred activity indicates how bad the activity fits the normal activity model given the observed events in the context.

Figure 1 is an illustrative example of an HMM used in E2A module. The nodes labeled with $A_j,\ j \in \{1, \ldots, M\}$ represent the (types of) activities defined in the insider mission scenario; $E_i,\ i \in \{1, \ldots, N\}$ $(N \gg M)$ represent the (types of) events reported by security sensors. An edge between two activity nodes represents the transition probability for the next activity after a given activity. On the other hand, an edge between an activity node and an event node represent the probability for that event being observed when the activity is performed. Note that the HMM is very sparse, because an activity usually only causes a few types of events being observed. Once the HMM is trained, for any given sequence of newly observed events, one can use the HMM to infer the underlying activity for each individual observation, as well as the suspiciousness of the inferred activity. The better the activity fits the model (in the context of other inferred activities), the less suspicious the activity is. On the other hand, when an activity does not fit the model well, it is considered suspicious, but not necessarily malicious. The suspiciousness values will be used by the CDMI module for further analysis.

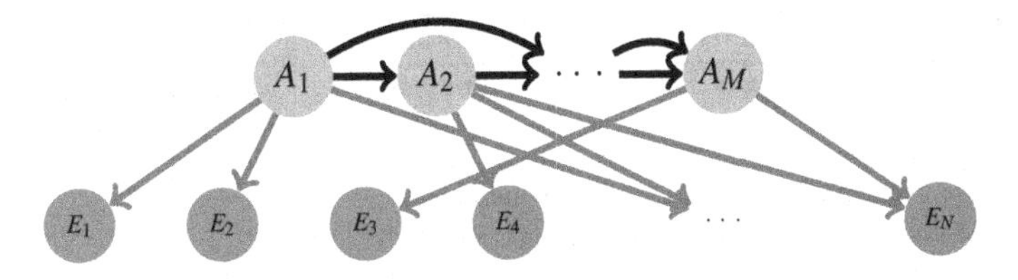

Fig. 1 An illustrative example of E2A hidden Markov model

3.2 Cross Dimension Mission Identification Module

The Cross-Dimension Mission Identification (CDMI) module processes the higher-level abstraction, i.e., the dimension specific activity level information to estimate the contribution that each set of hypothesized activities has made toward the completion of the insider mission. It is expected that tens of activity types will be used to represent hundreds of or more event types. CDMI aims at analyzing the activity suspiciousness value across insider dimensions to estimate the overall insider mission progress over time. An insider mission score, ranging between 0.0 and 1.0, will be produced to reflect the threat level of any ongoing insider mission.

Three major algorithms are developed to determine the mission score from the suspiciousness values of activities in different insider dimensions. Dynamic Activity Discovery (DAD) selects suspicious activities based on the E2A outputs, Intra-Dimension Fusion (IDF) aggregates the activity scores within each insider dimension to produce a completeness score for each dimension, and Cross-Dimension Fusion (CDF) takes the completeness scores and generates the final mission score.

The inputs to CDMI include the probability values (p_{ij}) that associate each event e_i to an activity a_j and the corresponding suspiciousness values (L_i) produced by the E2A module. Note that each event observable is now treated by CDMI as a potential insider activity with a suspiciousness value. The term 'suspiciousness' is emphasized because the goal is not to determine whether an event is truly an insider activity or not. Instead, the goal is to use the suspiciousness values to aggregate potential insider activities in different dimensions to determine a mission score.

The suspiciousness value, which is the log-likelihood, represents how much the corresponding individual event/activity deviates from the normal behavior given the contexts occurring before it. DAD further calculates the exponential weighted moving average (EWMA) of the suspiciousness values, to reflect how the sequence has been gradually deviating from normal behavior. The EWMA of log-likelihood is compared to a threshold derived based on the training set (i.e., the normal behavior). The events/activities that exceed the threshold will be used to produce the activity score for each activity type. The process of DAD is given in Algorithm 1.

The main objective of IDF is to evaluate how *complete* each insider dimension is given the suspicious activity level observed in each time window. A completeness score for a given dimension is determined by fusing the suspicious activities in the same insider dimension. The first step of this process is to determine a Suspicious

Algorithm 1 Dynamic Activity Discovery Algorithm

Given EWMA parameter α, log-loss threshold L, E2A probability p_{ij} and suspiciousness L_i
Set filtered log-likelihood $L_f(0) = L(0)$
for all Event e_i in the corresponding time window **do**
  EWMA log-likelihood $L_f(i) = \alpha L(i) + (1 - \alpha)L_f(i - 1)$
end for
Initialize Suspicious Activity Matrix M
for all Event e_i in the corresponding time window **do**
  if Filtered Log-likelihood $L_f(i) > L$ **then**
    Get the probability distribution vector $\mathbf{P} = (p_{i1}, p_{i2}, \ldots, p_{im})$;
    Append **P** to **M**
  end if
end for
return Suspicious Activity Matrix **M**

Activity Vector $\mathbf{V_A}$ by combining the suspiciousness values of the events in an observation window for each activity type. The combination process is based upon the Suspicious Activity Matrix **M**, and used a filtering mechanism as shown in Algorithm 2.

Algorithm 2 Suspicious Activity Vector Generation Algorithm

Given Suspicious Activity Matrix **M**, Threshold T, and parameters $\alpha_1 < \alpha_2$
for all Activity type a **do**
  Set $\mathbf{M}'(0, a) = \mathbf{M}(0, a)$
  for all Event e_i that has non-zero value **do**
    if $\mathbf{M}(i, a) < T$ **then**
      Set $\alpha = \alpha_1$ for less suspicious activities
    else
      Set $\alpha = \alpha_2$ for more suspicious activities
    end if
    $\mathbf{M}'(i, a) = \alpha \mathbf{M}(i, a) + (1 - \alpha)\mathbf{M}'(i - 1, a)$
  end for
  $\mathbf{V_A}(a) = 1 - (\prod_{i \in \text{observation window}}(1 - \mathbf{M}(i, a)))$
end for
return Suspicious Activity Vector $\mathbf{V_A}$

The Suspicious Activity Vector $\mathbf{V_A}$ represents the overall likeliness of an insider activity occurring in a time window. The filtering mechanism shown in Algorithm 2 is used to capture sudden surges of suspicious activities while exhibiting slow decays to maintain the lasting effects of insider activity across time windows. From here, the system evaluates the 'percent effort' spent in each activity type as compared to the overall effort within each dimension while accounting for the criticality of the activity types. The higher the percent effort (with lasting effect) is observed and/or the more critical the activity type is, the higher the 'completeness' score is for each dimension.

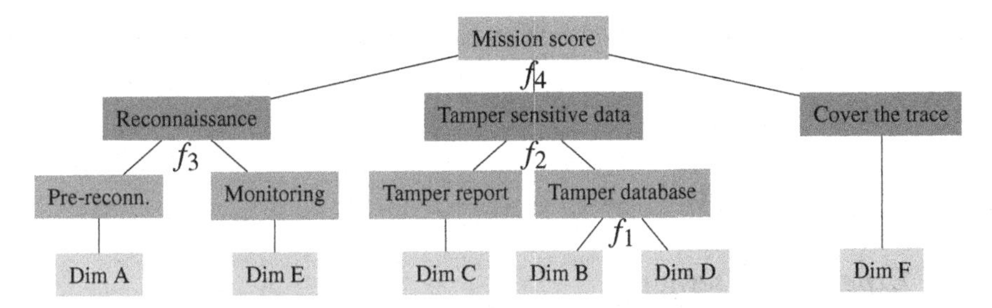

Fig. 2 Cross-dimension fusion structure

The design of the completeness functions for individual insider dimensions is based on the general framework of Ordered Weighted Average (OWA) [16].

To this end, CDF utilizes a hierarchical structure of fusion algorithms to combine the completeness scores of the various insider dimensions. Figure 2 shows the specific structure for the insider mission scenario described earlier. The hierarchical fusion structure consists of four fusion functions (f_1–f_4) and processes insider dimensions that behave similarly or tend to work together. Dimensions B and D both involve modification of the sensitive data; Fusing the two (f_1) gives an indication of the insider actions occurring on the database containing sensitive information. Alternative to tampering the database, the software agent can tamper the intelligence report of the sensitive information. Fusing the two types of tampering (f_2) shows the level of completeness with respect to the primary task of the insider mission, i.e., tampering the sensitive information. The design of f_1 and f_2 utilizes OWA framework to reflect that accomplishing either Dim B, D, or C is indicative to data tampering, but accomplishing more dimensions should still exhibit a higher overall mission score than just accomplishing one.

Similar concept is used to implement f_3 using OWA, since Dim A and E both reveal insider activity in learning about the sensitive data processing workflow process and system configurations. Dim F stands alone by itself to represent insider activities in covering evidences of data tampering. Function f_4 makes use of Sugeno Fuzzy Inference System [15] to integrate expert's knowledge for the final fusion and mission score generation. The Fuzzy system is designed so that not all of reconnaissance, data tampering and covering trace are needed to exhibit a high mission score. Based on expert recommendations, the fuzzy system emphasizes more on data tampering and covering trace. Particularly if sufficient and high confidence is shown for covering the traces of malicious activities, a sufficiently high mission score should be reached as activities in that dimension is not commonly observed.

The parameters used in the above DAD, IDF, and CDF algorithms are primarily derived based on qualitative recommendations from the domain experts for the specific insider mission scenario. Particularly, the weights used in EWMA as part of DAD, the weights used in OWA, and the fuzzy rules as part of CDF are determined by soliciting the relatively importance of and the relationship between the different activity types and insider dimensions. The threshold T and α_1/α_2 used in

IDFare designed to reflect how sensitive and how fast the system reacts to suspicious activities, respectively.

4 Experiments and Results

4.1 Experiment Design

A key obstacle in the way to insider threat research is the lack of real world data. There are two major challenges in obtaining such data. First, organizations are reluctant to share insider data due to business and security concern. Insider mission data usually contains sensitive information, such as organization security policies, sensor deployment, firewall configurations, etc. Moreover, in order to collect real insider data, all the network events should be monitored and logged, the process of which can incur expense that impact business bottom line. Second, not all the ground truth is known or tagged in real data. Users cannot tell which audit log entries are due to insider mission behavior, especially when the mission is at an early stage. Such knowledge is important for the analysis of insider's motivation and attacking strategy. In addition, it is challenging to obtain real data that reflects a variety of obfuscation techniques and colluding behaviors with different configurations of software agents. To address these challenges, this work elects to simulate insider mission process and generate the insider data set.

Insider event generation involves three distinct steps: First, each activity is decomposed into a partial order of event types. Second, the partial order of event types is verified to fulfill the activitys goal and to identify the corresponding constraints that must be satisfied. These constraints would clarify the data and control dependencies and invariants among the possible instantiations of the involved event types. Third, based on these data and control dependencies and invariants, a set of state-machines are used to automatically generate nondeterministic instantiations of the partial order of event types.

Normal background events are generated by a different set of state machines, each of which implements concurrent normal business operations. In order to build realistic normal behavior models, each workflow is decomposed into a partial order of activity types, which are mapped into a sequence of events. The insider mission events and normal background events are then interleaved to satisfy reasonable causality relationships between them, as well as taking into account that the malicious insider may attempt to hide the malicious events as much as possible.

This work considers an example of insider mission that aims at penetrating and potentially altering sensitive data, which involves a database managed by a DB administrator, a web application that allows Security Analysts to access and modify the sensitive data, and distributed file systems that store intelligence reports and review documents. The Security Analysts have the authority to directly update the data through the privileged accounts. The software agents could potentially possess the privileges to access a combination of the above victim systems. There are seven dimensions of activity categories, including those shown in Fig. 2 and one that collects irrelevant activities.

A number of event sequences are generated for HMM training to establish the baseline normal behavior, and to test a variety of colluding insider software agent behaviors. The normal background data is generated by analyzing the common business processes for intelligence organizations. A set of state machines are developed to reflect the business processes and used to generate such data. On the other hand, the insider attack data is constructed based on the insider mission scenario described above with built-in control dependencies. Both the normal business processes and insider mission processes are then simulated, with the event instances recorded in a sequential manner in the log files as the data sets to be used for training and testing. The data generation process is kept unknown from the insider mission detection algorithm development, so that the exact states and state transitions used to generate the simualted data is not known by the algorithm. Only the high level insider mission scenario and the type of business processes are known by both the data generation team and the algorithm development team.

Figure 3 shows an example of an event instance generated. Note that many of the attributes, e.g., `activity`, `user-id`, `insider-event`, and `obfuscation`, provide the ground truth information for verification purposes only, and are hidden from the insider mission detection algorithms. The attributes, `account` and `OS-account` provide information on whcih application accountand operating system account are used for the observed event, respectively. In addition to the accounts, the primary attributes used by the E2A/CDMI modules for insider mission detection are `type-id`, showing the event type, and the `asset-id`, showing the specific asset (e.g., a webpage, a file, an entry in the database) being accessed, along with the time stamps and IP/Port information.

```
<event-id="10"
  type-id="V0200"
  description="successfully login the TAS web UI"
  activity="D2"
  account="web1025"
  OS-account="a02005"
  user-id="a02005"
  asset-id="1.7.1"
  insider-event="no"
  obfuscation="no"
  start-time="2012/01/01 07:44:41"
  end-time="2012/01/01 07:44:41"
  source-ip="10.28.46.48"
  dest-ip="10.16.48.62"
  dest-port="22" />
```

Fig. 3 An example event instance

Each data set contains about 45,000–50,000 events. The trained HMM includes 32 states, 104 observable symbols, 466 transitions between states and 125 emissions from states to observable symbols. Additional sets of normal behavior are also generated for testing against the data sets containing insider activities. The insider test sets contain approximately 1,200 insider events each and are mixed with normal event behaviors. Individual event observables for insider activity and normal activity are not distinguishable. It is the state machines that cause the slight variation between the insider and the normal activities. Additional eight types of obfuscation techniques are also included in the insider test sets. They are defined as follows:

- Noise injection: inject noise into the insider mission process, such as conducting mission irrelevant events, accessing useless assets, etc.
- Event re-timing: change the time stamp of critical insider mission events to help disguise insider mission. One possible approach could be tampering the system clock.
- Self-throttling: lower the pace of mission execution by increasing time intervals between insider events.
- Event reordering: reorder insider events without dependency relationships to disguise insider mission.
- Activity splitting: repartition the insider mission or reorganize the original activities to get a new set of activities to fool detection modules such as E2A module.
- Leveraging equivalent event sequence (event renaming/event merging): replace a series of insider mission events with equivalent event sequence. For example, modifying a file in the file system can be achieved in two ways: either editing the file and saving the modified file into the file system or deleting the original file and creating a new file with the same file name with modified content.
- Removing traces: With escalated privilege automated agents can be configured to remove or modify mission critical event logs.

The performance of the mission identification system is evaluated based on the following metrics:

- False Positive Rate: the number of false positives divided by the total number of datasets each of which does not include an insider missions. The mission identification result for a dataset is false positive if it reports the detection of an insider mission even though the dataset does not contain an insider mission.
- False Negative Rate: the number of false negative results divided by the total number of datasets each of which does include an insider missions. The mission identification result for a dataset is false negative if it reports that there is no insider mission yet the dataset does include an insider mission in the ground truth.
- Precision: the number of true positive results divided by the total number of datasets identified as dataset including an insider mission. The mission identification result for a dataset is true positive if it reports that there is an insider mission and the dataset does include an insider mission in the ground truth.
- Recall (a.k.a. Detection Ratio): the number of true positive results divided by the total number of datasets that include an insider mission in the ground truth.

- Detection Time: the time period from the start of the first insider event to the time the insider mission is detected.

4.2 Experiment Results

Consider first the performance of E2A module. Ten datasets are used for testing. Table 1 shows the false positive rate (FPR), false negative rate (FNR), precision and recall using the suspiciousness value to determine whether an event is observed from an insider activity or not. In general, one would like to see high recall so that no insider events are dropped (for later mission identification tasks) and can tolerate a relative low precision (as later mission identification modules can mitigate this). Using the mean suspiciousness value from the training dataset as a threshold, one achieves very good recall but very poor precision. By adding standard deviation of the suspiciousness value, the precision is improved at the cost of reduced recall, while achieving reasonable false positive rate and false negative rate. On the other hand, using the maximum suspiciousness value from the training dataset sacrifices the recall too much. The weighted average between the maximum and the minimum suspiciousness values becomes usable only when the weight is leaned towards the minimum value. Indeed, the maximum value might be an outlier, and hence the weighted average can be too large to obtain a high recall. The above results recommend using the mean plus standard deviation approach, as commonly used in statistical control theory, to give low FNR and high Recall.

The poor performance of E2A module is expected, since this work builds upon the premise that differentiating legitimate from insider actions is not viable. However, the activity suspiciousness values produced by E2A help CDMI to analyze the insider activity levels across different dimensions, and thus to assess whether an insider mission is ongoing.

Table 1 E2A accuracy for insider event determination

Threshold	FPR	FNR	Precision	Recall
Max	0.00	0.64	0.94	0.36
Mean	0.60	0.03	0.02	0.97
Mean + stdandard	0.23	0.07	0.06	0.93
Mean + 2*standard	0.11	0.15	0.10	0.85
Mean + 3*standard	0.08	0.20	0.13	0.80
Min*0.0625+max*0.9375	0.00	0.64	0.94	0.36
Min*0.125 + max*0.875	0.00	0.64	0.94	0.36
Min*0.25 + max*0.75	0.00	0.64	0.94	0.36
Min*0.5 + max*0.5	0.02	0.49	0.25	0.51
Min*0.75 + max*0.25	0.13	0.12	0.09	0.88

Table 2 Detection accuracy

Configuration	TP	FP	TN	FN
1	5	0	5	0
2	5	0	5	0
3	5	1	4	0
4	5	0	5	0
5	5	0	5	0
6	5	0	5	0
7	5	0	5	0
8	5	0	5	0
9	5	0	5	0
10	5	0	5	0

To measure the robustness of the overall mission identification system, ten different sets of configurations are used. For each configuration, two training sets and ten testing sets, five with only normal events and five with mixed insider and normal events, are generated. Each dataset on average includes around 50,000 events. The detection results are shown in Table 2. True Positive (TP), False Positive (FP), True Negative (TN), False Negative (FN) are collected. As shown in the table, all insider missions are detected, with only one false positive reported. The false positive is due to the conservative setting of thresholds used in CDMI.

This excellent performance is achieved due to a few reasons. First, the objective here is to detect the existence of an overall insider mission, instead of differentiating insider events from the normal ones. The collective evidence, even with significant false negatives output from the E2A module, can still be indicative of the existence of an insider mission. In addition, the proposed system is built on an integrated data-driven and knowledge-driven approach. The definition of the activity types, the dimensions, and their relative importance (which translates to the parameters used in the fusion algorithms within the CDMI module) are derived by soliciting knowledge from domain experts for the specific insider mission scenario. The specific system developed here will not automatically work to detect a very different insider mission on a different organization, but the framework will apply. The results shown in this chapter suggest the promising outlook to detect insider mission using the novel integrated data and knowledge-driven approach, which is not conceivable in other existing works. The following shows additional results to demonstrate the robustness of the framework in the presence of missing observables and advanced obfuscation techniques.

For each configuration, the insider testing sets are reduced with each event independently dropped with a fixed probability. Five fixed probabilities, 5, 10, 15, 20, and 25 %, are used for different robustness tests. Table 3 shows the detection ratios for the cases with increasing dropping ratio. All insider missions are detected in this experiment even though 25 % of the events are dropped. This superior performance is

Table 3 Detection with missing observations

Drop probability	Detection ratio (%)	Average # alerts generated
0	100	122.69
0.05	100	102.70
0.1	100	98.74
0.15	100	94.20
0.2	100	93.51
0.25	100	93.27

due to that CDMI examines the suspiciousness values of activities in each time window, and raise an 'alert' as long as there is sufficient evidence suggesting an insider mission is ongoing in that time window. For a given test set, an insider mission is considered detected when the first alert is raised. As a result, as long as the removed events do not eliminate the possibility to raise alerts in ALL windows, the insider mission can be detected. Table 3 shows the averaged number of alerts generated for each case. While all insider sets are detected, not all time windows will generate alerts showing insider activities due to the dropped events.

Table 4 shows the detection ratios when different obfuscation techniques are used. As in the previous experiment, all insider missions are detected regardless of the obfuscation techniques used. The average number of alerts (i.e., time windows with insider mission detected) varies, however, when different obfuscation techniques are used. One interesting observation is that the 'event re-timing' and 'self-throttling' techniques result in significant different performance in terms of average number of alerts generated. This is because these two techniques spread the insider events over a longer time span, and thus more time windows see raised alerts.

Table 4 also shows the detection time when different obfuscation techniques are used. Other than 'event re-timing' and 'self-throttling', the mission identification system detects the insider mission around the same time if there were no obfuscation techniques. The longer detection time is due to the same reason as that for larger average number of alerts; that is, the spreading the same number of insider events

Table 4 Mission detection against obfuscation techniques

Obfuscation	Detection ratio (%)	Average # alerts	Average detection time
Noise-injection	100	61.9	125.5
Event-retiming	100	126.7	196.2
Self-throttling	100	126.7	196.2
Event-reordering	100	68.5	124.0
Activity-splitting	100	62.1	123.8
Equivalent-sequence	100	60.7	123.9
Trace-removing	100	53.1	124.1

over a longer period of time makes it more challenging to gather sufficient evidence to declare the existence of an insider mission. Obviously these types of evasion are difficult to detect due to the event sparsity.

5 Conclusion

Going beyond the classical intrusion detection, this work developed hierarchical data processing for insider mission identification by abstracting activities from lower level events, estimating level of suspiciousness, all of which have been evaluated for a final mission score that relies on both the data abstraction and domain knowledge. The emphasis is to show how one can reveal the insider mission while activities performed by automated software agents were hidden among the legitimate activities.

The integrated approach of data driven (E2A) and knowledge driven fusion of insider activity (CDMI) has been shown to be highly successful to differentiate cases where colluding autonomous agent activities are present versus those with no insider activity. Hierarchical fusion allows to account for the completion of individual insider dimension, driven by suspicious level of insider activities, and, thus, robust to obfuscation techniques attempting to hide the autonomous agent activities.

References

1. Ali, G., Shaikh, N.A., Shaikh, Z.A.: Towards an automated multiagent system to monitor user activities against insider threat. In: Proceedings of International Symposium on Biometrics and Security Technologies, pp. 1–5 (2008)
2. Bertino, E., Ghinita, G.: Towards mechanisms for detection and prevention of data exfiltration by insiders. In: Proceedings of the 6th ACM Symposium on Information, Computer and Communications Security, pp. 10–19 (2011)
3. Buford, J.F., Lewis, L., Jakobson, G.: Insider threat detection using situation-aware MAS. In: Proceedings of 11th International Conference on Information Fusion (2008)
4. Hu, Y., Panda, B.: Two-dimensional traceability link rule mining for detection of insider attacks. In: Proceedings of the 43rd Hawaii International Conference on System Sciences (2010)
5. Kohli, H., Lindskog, D., Zavarsky, P., Ruhl, R.: An enhanced threat identification approach for collusion threats. In: Proceedings of Third International Workshop on Security Measurements and Metrics, pp. 25–30 (2011)
6. Liu Y., Cobett, C., Chiang K., Archibald, R., Mukherjee, B., Ghosal, D.: SIDD: a framework for detecting sensitive data exfiltration by an insider attack. In: Proceedings of the 42nd Hawaii International Conference on System Science (2009)
7. Mathew1, S., Petropoulos, M., Ngo, H.Q., Upadhyaya, S.: A data-centric approach to insider attack detection in database systems. In: Proceedings of the 13th international Conference on Recent advances in intrusion Detection, pp. 382–401 (2010)
8. Maybury, M., Chase, P., Cheikes, B., Brackney, D., Matzner, S., Hetherington, T., Wood, B., Sibley, C., Marin, J., Longstaff, T.: Analysis and detection of malicious insiders. Technical report, MITRE (2005)

9. Parveen, P., Weger, Z.R., Thuraisingham, B., Hamlen, K., Khan, L.: Surpervised learning for insider threat detection. In: Proceedings of the 23rd IEEE International Conference on Tools with Artificial Intelligence, pp. 1032–1039 (2011)
10. Pfleeger, S.L., Predd, J.B., Hunker, J., Bulford, C.: Insiders behaving badly: addressing bad actors and their actions. IEEE Trans. Inf. Forensics Secur. **5**(1), 169–179 (2010)
11. Raissi-Dehkordi, M., Carr, D.: A multi-perspective approach to insider threat detection. In: Proceedings of IEEE Military Communications Conference, pp. 1164–1169 (2011)
12. Salem, M.B., Hershkop, S., Stolfo, S.J.: A survey of insider attack detection research. Insider Attack Cyber Secur. **39**, 69–90 (2008)
13. Santos, E., Nguyen, H., Yu, F., Kim, K., Li, D., Wilkinson, J.T., Olson, A., Jacob, R.: Intent-driven insider threat detection in intelligence analyses. Proc. IEEE/WIC/ACM Int. Conf. Web Intell. Intell. Agent Technol. **2**, 345–349 (2008)
14. Singh, S., Silakari, S.: A survey of cyber attack detection systems. Int. J. Comput. Sci. Netw. Secur. **9**(5) (2009)
15. Wang, L.X.: A Course on Fuzzy Systems. Prentice-Hall press, USA (1999)
16. Yager, R.R.: On ordered weighted averaging aggregation operators in multicriteria decision-making. IEEE Trans. Syst Man Cybern. **18**(1), 183–190 (1988)
17. Yang, J., Ray, L., Zhao, G.: Detect stepping-stone insider attacks by network traffic mining and dynamic programming. In: Proceedings of the 2011 International Conference on Advanced Information Networking and Applications, pp. 151–158 (2011)

A Game Theoretic Engine for Cyber Warfare

Allen Ott, Alex Moir and John T. Rickard

Abstract The nature of the cyber warfare environment creates a unique confluence of situational awareness, understanding of correlations between actions, and measurement of progress toward a set of goals. Traditional fusion methods leverage the physical properties of objects and actions about those objects. These physical properties in many cases simply do not apply to cyber network objects. As a result, systematic, attributable measurement and understanding of the cyber warfare environment requires a different approach. We describe the application of a mathematical search engine having inherent design features that include tolerance of missing or incomplete data, virtually connected action paths, highly dynamic tactics and procedures, and broad variations in temporal correlation. The ability efficiently to consider a breadth of possibilities, combined with a chiefly symbolic computation outcome, offers unique capabilities in the cyber domain.

1 Introduction

Game theory (1–15) is a mathematical theory of strategic behavior, in which a course of action (COA) consists of one or more individual moves taken by each player at a given stage of the game starting from their estimate of the current game state $S(k)$ at time k. A *game theory engine* is a computational device for advising a particular player as to the selection of future COAs based upon their estimate of the current state $S(k)$, given one or more evaluation functions $\varepsilon\,(S(j))$, $j = k + 1, k + 2, \mathrm{K}$

A. Ott (✉)
Distributed Infinity, Inc., 1382 Quartz Mountain Drive, Larkspur, CO 80118, USA
e-mail: aott@distributedinfinity.com

A. Moir (✉)
Distributed Infinity, Inc., 1230 N. Sweetzer Avenue, #302, Los Angeles, CA 90069, USA
e-mail: amoir@distributedinfinity.com

J. T. Rickard (✉)
Distributed Infinity, Inc., 4637 Shoshone Drive, Larkspur, CO 80118, USA
e-mail: terry.rickard@reagan.com

R.R. Yager et al. (eds.), *Intelligent Methods for Cyber Warfare*,
Studies in Computational Intelligence 563, DOI 10.1007/978-3-319-08624-8_10

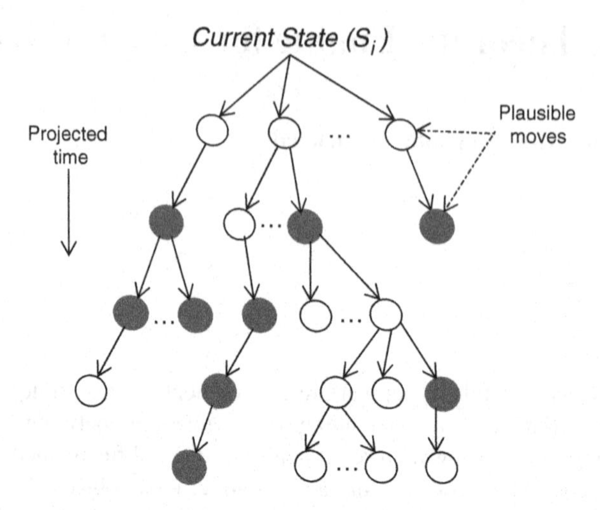

Fig. 1 Tree structure of a game theory engine output at a particular step of the game

that measure the change in the utility of future states that would result from a set of "moves" (i.e., actions taken) by himself and/or the other player(s).

The objective of a game theory engine is to identify for each player the set of feasible moves by all players from a given state, and to select the COA for a player that optimizes the sequence of future states with respect to his own assumptions at each step as the game proceeds. This optimal COA is also referred to as a "plan". The output of the game theory engine at each step of the game is represented as a tree with branches that are contingent upon the actions of all players. Figure 1 illustrates this structure in the case of a two-player game.

In general, each player in a game may have their own evaluation function, which in a two-player game we would denote by $\varepsilon_1(S)$ and $\varepsilon_2(S)$, respectively. In the simplest case, also known as a zero-sum game, the state S is commonly agreed by both players and assumed to represent the true state of the network, and the evaluation functions satisfy $\varepsilon_1(S) = -\varepsilon_2(S)$, i.e., one player's gain (loss) in value is equal in magnitude and opposite in sign to the other player's gain (loss).

In more complex and realistic cases, the current true state $S(k)$ of the network may not be available in full detail to one or more players. In such cases, each player may have their own unique (and perhaps only partially accurate) *estimates*, say $\hat{S}_1(k)$ and $\hat{S}_2(k)$, respectively, of the true state, while player 1 may have an estimate $\hat{S}_{12}(k)$ of the state perceived by player 2, and vice versa for player 2's estimate of the state $\hat{S}_{21}(k)$ perceived by player 1. Either player may assume the feasibility of certain moves by herself or her adversary that are in fact disallowed by the true state of the network. For example, a player may believe they know the password to a device and thus assume they can login and perform certain actions, when in fact they do not have the current password. The player may not be aware until a later time, if at all, that some of their moves were actually unsuccessful.

In addition to the potentially distinct estimates of state, the evaluation functions for each player may be different, so that for example in a two-player game, even if both players are in complete agreement on a common state S, the evaluation functions $\varepsilon_1(S)$, $\varepsilon_2(S)$, $\varepsilon_{12}(S)$ and $\varepsilon_{21}(S)$ may all be distinct. Thus, in addition to the true state, we may have to consider four different state estimates, as well as four different evaluation functions, at each step of a two-player game, and even more when additional players are involved.

Since a game theory engine is capable of considering a very large number of possible moves, it provides a natural mechanism for the modeling and analysis of cyber warfare offensive and defensive tactics. This is not to suggest that the human analyst can be replaced in this role. Instead, we consider this to be a useful tool to supplement the expertise of the human analyst by enabling the modeling of adversaries, the prediction of the efficacies of potential moves and the scoring of the vulnerability of a network (either one's own or an adversary's) to various cyber-attacks.

In this chapter,we describe a practical game theory engine denoted Themistocles that has been developed and employed over the past decade in cyber warfare analysis. At each time step k, Themistocles fuses past actions into a representation of the current state $S(k)$ of a network and the corresponding perceived states by all players. The latter states are functions of observed, inferred or hypothesized actions. Actions that could changethese state estimates may be observed directly (i.e., sensor data indicates the action), inferred indirectly from other observations, or hypothesized (neither observed nor implied by other objective evidence). All three types of actions involve varying degrees of uncertainty.

For each player, the feasible paths from their current state estimate and their estimate of their opponents' perceived state to a series of future states are constructed and scored with respect to the corresponding evaluation functions. The challenges are (1) to represent the evaluation functions of each player so that the scoring of each potential future state can be performed from the perspective of that player, and (2) to maintain a sufficient number of hypothesized future states to enable exploration of the full spectrum of path possibilities. In numerous formal cyber war games, Themistocles has demonstrated a capability to generate recommended COAs that met with the approval of human experts monitoring the games as being consistent with their best judgment of strategies and closely predicted the actions and tactics of human agents.

The remainder of this chapter is organized as follows. Section 2 describes the structure and algorithms of Themistocles. Section 3 presents examples of Themistocles' employment in cyber warfare scenarios. Section 4 concludes.

2 Themistocles Structure and Algorithms

Themistocles is comprised of four major software components, as illustrated in Fig. 2. The main processing flow is contained within the Search and Move Generation components. The Scheduler manages time within the game and initiates processing

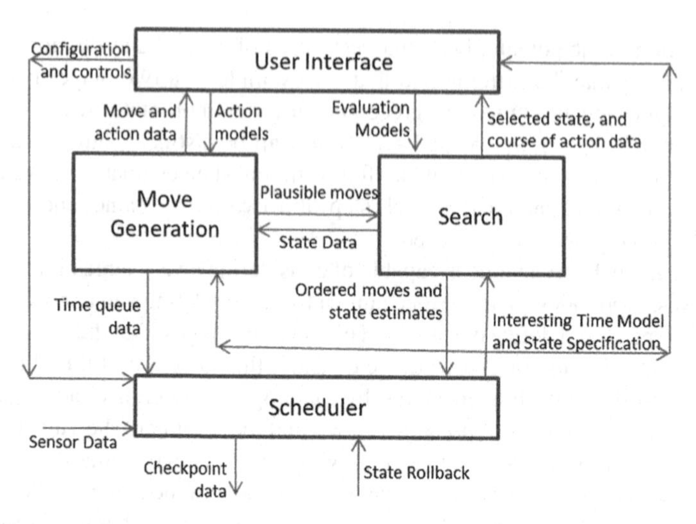

Fig. 2 Diagram of Themistocles components and their inputs/outputs

within the other components. The User Interface manages interaction with the user, including user controls and displays. This section first lays out the definitions required to specify system components and variables in Themistocles and then describes each of these in turn.

2.1 Definitions

The following definitions further detail the structure of Themistocles as a game-theoretic engine for modeling and analyzing cyber warfare scenarios.

Domain. The operating domain of Themistocles is a network of digital devices that are potentially capable of communicating with one another over this network. The devices may or may not be operable, and a given device may or may not be accessible by one or more players in a game.

State. The state in Themistocles is defined as the set of all variables and their associated values needed to characterize the system situation at a given point in time. In all cases, templates may be used to configure the system for ease of setting up a game. Global state variables characterize the overall situation on the network, while local state variables describe the situation on each device. Local state variables describe devices' pertinent attributes, operational status and accessibility. The number of local state variables is device specific, but may range upwards of 100 variables of different types. Examples of these include the on/off state of the device, whether it is a server, the operating system (O/S), the root password, as well as subjective variables such as the degree of belief in current suspicious activity on the device, etc. Global state variables include subjective variables such as the *work cost* of each prospective move,

Table 1 Partial list of local state variables for a network computer

Cyber object state variable	Value type
Admin password known	Boolean
BufferOverflow vulnerability	Boolean
Database service	Numerical
FTP service	Boolean
HTTP service	Boolean
Has account	Boolean
Has password file	Boolean
isCritical	Boolean
isHostKnown	Boolean
isHostUp	Boolean
isServer	Boolean
Patch version	Numerical
Physically accessible	Boolean
Physically at machine	Boolean
Root kit installed	Boolean
Needs investigation	Boolean
Key logger installed	Boolean
Suspicious activity seen	Numerical

Table 2 Partial list of global state variables for a network

Global state variable	Value type
Danger level	Numerical
Paranoia	Numerical
Political risk	Numerical
Risk	Numerical
Work	Numerical
Number of victims	Numerical
Number of decoys	Numerical

the *cumulative work cost* for each player, the *risk of discovery* for each prospective move for a player (usually only factored into the evaluation functions ofred players) and the cumulative *paranioa* of a blue player (used to determine the eligibility of certain drastic moves such as system restores).

Table 1 provides a listing of some typical local state variables for a network computer, while Table 2 provides a listing of some typical global state variables for the network. Each of these state variables has a set of sub-values corresponding to its true state and its estimated state by each player in a game, both for themselves and for the other players. Thus in a two-player game, there will be 5 sub-values for each state variable, i.e., truth state, player 1's own estimate of the state values and his estimate of player 2's assignment of state values, and similarly for player 2.

Table 3 Partial list of typical moves available to network defender/attacker

Defender move name	Move effect	Opponent observables
AnalyzeSystemLogs	Logs viewed, get info such as a login record or service installation	None
AnalyzeDataLogs	Database viewed, learn whether data was modified, deleted, or added	None
IP filter	Blocks a given set of IP addresses at the firewall	SYN flood stops working
InvestigateShutdown	Determine if a host shutdown was legitimate	None
NotifySecurity	Security team is on alert	More extreme counter moves
RestoreSystemFromBackup	Deletes all data and software, puts it back into standard start state	Lost connection, lost malicious services, lost backdoor
Attacker move name	Move effect	Opponent observables
Modify data	Corrupt data in a database	Tripwire alert on modified files
Port scan	Determine IP's of host's on a subnet and the services they offer	TrafficAnalyzer alert (i.e., Snort)
SetupBot	Take over a machine for later attacks	None
SQL injection	Gain root privileges by embedding string literal escape characters into the login command	None
SYN flood	Distributed denial of service (DDoS) by sending thousands of TCP SYN packets to a single machine	Service unusable or network slow

Move. A move is a member of a relatively small set of steps (i.e., O (10) actions) that a player can execute on a given device (e.g., a login). Moves have prerequisites (e.g., a device must be turned on, the player must know the login password, and the paranoia level makes the move eligible) and effects. The effects include a work cost associated with the move, which is expert-assigned, and changes in local or global state variable values (which in turn can add to the cumulative work cost for a player). Table 3 provides a listing of typical moves, along with their prerequisites and effects. **Move generation**. Move generation is the process of creating a set of *feasible* moves for a particular player, given the current state of the system. This set also has relatively small cardinality in most instances, since a particular player typically has access to only a fraction of the devices on the network. As well, the values of both local and global state variables can further prune the set of feasible moves.

Objective function. An objective function is one of a set of utility functions for each player whose independent variables are local and/or global state variables of the system. These state variable values are mapped into negative or positive integer scores, with the sign depending upon whether a given state of the system confers negative or positive benefit to the player. The scores are expert-assigned on a relative scale on the interval $[-5 \times 10^4, 5 \times 10^4]$, and reflect the cost/benefit of system state variable values. Thus a score of 1000 (-1000) reflects 10 times the benefit (cost) of a score of 100 (-100). A weighted average of these scores is calculated over all of the objective functions for each player, and the result is mapped via a sigmoid function to the interval $[1, 10^5]$ to produce an overall score for a given system state, from that players' perspective. This range is chosen in order to provide adequate dynamic range to the normalization steps that map COA state scores back into *utility* values, the latter residing in the unit interval [0, 1]. Table 4 presents a typical set of objective functions for a blue player (defender), while Table 5 presents a typical set for a red player (attacker).

In addition to the state-related scores calculated from these objective functions, there are also work costs (with negative values) that reflect the time and expense associated with a given move. These costs are cumulated over sequences of moves, and the cumulative work cost upon arriving at a given state is deducted from the overall score associated with that state. In addition, there are scoring-related state variables such as the *danger level* for the network, a number in [0, 1] that determines how the scores of individual players' COAs are combined into a joint COA score, with a value of 0.5 giving equal weight, a value of 0 assigning all weight to the red players' score(s) and a value of 1 assigning all weight to the blue players' score(s).

Utility. Utility is a normalized score associated with each state involved in a particular COA, i.e., with each state resulting from a sequence of moves by the players in the game. The normalization is with respect to all feasible moves deriving from the current system state. Thus the utility of the successive states in a COA monotonically

Table 4 Blue player objective function descriptions

Objective function	Description
Preserve availability	Adds points for each host under supervision if the host is up and working properly
Investigate suspicious activity	Adds points for states that provide information about a host that has gone down or is non-functional, even if it isn't fixed
DoS defense	Adds points for maneuvers to stop a denial of service, such as blocking IP addresses, ports, or applying patches
Worm defense	Adds points for applying patches; deducts points for non-critical ports being open, deducts points for each host infected
Submit weekly report	Adds points for successfully uploading data to a database on a weekly basis
Minimize work	Deducts points for executing moves that utilize administrator time/energy

Table 5 Red player objective function descriptions

Objective	Description
Corrupt database	Adds points for modifying data on any database
Corrupt web server	Adds points for modifying data on any web server
Cover tracks	Adds points for removing log entries, software installations, etc. that result from an attack and could lead to being caught
DoS host	Adds points for preventing network access to any host
Gain server root account	Adds points for obtaining a username/password on a server
Minimize risk	Deducts points for executing moves that have risk
Poison DNS	Adds points for modifying host files to point to one of your own servers
Remote reconnaissance	Adds points for mapping an opponent's network and determining what services and vulnerabilities are there
Setup bots	Adds points for getting root privileges on remote machines
Steal data	Adds points for exfiltrating data from any host
Steal server data	Adds points for exfiltrating data from any server

decreases as the depth of the COA increases, and a COA is terminated when the utility of its leaf node falls below a cutoff threshold. This process is further described in the Search component below.

Action Queue. The Themistocles Action Queue manages the execution of all moves. It has two primary functions: (1) to test the effects of each move, and (2) to manage the time clock of the game. When a prospective feasible move is added to the action queue, the resulting state change is calculated and the game time is advanced to the next *interesting time*. The latter time is the minimum increment of time until the move generates an observable event, or until the move completes, or until a *pass time* is reached (i.e., the maximum increment of time permitted in the game scenario). Once a particular move has been added to the Action Queue, the overall utility of the resulting state is calculated and stored. Following this, the move is removed from the Action Queue and the next feasible move is added, with this process repeated until the respective utilities of all feasible moves have been calculated.

With these definitions, we now proceed to describe the four components of Themistocles in Fig. 2 and their interactions.

2.2 Search Component

The search engine is the core of the Themistocles software. The search engine performs the selection and evaluation of prospective feasible moves recursively over time. The output of the engine at each step in the game is data describing the prescribed COA and the states corresponding to each evaluated COA.

The Themistocles search process is a tree-based search designed quickly to produce an initial prescribed COA, using a leaf node score cutoff threshold in combination with a maximum tree depth, and then successively to refine the prescribed COA by deepening the search via iterative reductions in the cutoff threshold, as elaborated upon below. Each player performs his own tree search at each step of the game, based upon his own estimates of the system state and his estimates of the system state assumed by the adversary(s).

The root node of a player's search tree corresponds to the current state of the system $S(k)$ at time step k in the game. Starting with the root node, the tree is constructed in a partially serialized manner by considering the collection of child nodes $N^n_{i,j}$ at levels $n = 1, 2, K$ that would result from the feasible move sets of a given player and her adversary(s), where the superscript n refers to the depth of the tree, i indexes the child nodes at a given level n and j indexes the child nodes of parent node i ($n = 1$ corresponds to the children of the root node, for which only a single parent node exists, i.e., $i = 1$ only for this level). Each node in the tree corresponds to a prospective future system state $S^p_{ij}(k+n), n = 1, 2, K$ for player p, where these states are ordered in a time sequence dictated by the move effects.

A utility value $V\left(S^p_{ij}(k+n)\right)$ for player p is calculated for each child node, and these values are then normalized to sum to unity by dividing each child node value by the sum over all child node values, resulting in a normalized utility value $V'\left(S^p_{ij}(k+n)\right)$, i.e.,

$$V'\left(S^p_{ij}(k+n)\right) = \frac{V\left(S^p_{ij}(k+n)\right)}{\sum\limits_j V\left(S^p_{ij}(k+n)\right)}. \tag{1}$$

For $n = 1$, i.e., the first level of child nodes from the root node, these values are equated to the corresponding nodes' game score denoted by $P^p_{ij}(k+1)$, i.e., $P^p_{1j}(k+1) \equiv V'\left(S^p_{1j}(k+1)\right)$. For $n > 1$ the $V'\left(S^p_{ij}(k+n)\right)$ are multiplied by the parent node's game score $P^p_{i'i}(k+n-1)$ (where i' is the index of the grandparent node), resulting in a game score for node $N^n_{i,j}$ given by

$$P^p_{ij}(k+n) = P^p_{i'i}(k+n-1)V'\left(S^p_{ij}(k+n)\right) \tag{2}$$

If $P^p_{ij}(k+n) > P_{cutoff}$ for a given child node $N^n_{i,j}$, then node $N^n_{i,j}$ becomes a parent node for a follow-on set of moves that deepens the tree. Thus the score of the terminal node for a given COA(path) through the tree decreases as the tree depth increases. Some moves may be generated with an associated probability, in which case the child node utility value prior to normalization is also multiplied by its (expert assigned) probability of occurrence.

The above process is continued until no parent node has a child node with a score above the cutoff threshold, at which point these parent nodes become the leaf nodes of the tree. This search for a particular player is illustrated in Fig. 3, where the white nodes represent the states resulting from that player's move choices and the dark

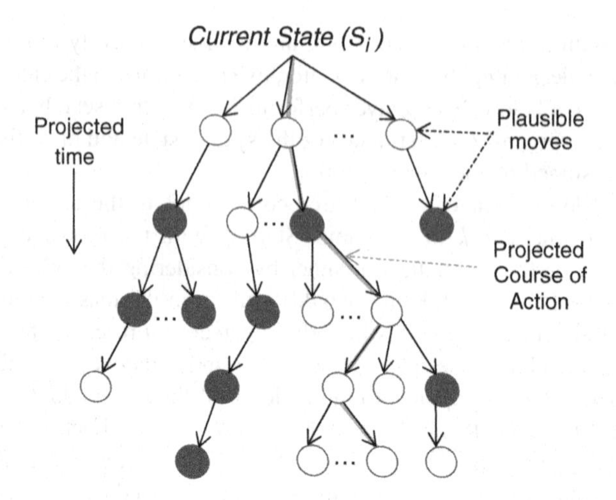

Fig. 3 Search tree from a given current state, showing projected moves by both players in a two-player game

nodes represent the states resulting from the adversary's move choices. Note that the search tree admits a combination of these states in any path through the tree.

The COA is scored based upon knowledge engineering defined abstract parameters specifying expected effects of moves and weights of objective functions for a given player evaluation model. Themistocles instantiates the moves and effects of each move and calculates the COA score at each node based upon the accrued move effects and weighted objective functions for each player. The path terminating at the highest scoring leaf node is selected as the prescribed COA and is then executed in an autonomous game simulation. In an interactive game, the top three scoring COAs are presented to the user, who then selects the one of his choice.

2.3 Move Generator

A move is defined to have the following characteristics:

- A list of preconditions.
- Effect on state upon initiation.
- Effect on state upon completion.
- A list of conditional effects during execution.
- Timing information for the entire move and each effect.
- A list of possible outcomes and probabilities for each effect.

In addition to explicitly modeling timing effects and stochastic move outcomes, the move set differs from traditional game move sets in another fundamental way. In a traditional game, two opposing players alternate moves. In most real-world domains such as cyber-warfare, this is simply not the case. Each player has the option of

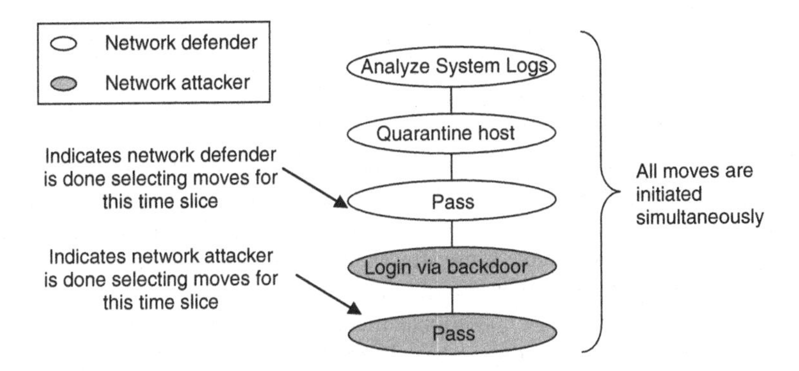

Fig. 4 Snippet of partially-serialized game tree

choosing multiple moves that are executed simultaneously. In fact, both players will frequently be executing multiple actions at the same time.

To accommodate this, the search utilizes an untraditional approach to tree construction where all simultaneously chosen moves are serialized, such that they are listed in order in the tree despite the fact that they will be executed at the same time. This is accomplished by introducing a new move type: *Pass*. A pass indicates that the player will not be choosing to begin any additional actions until the next interesting time. All moves chosen before a pass are interpreted as beginning at the same simulated time. Figure 4 shows an example of a small subset of a partially serialized game tree.

When there are no players left to choose a move at a particular interesting time, the serialized move tree is parsed and the scheduler is notified of the postulated move selections for each player and the time queue entries for each action chosen.

While each move has a duration and set of possible outcomes associated with it, both players may or may not be aware of these outcomes. Awareness of the state of the network is based on available resources, and may be contingent on making moves to gain information. Even when a move produces an observable, such as a message logged by a deployed Intrusion Detection System (IDS), the players may not be in a position to see the observable without further action. In the event that a player can see the observable the system will give that player a chance to respond to the observed event.

Note that both players are not necessarily given the option of moving during a particular slice of time. Players only move if one of the defined events occurs such that they are aware of it. Thus, if the defender completes a move, the defender will have the option to choose more moves but the attacker will only have that option if the event has produced an observable to the attacker.

For added realism, there are three types of dynamic environmental moves included in any game: pure, scheduled, and consequential. Pure environmental moves include elements of the environment that have unknown or dynamically changing attributes. Schedule driven moves have scheduled preferential occurrence (such as circadian rhythm driven actions). Consequential environment effects simulate unexpected results tied to a specific state change or move execution. The actual state effects can be modeled that same way as player moves, using the action model. However, the triggering effect can be different.

Pure environmental actions are simulated using a Bernoulli or Poisson distribution tied to specific environmental factors. This technique was used effectively in training programs for noise generation. This simulation takes into account the state context so that the outcomes will make sense; for example, actions will not occur from an object that isn't capable of the proposed function.

For schedule driven factors, conditional effects are applied based upon the time of occurrence. The time simulation can be selected based upon the particular environmental factor being modeled. For example, this technique was used in space network operations to model environmental factors such as thunderstorms, which in some areas are highly dependent upon the time of day.

2.4 Scheduler

The Scheduler is responsible for maintaining the time queue of "interesting times". These interesting times are provided by move postulation, including start and end time, sensor data input, and user input.

When a move is executed, the scheduler inserts the selected moves into the action queue, advances time to the next interesting time, decides which player next has a turn, and calls move generation to provide that player with all available moves. The move start times are the times from the serialized move tree after composition of all projected player move selections. The move completion times are the times at which the last move effect is complete whether or not there is an observable state effect.

A move may have zero or more expected impacts to state and zero or more observables for any player as a result of the move execution at specified times during its execution. A Pass move has the simplest time data, having no effects on state times but a completion time. Any time a move completes, whether or not it had an impact on state, the Scheduler returns control to the move generator to determine whether or not another move of any type by any player should be initiated. Move generation augments the serialized move tree with moves selected for any player followed by a pass move.

The Scheduler maintains two interesting time queues. The primary simulated game time queue is the game action queue. The game action queue is maintained for all actions derived from human or computer selected moves, environment model moves, and sensor data reports. The secondary queue is the search action queue. This queue is maintained for all actions derived from projected moves by the search engine during a course of action evaluation. Any time an interesting time from the game action queue implies a change of state, the scheduler initiates a course of action evaluation by copying the game action queue to the search action queue, and initiating a search with the search action queue. The search action queue is then managed with projected moves and state changes selected by the search engine and driven by the search action queue.

When a course of action path has been evaluated to the point where the cutoff score has been reached, then the scheduler will back up the search action queue to the time of the last state that was not completely evaluated (paths with scores higher than the cutoff still exist), and reinitiate the search engine. When all paths have satisfied the cutoff, then the Scheduler will return to the game action queue and send results of the latest state search to the appropriate player. A computer (automated) player will always select the top scoring move(s), a human player can select moves indicated by the search results and/or any other set of valid moves. This process repeats continuously until there are no further actions in the game action queue. A condition where no further actions exist in the game action queue occurs when the game has been updated comprehensively but a human player has yet to finish move selection or when a predetermined time limit has been reached.

The scheduler may run Themistocles much faster than real time or much slower than real time. The scheduler runs Themistocles as fast as possible between the game action queue entries. If the primary (game action queue) processing is provided solely by computer input and not held for sensor data input, then Themistocles can run many times real time. If the cutoff score is set very low, requiring deeper and broader analysis (search action queue), or the moves in the primary (game action queue) are defined in very fine grained time or with operator delay, then Themistocles can run many times slower than real time.

The Scheduler also manages checkpoint retention, saving the state of the game at specified times. The checkpoint initiation can be based upon a change in state trigger, a simulation time trigger, or an operator trigger. As requested by the user, the Scheduler restores the game to a full fidelity image at a specified time from a stored checkpoint file.

2.5 User Interface

The User Interface (UI) leverages JAVA graphics packages to provide a graphical representation of the state. It displays sufficient information that the observer may understand the status of the game and the state of the target network. The UI may present multiple views from the perspective of individual players including state, resources, and moves available. In human mode, the player has an end turn button that basically completes the move selections for that interesting time and triggers the action queue to insert the pass move for the active display player. The UI also supports configuration of the game simulation including checkpoint and rollback.

The User interface connects to the game server using JAVA Remote Method Invocation (RMI) so that a client can be run on any other (potentially remote) machine. As well, any interface could be created and connected to the game server through the RMI.

3 Examples

This section presents an example of Themistocles' use in a realistic cyber warfare game. We first describe the network environment being defended, including the security measures and policies in place. We next present a scenario involving the red and blue players' objectives. We then describe the red players' attack moves and the resulting observables that are generated for the blue player by these actions. We step through the sequence of top-scoring COAs generated by Themistocles in the context of the moves undertaken by the red and blue players. We conclude with a summary of the results.

3.1 Cyber War Game Environment

We consider a network of 10 regular workstations that are used on a daily basis, as shown in the Themistocles screen capture of Fig. 5. Two additional workstations are used for system administration. There is also an internal web server used for organizational data sharing, an internal database server holding proprietary data shareable only with employees and an internal email server. At the network interface, there is a firewall with virtual private network (VPN) support. Outside the firewall, in the network demilitarized zone (DMZ), there is a network intrusion detection system (NIDS) machine with backup and archiving capabilities and a web server for sharing data with remote employees. An unknown number of external machines have valid access to the network.

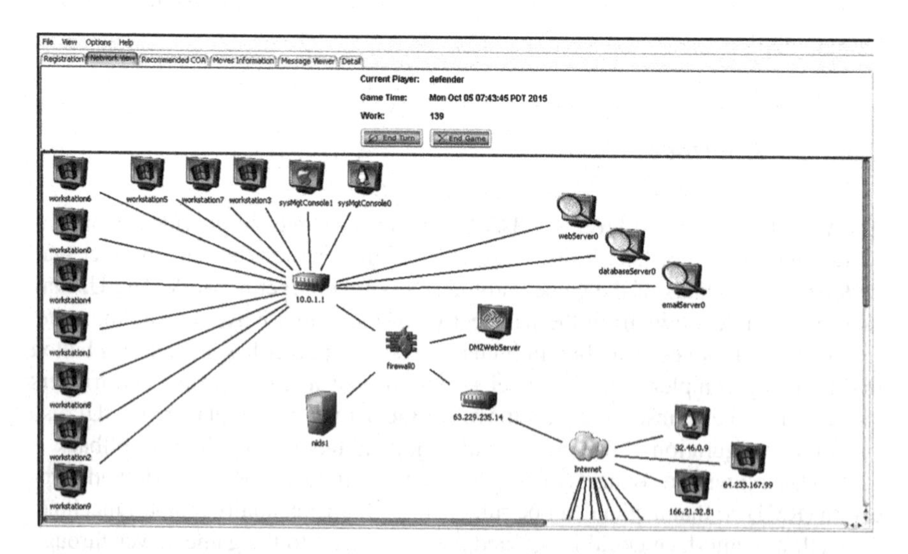

Fig. 5 Themistocles screen shot showing the cyber warfare game environment

Certain security measures and policies have been implemented on the network. To wit, the fixed mission of the network allows the firewall to be configured to restrict incoming and outgoing traffic only to those services needed. Source information comes only via VPNs. The operating system (O/S) configurations have been stripped of unneeded functionality and services. All application software on the network machines is under configuration management. Finally the external IDS detects any attempts to violate the firewall rules.

3.2 Attack Scenario

The objectives of the red player in the game are to steal information from any computer on the network and to maintain access to the network for future attacks, while minimizing the chance of being discovered. We assume that the red player knows the IP address of the external web server in Fig. 5.

The objectives of the blue player are to maintain the availability of all servers while minimizing the amount of effort and resources required to do so, and to gather intelligence on any attacks in progress.

The attack proceeds as follows. Using a bot to help avoid attribution, the red player scans the DMZ web server for vulnerable services running on the machine. An exploit is used to gain root privileges on this server, ensuing with a backdoor installation. Using this backdoor, the red player maps the internal network. From this point, the red player attempts the exfiltration of data and the installation of root kits on any and all network devices that are exploitable. When the red player observes the failure of an attack against any particular host, a new host is selected for attack.

As a result of the red players' attack, the blue player observes the following on their control workstations:

- The IDS detects the port scan of the external web server. Due to background network activity, this observation does not lead to any response.
- A tripwire detects that new software has been installed on the external web server (i.e., the backdoor software).
- The IDS detects heavy download traffic to the external web server as the red player performs data exfiltration.
- A tripwire detects the upload of the rootkit software.

With these preliminaries, the blue player employs Themistocles to aid in her selection of moves to counter the attack in progress.

3.3 Themistocles Recommended Courses of Action

Table 6 shows the sequence of moves generated by Themistocles as the game proceeds from attack initiation to its conclusion. To illustrate the game, we will examine the first couple of COAs recommended by Themistocles for the blue player at the conclusion of the preceding sequence of moves by the red player.

Table 6 Sequence of moves by red and blue players

Move	Owner	Machine	Function
SetupExternalProxy	Attacker	externalHostX	
FTPScan	Attacker	externalHostX	DMZWebserver
AnalyzeSystemLogs	Defender	nids1	
Exploit FTP	Attacker	externalHostX	DMZWebserver
UploadHostileSoftware	Attacker	externalHostX	DMZWebserver
InstallBackdoor	Attacker	externalHostX	DMZWebserver
InstallNIDS	Attacker	sysMgtConsole	
LoginViaBackdoor	Attacker	externalHostX	DMZWebserver
ScanSubnetForVulnerabilities	Defender	sysMgtConsole	10.0.1.*
PingSubnetInternal	Attacker	DMZWebserver	10.0.1.*
PortScanSubnetInternal	Attacker	DMZWebserver	10.0.1.*
ExploitFtp	Attacker	DMZWebserver	Web server
UploadHostileSoftware	Attacker	DMZWebserver	Web server
HardenSystem	Defender	Web server	
InstallRootkit fails	Attacker	DMZWebserver	Web server
ExploitFtp	Attacker	DMZWebserver	DB server
ExfiltrateData	Attacker	DMZWebserver	DB server
DeployHoneypot	Attacker	DB server	
ModifyData	Attacker	DMZWebserver	DB server
ExfiltrateData fails	Attacker	DMZWebserver	DB server
ExploitFtp	Attacker	DMZWebserver	e-mail server
UploadHostileSoftware	Attacker	DMZWebserver	e-mail server
ApplyPatches	Defender	e-mail server	
InstallRootkit fails	Attacker	DMZWebserver	e-mail server
InstallRootkit	Attacker	externalHostX	DMZWebserver
Quarantine	Defender	DMZWebserver	

Given the red players' objectives, he begins with an FTP scan on the DMZ web server followed by a Pass. At the blue players' first turn, she observes the FTP scan and Themistocles analyzes the feasible subsequent moves and selects the highest-utility COA as shown in Fig. 6. Note that this COA includes moves by both the blue and red players, from the blue players' perspective of their own and the red players' scoring of these moves. The normalized utility of each successive move is shown in the next to last column. The depth of a COA tree is limited by setting the cutoff utility at 5×10^{-5}. Thus the COA in Fig. 6 represents the deepest COA having the highest utility above this cutoff threshold.

Referring back to the actual game moves in Table 6, the blue player elects to analyze the system logs as recommended for the next step by Themistocles. The red player then counters with the FTP exploit, uploading and installing the backdoor on the DMZ web server. When the blue players' next turn comes, Themistocles

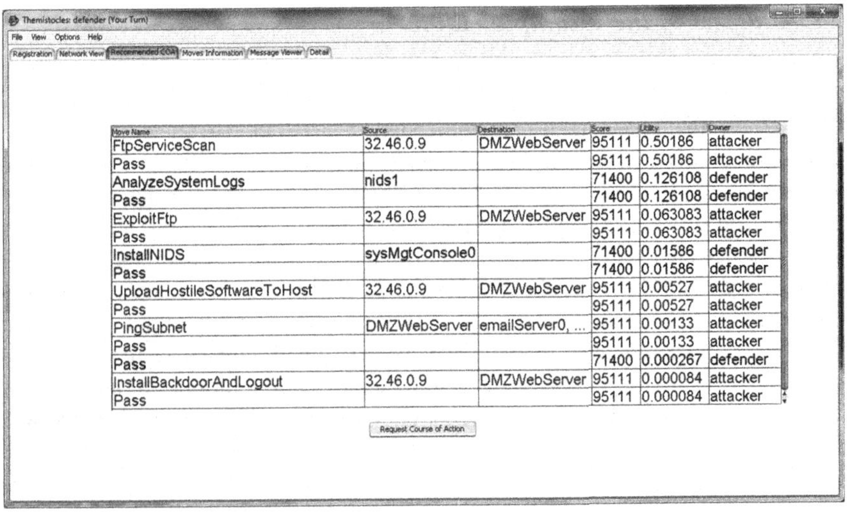

Themistocles: defender (Your Turn)

File View Options Help

Registration | Network View | Recommended COA | Moves Information | Message Viewer | Detail

Move Name	Source	Destination	Score	Utility	Owner
FtpServiceScan	32.46.0.9	DMZWebServer	95111	0.50186	attacker
Pass			95111	0.50186	attacker
AnalyzeSystemLogs	nids1		71400	0.126108	defender
Pass			71400	0.126108	defender
ExploitFtp	32.46.0.9	DMZWebServer	95111	0.063083	attacker
Pass			95111	0.063083	attacker
InstallNIDS	sysMgtConsole0		71400	0.01586	defender
Pass			71400	0.01586	defender
UploadHostileSoftwareToHost	32.46.0.9	DMZWebServer	95111	0.00527	attacker
Pass			95111	0.00527	attacker
PingSubnet	DMZWebServer	emailServer0, ...	95111	0.00133	attacker
Pass			95111	0.00133	attacker
Pass			71400	0.000267	defender
InstallBackdoorAndLogout	32.46.0.9	DMZWebServer	95111	0.000084	attacker
Pass			95111	0.000084	attacker

Request Course of Action

Fig. 6 Highest-utility COA for both players upon observation of the red FTP scan

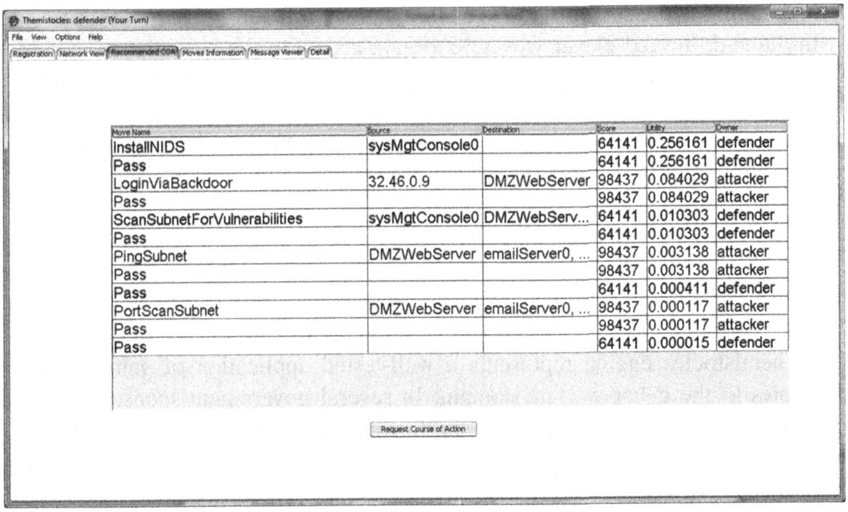

Themistocles: defender (Your Turn)

File View Options Help

Registration | Network View | Recommended COA | Moves Information | Message Viewer | Detail

Move Name	Source	Destination	Score	Utility	Owner
InstallNIDS	sysMgtConsole0		64141	0.256161	defender
Pass			64141	0.256161	defender
LoginViaBackdoor	32.46.0.9	DMZWebServer	98437	0.084029	attacker
Pass			98437	0.084029	attacker
ScanSubnetForVulnerabilities	sysMgtConsole0	DMZWebServ...	64141	0.010303	defender
Pass			64141	0.010303	defender
PingSubnet	DMZWebServer	emailServer0, ...	98437	0.003138	attacker
Pass			98437	0.003138	attacker
Pass			64141	0.000411	defender
PortScanSubnet	DMZWebServer	emailServer0, ...	98437	0.000117	attacker
Pass			98437	0.000117	attacker
Pass			64141	0.000015	defender

Request Course of Action

Fig. 7 Highest-utility COA for both players upon observation of the backdoor installation

generates the highest-utility COA shown in Fig. 7, whereupon the blue player follows the recommended next step and installs a NIDS on an internal router.

The red players' next step is to login via the backdoor software he has installed, as shown in Table 6. The game proceeds as shown by the moves in this table to the final step where the blue player quarantines the DMZ web server.

3.4 Analysis of the Example Game

The blue player began seeing suspicious behavior from alerts by its external NIDS early in the game. Themistocles recommended an installation of an additional NIDS on an internal router in order to catch malicious network activity if the red player had gotten inside the defenses. A vulnerability scan was recommended to determine if any unauthorized services had been installed. The backdoor installed on the DMZ web server was detected, and this forced the blue player to make an important decision—immediately remove the backdoor and make the DMZ web server unavailable to its regular users, or keep the server running and use this as an opportunity to identify the attacker. In keeping with the blue players' mission, Themistocles recommended the latter by deploying a "honeypot," while making sure to turn off all unnecessary services on the internal web server being attacked. The red player notices the web server is no longer available to attack, so heads for the database server. This leads the blue player to gather information on the red player, but the latter figures out that it is a honeypot when a data corruption attempt doesn't succeed. This is because honeypots do not have real data on them, so modifications aren't written to disk. The red player moves on to the e-mail server and at this point the blue player decides this is getting too aggressive and decides to shut out the red player by applying patches and quarantining the server to remove all malicious software.

In the end, the red player was able to gain access to the blue players' internal systems and map the entire internal network, but was not successful in stealing any data. One rootkit was successfully installed on the external DMZ web server, but that was a risk the blue player desired to take in order to gather more information on the red players' identity.

4 Conclusion

The Themistocles engine represents a well-tested application of game-theoretic principles to the cyber warfare domain. In several government-sponsored formal cyber war games, Themistocles has been shown to generate COAs for both offensive and defensive cyber warfare scenarios that are consistent with the move choices of independent experts monitoring the game.

Future work in this area will include the fuzzification of move scores to both type-1 and interval type-2 membership functions and the use of hierarchical linguistic weighted power means for the aggregation of COA scores (16–18). This will enable us to take account of the inherent imprecision associated with the costs/benefits of individual moves, and to employ a perspective ranging from the most pessimistic to the most optimistic on the aggregations of these scores.

References

Katz, A., Butler, B.: "Game Commander"-Applying an architecture of game theory and tree look ahead to the command and control process. In: Proceedings of the Fifth Annual Conference on AI, Simulation, and Planning (AIS94). Florida (1994)

Samuel, A.L.: Some studies in machine learning using the game of checkers. IBM J. Res. Dev. **3**(3), 211–229 (1959)

Tesauro, G.: TD-Gammon, a Self-Teaching Backgammon Program, reaches master-level play. Neural Comput. **6**(2), 215–219 (1994)

Hsu, F., et al.: Deep thought. In: Marsland, T.A., Schaeffer, J. (eds.) Computer, Chess, and Cognition, pp. 55–78. Springer, New York (1990)

Hamilton, S.N., Hamilton, W.L.: Adversary modeling and simulation in cyber warfare. International Information Security Conference (2008)

Dudebout, N., Shamma, J.S.: Empirical evidence equilibria in stochastic games. In: 51st IEEE Conference on Deicsion and Control, December 2012 (2012)

Gopalakrishnan, R., Marden, J.R., Wierman, A.: An architectural view of game theoretic control. ACM SIGMETRICS Perform. Eval. Rev. **38**(3), 31–36 (2011)

Marden, J.R., Arslan, G., Shamma, J.S.: Connections between cooperative control and potential games. IEEE Trans. Syst. Man Cybern. Part B Cybern. **39**, 1393–1407 (2009)

Chen, L., Low, S.H., Doyle, J.C.: Random access game and medium access control design. IEEE/ACM Trans. Networking **18**(4), 1303–1316 (2010)

Chandra, F., Gayme, D.F., Chen, L., Doyle, J.C.: Robustness, optimization, and architectures. Eur. J. Control **5–6**, 472–482 (2011)

Chen, L., Li, N., Jiang, L., Low, S.H.: Optimal demand response: problem formulation and deterministic case. In: Chakrabortty, A., Ilic, M. (eds.) Control and Optimization Theory for Electric Smart Grids. Springer, New York (2012)

Carmel, D., Markovitch, S.: Learning and using opponent models in adversary search. Technical report CIS9606 (1996)

Meyers, K., Saydjari, O.S., et al.: ARDA cyber strategy and tactics workshop final report (2002)

Hamilton, S.N., Miller, W.L., Ott, A., Saydjari, O.S.: The role of game theory in information warfare. The information survivability workshop (2001)

Hamilton, S.N., Miller, W.L., Ott, A., Saydjari, O.S.: Challenges in applying game theory to the domain of information warfare. The information survivability workshop (2001)

Rickard, J.T., Aisbett, J.: New classes of threshold aggregation functions based upon the Tsallis q-exponential with applications to perceptual computing. Accepted for publication in IEEE Trans. Fuzzy Syst. (2013)

Rickard, J.T., Aisbett, J., Yager, R.R., Gibbon, G.: Linguistic weighted power means: comparison with the linguistic weighted average. In: Proceedings of FUZZ-IEEE 2011, 2011 World Congress on Computational Intelligence, June 2011, pp. 2185–2192. Taipei, Taiwan (2011)

Rickard, J.T., Aisbett, J., Yager, R.R., Gibbon, G.: Fuzzy weighted power means in evaluation decisions. In: Proceedings of World Symposium on Soft Computing, Paper #100, May 2011. San Francisco, CA (2011)

References

Mission Impact Assessment for Cyber Warfare

Jared Holsopple, Shanchieh Jay Yang and Moises Sudit

1 Introduction

Cyber networks are used extensively by not only a nation's military to protect sensitive information and execute missions, but also the primary infrastructure that provides services that enable modern conveniences such as education, potable water, electricity, natural gas, and financial transactions. Disruption of any of these services could have widespread impacts to citizens' well-being. As such, these critical services may be targeted by malicious hackers during cyber warfare. Due to the increasing dependence on computers for military and infrastructure purposes, it is imperative to not only protect them and mitigate any immediate or potential threats, but to also understand the current or potential impacts beyond the cyber networks or the organization. This increased dependence means that a cyber attack may not only affect the cyber network, but also other tasks or missions that are dependent upon the network for execution and completion. It is therefore necessary to try to understand the current and potential impacts of cyber effects on the overall mission of a nation's military, infrastructure, and other critical services. The understanding of the impact is primarily controlled by two processes: state estimation and impact assessment. State estimation is the process of determining the current state of the assets while impact assessment is the process of calculating impact based on the current asset states.

Cleared for public release on 23 Apr 14, #88ABW-2014-1911

J. Holsopple · M. Sudit
Center of Multisource Information Fusion, CUBRC, Inc., Buffalo, NY, USA
e-mail: holsopple@cubrc.org

S. J. Yang (✉)
NetIP Lab, Department of Computer Engineering, Rochester Institute of Technology, Rochester, NY 14623, USA
e-mail: jay.yang@rit.edu

R.R. Yager et al. (eds.), *Intelligent Methods for Cyber Warfare*,
Studies in Computational Intelligence 563, DOI 10.1007/978-3-319-08624-8_11

In this chapter, we consider the case of a military computer network that could be subjected to external and insider attacks through various physical and virtual vulnerabilities. The goal is to provide an estimate of the Nth-order impact of cyber threats while the missions are in operation. This is accomplished by a tree-based structure, referred to as a *Mission Tree*, which models the relationships between various missions, tasks, and assets. The relationships are modeled using Order Weighted Aggregators (OWAs),which provide for a diverse set of relationship types. The Mission Tree is different from other methods of modeling mission relationships in that it is capable of providing a quantitative estimate of impact by propagating the impacts "up", from the leaves to the root, through the tree.

Another key aspect of impact assessment is that missions or tasks will change during the course of warfare. This chapter will also explore how to dynamically change the mission tree to account for scheduled or non-scheduled changes and how those dynamic changes can affect how one performs impact assessment. This chapter will present a novel approach to address all of these aspects.

To ensure consistency of vocabulary, Sect. 2 provides a set of definitions, followed by a review of the existing methods for mission planning and assessment. Sections 4 and 5 discuss the mission tree structure and how it is used for mission impact assessment. Several examples and simulation results will be presented in Sect. 6 to demonstrate the utility of the mission tree.

2 Definitions

It is necessary that a state estimation process be executed first to determine asset damage, which is then fed into the mission impact assessment process. However, before discussing these processes in detail, it should be noted that there is no common vocabulary defined for mission impact assessment. In fact, vocabulary used for mission planning and impact assessment within one organization can sometimes conflict with the vocabulary used by another organization. Therefore, a set of definitions for various mission planning and impact assessment concepts is provided to maintain consistency throughout the chapter. The reader should take careful consideration of how these definitions differ from their organization's to maximize their understanding of the concepts presented in this chapter.

While this book focuses on computer security and cyber warfare, the concepts presented in this chapter are intended to be generic enough such that they are applicable to other application domains. Each definition given below will also provide a short description of how it applies specifically to cyber warfare.

Situation awareness—the state of knowledge that results from a process [1]. Situation awareness involves the detection and assessment of the threats to a computer network.

Situation assessment—the process that provides outputs that can be analyzed to gain situation awareness [2]. The situation assessment process is typically comprised

of a combination of sensors and data aggregation software manually analyzed by computer security experts.
Environment—the specific location in which events are being monitored and assessed. The protected computer network comprises most of the environment with respect to cyber warfare, but it can contain other physical or virtual elements such as buildings, rooms, or other networks.
Entity—something that has a distinct, separate existence, though it need not be a material existence [2]. Given this very general definition, it is important to define the granularity at which an entity is being defined for an application.
Object of Interest (OOI)—an entity or mission in the environment that should be monitored. In cyber warfare, this will generally refer to physical and virtual entities such as hosts, switches, services, and even communication links. The actual OOIs for a given network will depend heavily on how much is known about the network and what is considered critical or important enough to monitor.
OOI State—the condition of the OOI defined relevant to the application domain. An OOI may be in multiple states (e.g., *damaged* and *limited operations*).
Situation—a collection of activities and their effects on the environment at a given time.
Activity—something done as an action or a movement. They are composed of entities/groups related by one or more events over time and/or space [2].
Event—an occurrence that affects the environment [2].
Observable—one or more attributes of an event or object from a sensor or some form of intelligence.
Mission—a process by which a goal is intended to be achieved. It is possible that missions can be contained within other missions. It should be noted that we use the term mission in a very general sense throughout this chapter to represent any type of task that must be executed to achieve a desired goal.
Asset—an OOI that supports one or more missions whose state can be determined by a situation assessment algorithm. It should be noted that the granularities by which assets are defined with respect to the mission tree vary by application as well as the existing pre-processing mechanisms in place. We discuss this issue later in the paper.
Role—a function that an asset performs.
Impact—a quantitative assessment of how much a mission is affected by a given activity or situation. For consistency throughout this chapter we will use the term "impact" to imply "negative impact", unless otherwise specified.
Damage—a quantitative assessment corresponding to the state(s) a given asset is in with respect to its ability to perform a given role.

3 Mission Impact Assessment: A Brief Background Review

Mission impact assessment is not a new concept; however, it has traditionally been a manual approach. Grimaila and Fortson [3] argue the importance of fast and accurate damage assessment, especially in "cyberspace where attacks can occur in milliseconds and may have a greater impact due to the complexity and interconnectedness of the information infrastructure. A failure to immediately detect, contain, remediate, and assess the impact following a cyber attack may result in other unforeseen higher order effects that may not be immediately apparent." The approach for impact assessment considered in this chapter can not only be used in the forensic phase of a cyber attack (i.e., after it has already occurred), but also during the attack. The ability to identify impacts that may not be "immediately apparent" *during* an attack can greatly improve the incident reports and enhance mitigation strategies by protecting assets that are not directly attacked. Grimaila and Fortson [3] also mention that a hurdle to fast and accurate damage assessment is the lack of asset documentation. This is problematic because the only way to estimate the direct and indirect impacts is to have an understanding of how assets interact with each other and are used for various missions and tasks. This understanding is also critical to the approach discussed in this chapter. In the past, the important components or concepts of a mission have been described in written language and/or diagrams. In some cases, including the use of cyber infrastructure to support military missions, the missions could be "implied" or undocumented by those who are assessing the health of the networked computing and storage systems. As such, modeling the mission in such a way that is understood by computers has been a daunting task.

Muccio and Kropa [4] describe cyber mission assurance as a four step process: (1) Prioritize mission essential functions, (2) map critical cyber assets, (3) vulnerability assessment of mission essential functions, and (4) mitigation of vulnerabilities and risks. This chapter will primarily focus on (2) and (3) to aid in (4). The prioritization of assets is also critical to mission impact assessment because it will determine the level of detail by which the models need to be developed.

Computer-aided mission impact assessment is part of a Decision Support System (DSS), which is a computer system/application that assists an analyst in evaluating the health of a system, task, or mission. A DSS can be traced back to 1982 when Ben-Bassat and Freedy [5] outlined the formal requirements for a generic DSS system that assessed the threat probabilities on various aspects of a given system.

Musman, et al. [6] discuss the evaluation of cyber attack impact on missions. They argued how critical it is to have mission models and descriptions stronger than what is available today. Their approach uses Business Process Modeling Notation (BPMN) and utilizes multiple information sources to develop the models necessary for impact assessment. Using cyber incident reports, they manually modify the mission model to produce new estimates, though the authors admit that this is a shortfall and are considering various alternatives to provide a faster impact estimate that is critical to effective impact assessment.

In the past decade or so, research has focused on modeling the mission dependencies to help facilitate computer-assisted analysis of current missions. D'Amico et al. [7] focused their research specifically towards computer networks by creating an ontology of the mission dependencies. Their approach focused on modeling how *cyber assets* provide a *capability* for each *mission*. While this approach provided the necessary modeling capability of mission relationships, it still required a graphical analysis of the events to determine what was affected since it did not provide a numeric estimate for mission impacts.

Jakobsen [8] proposed the use of dependency graphs for cyber impact assessment. He also proposed the use of an "Impact Factor" to represent the capability of an attack affecting a given asset and the "Operational Capacity" that indicates the level to which the asset is compromised. These variables are similar to the impact scores and state estimation approaches discussed in this chapter. In addition, there has also been a wide array of other approaches to identify cyber attacks and their effects on computer networks, e.g., [9].

4 Asset State Estimation

The first step to determining mission impact is to determine a quantitative value of the damage to the objects of interest (OOIs) in the environment at a given time, which we will refer to as state estimation. The damage is calculated using inputs from an observable correlation process, which has grouped together observables in a meaningful way for a certain situation.

In the case of cyber warfare, the damage to an OOI corresponds indirectly to the level at which the information from hosts, services, and communication links can be trusted by the blue team. This section explores different ways that $d_i(t)$, the damage to an object of interest i at time t, can be calculated and used by the mission tree.

The state estimation process is not restricted to any single method of calculation. However, the following requirements are imposed for each value of $d_i(t)$:

1. Range of $d_i(t)$ is [0,1]
2. $d_i(t) = 0$ indicates that the OOI is operating normally and has no damage.
3. $d_i(t) = 1$ indicates that the OOI is unable to perform or be trusted for any of its tasks
4. $d_i(t) \in (0, 1)$ indicates that the OOI has some damage to it, but is still able to operate in a limited state. The extent to which the operations are limited tends towards the appropriate end of the range.

These requirements ensure that the damage scores are as consistent as possible across state estimation algorithms for different entity types. This in turn allows multiple state estimation algorithms to be used and potentially combined in various ways (see multiple state-space estimation, Sect. 4.4) to estimate $d_i(t)$ for a various types of OOIs.

4.1 Simple State Estimation

Simple state estimation is a direct calculation or assignment of $d_i(t)$ based on a set of rules or equations. As long as the rules or equations have known upper and lower

Table 1 Example rules for simple state estimation

Rule (Weight)	Criteria	Value
Firewall ($w_1 = 0.3$)	Traffic allowed	1 (= mx_1)
	Traffic forbidden	0
Attack type ($w_2 = 0.3$)	Reconnaissance	1
	User privilege escalation	2
	System (Root) privilege escalation	3 (= mx_2)
Connectivity ($w_3 = 0.4$)	Not connected	0
	Connected	1 (= mx_3)

bounds to the values, it can easily be normalized into the [0,1] interval. For example, consider three rules (Firewall, Attack Type, and Connectivity) that determine the current damage score of the state. Based on the maximum value of the score for each rule, each score can be normalized and combined with the other rules using a combination function such as weighted sum. In general, for R rules, each rule, r, will have a maximum possible value, mx_r, as well as a value, v_r, assigned to it. If we assign a weight, w_r, to each rule, we can calculate the weighted sum as:

$$d_i(t) = \sum_{r=0}^{R-1} W_r \frac{v_r}{mx_r}$$

The rules, criterion, and values are summarized in Table 1 which can be determined through various means applicable to each rule. For example, the firewall rule can utilize the firewall configuration to determine if certain IP addresses, protocols, and ports would be allowed. The attack type can be discerned by an alert aggregation tool that categorizes observables. The connectivity can be analyzed by using a known model of the network and routing tables to determine connectivity.

In this example, suppose a Reconnaissance observable is received and it reflects a penetration through the firewall to a connected target. In essence, this models a successful reconnaissance action on a target. Using a simple weighted sum, the aggregated asset damage score is as follows:

$$\begin{aligned} d_i(t) &= W_1 \left(\frac{1}{mx_1}\right) + W_2 \left(\frac{1}{mx_2}\right) + W_3 \left(\frac{1}{mx_3}\right) \\ &= (0.3) \left(\frac{1}{1}\right) + (0.3) \left(\frac{1}{3}\right) + (0.4) \left(\frac{1}{1}\right) = 0.8 \end{aligned}$$

In the equation above, it should be noted that each rule is normalized by its maximum value to obtain a final value on [0,1]. Also, while we have just presented the weighted sum as an example, any method to combine the values could be acceptable provided that the damage score calculations make sense for the rules.

4.2 Unbounded State Estimation

Recall that $d_i(t)$ must have a value on the [0,1] interval. However, some impact equations or rules do not have an upper bound. A simple example of this would be a score that increases by one every time a successful attack targets an OOI. In order to resolve this, one can simply define a maximum value corresponding to a value that is, for all intents and purposes, a value high enough that analysts would need to act upon. Once the maximum value is known, the damage score can be normalized into the [0,1] interval by saturating all values to a maximum then dividing by the maximum score. It should be noted that when exponential or special polynomial functions are used, the assessed damage scores can be close to the extreme values given the nonlinearity nature of the functions.

4.3 Single State-Space Estimation

This is a variant of the simple state estimation where, instead of the direct calculation, a qualitative state is determined then converted into a quantitative value. Such an approach provides a descriptive, language-based state for each OOI.

In cyber warfare one can consider the "Red" state space for an OOI, corresponding to the level of control or knowledge the red team may have on an OOI. To demonstrate this approach, consider the following five mutually exclusive states in the order of increasing severity:

- **Normal**—there is no indication of malicious or suspicious activity affecting the OOI. This is the default state for each OOI.
- **Attempted**—there has been at least one malicious or suspicious activity targeting the OOI, however, the targeted attacks have been unsuccessful.
- **Discovered**—there has been at least one successful attack targeting the OOI, however, the targeted attacks have only yielded information about the OOI.
- **Partially Compromised**—there have been successful attacks targeting the OOI that have given some control of it to the red team, however, the red team does not completely control the OOI.
- **Compromised**—there has been at least one successful attack targeting the OOI which has given the red team complete control of the OOI, thus is cannot be trusted by the blue team.

The determination of the qualitative state can be accomplished by establishing rules using various elements in the cyber environment such as, but not limited to, network connectivity, firewall rules, routing tables, sensor locations, and known vulnerabilities and services.

Once the state for an OOI has been determined, one can define a simple lookup table to calculate the damage score. An example lookup table is shown in Table 2. This example assumes a linear increase in damage score for each given state; however,

Table 2 Example states and damage scores for single state space estimation

State	$d_{i(t)}$
Normal	0.0
Attempted	0.25
Discovered	0.50
Partially compromised	0.75
Compromised	1.00

the values can be assigned logarithmically, polynomially, exponentially, or by some other method that adequately describes how the relative damage score increases:

4.4 Multiple State Space Estimation

This is an extension of single state-space estimation where an OOI can be evaluated across multiple state spaces. This method requires the combination of the $d_i(t)$'s for each state space. For example, suppose that, in addition to the Red State Space described above, there is an Operational State Space that defines the following four states: Operational ($d_i(t) = 0.0$), Maintenance (0.3), Degraded (0.7), Non-operational (1.0). It is therefore possible for an OOI to be in both the Degraded (0.7) and Partially Compromised (0.75) states simultaneously. Using this method, a new index, j, is added to the damage score, $d_{i,j}(t)$, to represent the state space the damage score is relevant to.

A combination function is needed to combine $d_{i,j}(t)'s$ into an overall damage score. Specifically, let J be the set of state-spaces, acombination function, $\bigoplus$, is such that $d_i(t) = \bigoplus_{j \in J}(d_{i,j}(t))$.

There are a few different combination functions that can now be considered:

4.4.1 Worst State

The simplest combination method is to use the worst state, which will be indicative of the highest damage score for a given state-space. This is defined by the max function:

$$d_i(t) = max_{j \in J}(d_{i,j}(t))$$

4.4.2 Weighted Sum

A different approach is to assign weights to each state space and combine them using a weighted sum. This approach only tends to make sense in a situation where some state-spaces may be more important than others when it comes to overall damage

Table 3 Example asset damage score combinations comparing DST with modified DST

Asset	$d_{i,1}(t)$	$d_{i,2}(t)$	$d_{i,3}(t)$	DST	Modified DST
1	0.1	0.2	0.3	0.001	0.3
2	0.1	0.1	0.8	0.047	0.8
3	0.1	0.7	0.8	0.509	0.903
4	0.9	0.7	0.8	0.988	0.988

and when all state spaces should have high damage scores to indicate a high impact.

$$d_i(t) = \sum_{j \in J} W_j d_{i,j}(t)$$

4.4.3 Modified Dempster-Shafer

Dempster-Shafer Theory (DST) [10] allows one to combine evidence from different sources to determine a degree of belief taking all pieces of evidence into account. DST uses a frame of discernment to define the possible states and mass functions (also referred to as basic probability assignment functions) to define the belief estimate from a given source. If we consider a source to be synonymous with a damage assessment for a given state space and a frame of discernment to be {D,N}, where D represents the belief that the asset is unable to perform its role ($D = d_{i,j}(t)$) and N represents the belief that the asset is able to perform its role ($N = 1 - d_{i,j}(t)$), we can use DST to combine the assessments into a singular assessment of the asset damage taking into account its damaged state for all state space.

One of the advantages of using DST to combine scores is that "high" scores can be combined into an even higher score. Likewise, "low" scores can be combined into an even lower score. This latter behavior is undesirable for calculating impact, so a piece-wise combination method can be used to combine the state spaces:

$$d_i(t) = \begin{cases} max_{j \in J}(d_{i,j}(t)) & , \textit{if all } d_{i,j}(t) < 0.5 \\ \bigotimes_{j \in J, d_{i,j}(t) \geq 0.5}(d_{i,j}(t)) & , \textit{if } \exists\ d_{i,j}(t) \geq 0.5, j \in J \end{cases}$$

With Dempster-Shafer a "high" score is greater than 0.5, and a "low" score is less than 0.5. As such, the above combination method uses the maximum score when all damage scores are low. This ensures that the final combined value will be at least as bad as the worst state. If at least one score is greater than 0.5, we use Dempster-Shafer combination to combine the high scores. This ensures that the combined score will be even higher than any single damage score.

Table 3 illustrates various combination scores for assets assessed across three different state spaces. It should be noted that in all cases, the modified DST score was at least as high as the DST score. Asset 1 had all low damage scores, and the traditional DST calculation drove the final score even lower. This is undesirable because at a minimum we want to consider the lowest value. The modified DST calculation accounts for this by simply taking the maximum of the numbers, which is 0.3. Also note the disparity in the damage scores for assets 2 and 3. Having at least one state space with a high damage score should be of concern, so we ignore the low estimates and only combine the high ones. Note that asset 4's combined damage score is higher than any single state space.

4.5 Considerations for Damage Degradation over Time

For general mission impact assessment, it may be sensible for the impact or damage to decrease simply due to time. An example of this is an asset who was damaged, but has not received any impact changes for such a long time that the impact or damage is no longer relevant. As such, it may make sense to define an aggregation function that linearly, exponentially, or logarithmically discounts the impact or damage as time progresses, then "resets" itself once an event occurs to change the score. For example, we can define the following linear function such that t' is the last time state estimation changed the damage score due to an event and λ is the rate at which the damage degrades:

$$d_i(t) = \max(0, d_i(t') - \lambda(t - t')d_i(t'))$$

However, this concept could potentially lead to false negatives in the cyber domain, where "low and slow" attacks are very common. In a low and slow attack, a hacker can sometimes wait months between attacks on the network. The idea is that even if sensors or analysts do pick up their activity, the events will be too far apart to correlate with each other. In addition, if a hacker has successfully setup a backdoor undetected, the backdoor may be there long enough that it is also present in backups and may continue to be present even if a server or host is restored from a backup if the attack is detected. As such, we recommend that each "high" damage score be taken into consideration and if it is determined to be a false positive, the state estimation software should be able to allow for manual overrides to reduce the damage to the correct level. Automatic degradation may unnecessarily discount the damage if it goes unresolved over time.

5 Mission Tree Methodology

A mission tree is a tree-structure that captures the relationships between missions and assets. The mission tree is currently implemented as the impact assessment capability for Future Situation and Impact Awareness (FuSIA) [11]. As mentioned in Sect. 2, a "mission" is generically defined as any sort of task that must be performed, and an "asset" is a resource required for the execution of at least one mission. The mission tree takes the damage scores calculated by state estimation and propagates them through the mission tree to provide mission impact estimates indicating the current health of a mission. The mission tree can also be used with various predictors or "what if scenario" interfaces to estimate future mission impact. The mission tree is intended to provide an estimate of mission impact, which is intended to trigger a deeper analysis for command and control activities. The intent of providing these estimates is to make more efficient use of an analyst's time since the impacts are a fast indicator of whether one or more events are adversely affecting a mission.

For example, a situation assessment tool such as FuSIA [11] will be able to suggest that a computer on a network has been compromised by a computer hacker using a Single State-Space Estimation. However, in order to determine how that impacts the other missions being executed, one must know not only how that computer supports each mission, but also be able to combine that information with the current state of other assets supporting that mission. In addition, that mission could be used to support other missions. The mission tree models these relationships and provides for various means of aggregating the data to provide a quick estimate of how far reaching the impacts are across other missions.

A static Mission Tree implementation would represent the state of the missions only at a specific point in time. In many applications, certain attacks may affect the mission more critical than others at different timeframes. This is even more challenging if new missions emerge over time, assets are re-assigned to assist in other missions, certain missions must be executed within a given timeframe, or the criticality of a mission or asset changes over time.

These dynamic and temporal changes must be taken into account in order to perform an accurate impact assessment in a timely and accurate manner. In Sect. 6, we will demonstrate how the mission tree is developed to incorporate dynamic and temporal mission changes through a fictional military computer network example.

Figure 1 shows the basic Mission Tree structure. A Mission Tree consists of three different types of nodes—assets, aggregations, and missions. Asset nodes must always be leaf nodes. An aggregation node is a node that performs a mathematical function to calculate the combined impact of all of the children nodes. Finally, the third node type is a mission node, which represents any type of task that needs to be executed either as the primary mission (the root node) or in support of another set of missions.

5.1 Asset Nodes

An asset node is defined by a 3-tuple (i,e,c) for a role r where i is the damage score for asset e in support of the parent mission with a criticality c. The criticality is used to describe the importance of a given node to the parent mission. An asset node must always be a leaf node and have an aggregation node as a parent.

Let $\mathbf{s}_{a,r}$ be a vector of probabilities such that $\mathbf{s}_{a,r}(\mathbf{i}) = \mathbf{p}_i$ where p_i is the probability that the function of the asset a is in state i. with respect to its role r. For example,

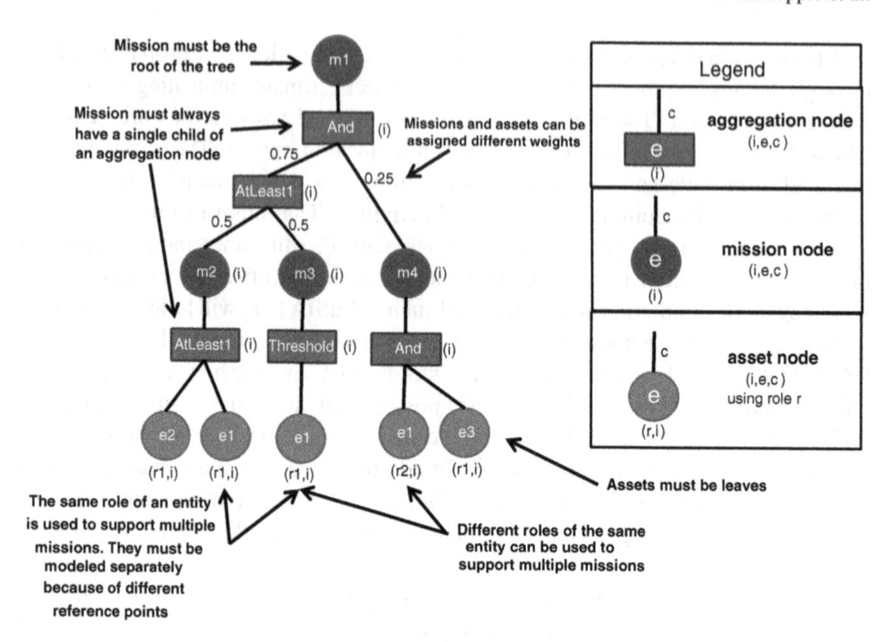

Fig. 1 Mission tree structure

$\boldsymbol{S_{a,r}} = (p_{normal}, p_{attempted}, p_{discovered}, p_{partially\_comp}, p_{comp})$ corresponds to the set probability that role r for asset a is in the Normal, Attempted, Discovered, Partially Compromised and Compromised states, respectively. $\boldsymbol{S_{a,r}}$ should be defined for every mutually exclusive set of states. If an asset only performs a single role, the r index may be removed.

Let $\boldsymbol{d}^*$ be a vector of the "numerical damage" for a given state. For example, we can define d such that $\boldsymbol{d}^* = (\boldsymbol{d}^*_{normal}, \boldsymbol{d}^*_{attempted}, \boldsymbol{d}^*_{discovered}, \boldsymbol{d}^*_{partially\_comp}, \boldsymbol{d}^*_{comp}) = (0, 0.2, 0.4, 0.6, 0.8)$. The $\boldsymbol{d}^*$ vector should be defined for every mutually exclusive set of states. We can therefore define $\boldsymbol{i} = \boldsymbol{d_{a,r}} = \boldsymbol{s_{a,r}} \bullet \boldsymbol{d}^*$ to be the damage score for an asset's role.

Using this definition, an entity can be represented across multiple missions by multiple asset nodes referencing the same entity. For example, if the network service "SSH" is an entity that supports three missions, three asset nodes referencing the SSH service are created. As a result, any change to the damage of the entity will trigger the recalculation of impact from each of these three asset nodes.

It is important to note that asset nodes merely *reference* the actual asset instance. So in essence, an asset node represents the damage score of the asset instance *with respect to* its parent mission. The reason for this distinction is to preserve a tree structure for an easier understanding of the impact score calculation process.

5.2 Mission Nodes

A mission node is also defined by a 3-tuple *(i,e,c)* such that *i* is the impact score for mission *e* in support of the parent mission with a criticality *c*. The criticality is used to describe the importance of a given node to the parent mission. Every mission node must contain a single parent (except for the root) and a single child aggregation node. For a mission node, *i= i(child aggregation node).*

Recall that we generically refer to a mission as any task that needs to be completed to accomplish a given goal. Organizations may use strict mission structures comprising of multiple layers of missions, sub-missions, tasks, etc. However, this structure is not consistent across all organizations. So, for the sake of the mission tree, each of these elements are considered to be a "mission" since they are functionally treated the same for impact calculations. As a result, the depth of the mission tree can vary between organizations. A mission type is assigned to each mission to allow the user to distinguish which mission, sub-mission, task, etc. is being modeled.

5.3 Aggregation Nodes

Along with the tree-structure, the aggregation node enables the mission tree to not only model various relationships between its child assets, missions, or other aggregation nodes, but to also "propagate" impacts up the mission tree. The aggregation node enables mission-asset relationships through the definition of various functions that model various asset behaviors. An aggregation function shares mathematical similarities to a combination function used for state estimation. However, the combination of state spaces typically utilizes a single function, whereas a mission may utilize multiple functions forming more complex relationships aggregated together in a way meaningful to the mission-asset relationships.

Assets may have redundant behavior for missions, so "at least 1" of them may need to be un-impacted for the parent mission to execute. Assets could also have complementary behavior, meaning that "all" assets must be functional for mission execution. The aggregation nodes also provide for "at least N" relationships as well as "threshold" nodes. These various relationships all help to provide for an intelligent propagation of impact scores "up" the mission tree based on the estimated damage to each asset. The use of multiple aggregation nodes also allows for more complex relationships to be modeled.

An aggregation node calculates the combined impact of all children nodes and is defined by a 3-tuple *(i,e,c)* where *e* is an aggregation function with criticality *c* and *i=f(e)*. Every aggregation node must contain a parent that is either a mission node or an aggregation node. While an aggregation function can be any type of function, we adopt Yager's aggregation functions [12] due to their flexibility in defining various logical and mathematical relationships.

Yager's aggregation functions use a weighting vector multiplied with a *sorted* vector to perform various mathematical functions, such as maximum, average, and minimum. Due to their flexibility in function definition, they were chosen as the primary calculation means for the aggregation functions.

Each aggregation node is defined by a vector of weights, w, and a sorted vector, v_s, for Yager's aggregation calculations. The sorted vector is a vector sorted in descending order of all i^*c(impact multiplied by criticality) values defined by each child. The dot-product of each vector yields the impact score, i, for the aggregation node.

True Maximum is defined such that $w = [1, 0, \ldots 0]$. A true maximum calculation is equivalent to an "And" relationship between all of the children. This indicates that all of the children have complementary roles and must be fully functional to complete the mission, so the mission is "only as strong as its weakest link."

Weighted maximum is defined such that $w = [z, |ch|/(z-1), |ch|/(z-1), \ldots]$, $z <= 1$, where z should be close to one and $|ch|$ is the number of children. This can be used when it is highly desirable for all children to be fully functional, but the mission isn't completely non-functional when a child goes down.

True Minimum is defined such that $w = [0, \ldots, 0, 1]$. A true minimum calculation is equivalent to an "Or" relationship between all of the children. This indicates that all of the children are performing redundant roles. Thus only a single child needs to be operational for the mission to execute.

Weighted Minimum is defined such that $w = [|ch|/(z-1), \ldots, |ch|/(z-1), z]$. This can be used when the degradation of the functionality of the child will minimally impact the mission.

Average is defined with $w = [1/|ch|, \ldots, 1/|ch|]$. The average aggregator essentially captures a "consensus" between the children. However, it has not been found to be applicable to impact assessment calculations, so the use of this aggregator is not recommended.

It is also possible to model "At-Least-N" relationships that capture the situation where a minimum number of children are required to complete a mission. The "At-Least-N" operator is defined such that $w = [0, \ldots, 0, w_1, \ldots, w_n]$ where $w_n > \ldots > w_1$ and n is the minimum number of objects that must be operational. These are the primary aggregation functions used for the mission tree, and are mostly applicable to asset damage scores calculated by Single State Estimation. However, other aggregation nodes may be chosen dependent upon the state estimation technique used. For example, an impact score that grows exponentially may be skewed towards the higher end of the numbers. One could design an aggregation node that better distributes the scores evenly between 0 and 1. Finally, the modified Dempster-Shafer combination discussed in Sect. 4.4.3) could also be used as an aggregation function. This will allow the impacts to "grow" if multiple assets are impacted, thus being a good function for non-redundant assets. So in essence, the aggregation nodes can be used not only to model asset relationships, but to also change state estimation values as deemed necessary. Table 4 shows two examples of "at-least-2" operators.

Table 4 Two examples for "At-Least-2" operations

Sorted input	0.9	0.9	0.1	
Weight	0	0.75	0.25	Output
	0	0.675	0.025	0.7
Sorted input	0.9	0.6	0.4	
Weight	0	0.75	0.25	Output
	0	0.45	0.1	0.55

It should be noted that each aggregation node is defined independent of the number of its children. As such, when children are added or removed from an aggregation node, the new calculation vectors are automatically recalculated.

The threshold aggregation node is an example of one not using OWA. It is also a special aggregation node in that it can contain only a single child. A threshold aggregation node is a simple function that defines a minimum threshold of an impact value before it should be propagated upwards. This function simply returns a value if it is greater than the threshold, but returns 0 otherwise. This function can be important depending on the state estimation techniques used for the assets. If the "low" values for a state estimation technique are not important enough to propagate through to the parent missions, a thresholding node may be used.

These are the primary aggregation functions used for the mission tree, and are mostly applicable to asset damage scores calculated by Single State Estimation. However, other aggregation nodes may be chosen dependent upon the state estimation technique used. For example, an impact score that grows exponentially may be skewed towards the higher end of the numbers. One could design an aggregation node that better distributes the scores evenly between 0 and 1. Finally, the modified Dempster-Shafer combination discussed in Sect. 4.4.3 could also be used as an aggregation function. This will allow the impacts to "grow" if multiple assets are impacted, thus being a good function for non-redundant assets.

So in essence, the aggregation nodes can be used not only to model asset relationships, but to also change state estimation values as deemed necessary.

5.4 Calculating Impacts

A tree structure was chosen to represent the mission relationships because it naturally lends itself to "bottom-up" calculations. The calculations are performed using an optimized depth-first traversal of the mission tree to eliminate unnecessary calculations. For example, if a single asset's damage has changed, it is only necessary to recalculate the nodes on that particular branch. In addition, if the value on a node of that branch does not change, it is not necessary to continue the calculation since the parent nodes will not have changed.

After all of the calculations are performed on the mission tree, each asset and mission node has an impact score calculated for it, which corresponds to the fused effect the child nodes have on that node's ability to perform its task.

Initially we will assume the calculations to be specific to a given time t. This will ensure that the data structure is static for the analysis. For a given set of events at time t, we have one or more state estimation algorithms to determine the states of different assets within the environment.

When the calculated value of a node changes from its previous assessment, its parent is notified of the change, at which time the parent recalculates its new impact score. These recursive calculations continue until a calculated value does not change. After all of the damage scores have been determined, the parent aggregation nodes for all nodes whose damage scores have changed are then recalculated.

Due to the structure of the mission tree, every mission node has at most 1 child, which is an aggregation node. The value of this aggregation node also represents the impact score for the mission node.

5.5 Handling Mission Changes

Recall that we initially assumed the calculations for the mission tree to be performed for a single point in time. While this assumption was initially made for simplicity, it may not be a practical assumption in most applications, especially in the cyber domain where the life cycle of a cyber operation can be short. Over time, missions can be added or removed, assets may only be available at certain times, and missions may only be able to be executed within a given timeframe. As such, the assumption of the Mission Tree being defined for a specific time t may not be valid as temporal considerations would need to be taken into account. Therefore, the mission tree must be able to evolve with these changes.

In this section we will consider the different types of changes the Mission Tree can accommodate and how they can be triggered.

5.5.1 Types of Mission Tree Changes

There are various ways in which a Mission Tree can change. At first glance, how to handle the changes may seem as obvious as changing a node or a value within the tree, but there are subtle considerations to properly and efficientlymake the changes. In this sub-section, we will describe the different ways a mission tree can be modified and the considerations when making the change.

Adding an Asset

A new asset is typically created when a new resource becomes available for a given mission. The physical addition of the asset is as simple as adding the node as a child to an aggregation node. However, the following must also be taken into account:
Criticality—Determine how critical the asset is to the mission, relative to the other assets available.
Relationship to other assets—Determine whether the asset performs the same role as other assets supporting the mission. If it does, it should be added as a child underneath an "Or" or "At-Least-N" aggregator to indicate that it is a redundant asset. If the asset performs a unique and critical role to the mission, it should be added as a child to an "And" aggregator to indicate that this is a required asset. This is critical in ensuring that the aggregation nodes are correctly defining the mission dependencies. The aggregation nodes may need to be re-worked if the asset introduces a new mission dependency.

Adding a Mission

The considerations for adding a mission are very similar to adding an asset. However, unlike asset nodes, mission nodes are not leaves, we must also consider the tree structure "below" the mission.
When adding a mission, one must also consider which assets support the mission and also how those assets work together to perform that mission. This analysis will form the necessary structure for the aggregation nodes to capture these relationships.

Removing an Asset or Mission

When an asset or mission is removed, all of its children are also removed. As a result, careful considering must be taken into account so as to not make existing missions unable to perform their tasks. The impact of such node removals can become immediately evident when they are removed from the mission tree, which can be a benefit to the mission planner to assess the potential problems with removing a mission or asset. The obvious indicators of problematic removals are aggregation nodes with no children. Such nodes indicate that there is a specific dependency that is no longer modeled – and thus may indicate that a mission would be unable to perform. In addition, due to "At-Least-N" aggregators, it is possible that an aggregation node may not have enough children. These potential problems can be brought to the attention of the analyst so that they can be resolved immediately.

Re-Assigning an Asset

As missions evolve, assets may need to be re-assigned for various reasons. As such, the re-assignment of an asset is necessary to model. However, it is simply equivalent to the removal and subsequent addition of the mission and/or asset.

5.5.2 Change Triggers

We define a change trigger, as an event that causes a change to the mission tree. When a change is triggered, a list of actions, A, is executed to change the mission tree. The list of available actions were described in Sect. 5.5.1. Triggers are a critical inclusion to the mission tree in order to handle any changes to the mission tree, whether they are unplanned or planned.

Functional Trigger—These are one-time changes to the mission tree. As business or a particular task evolves, new missions may need to be created or existing missions are deemed unnecessary. These triggers are typically manual changes that must be made in order to accommodate changes that did not have a predictable time at which they became effective.

Absolute Temporal Trigger—These are one-time changes that are triggered by a single point in time. These changes typically represent a predictable change to the mission tree, such as a deadline for a given mission. When deadlines for missions have occurred, the mission is permanently removed from the mission tree. In addition, known or planned tasks in support of a mission can be created at the given point in time.

Cyclical Trigger—These changes are characterized by a predictable and cyclical change in the mission definition. These changes are typically caused by a business cycle, where certain assets may be more critical during normal business hours. In addition, due to resource availability, assets may only be available within certain timeframe, so there are only specific periods of time at which the assets are able to affect the mission. These changes result in a cyclic change of the mission tree.

5.5.3 Managing Mission Tree Changes

Depending on how frequently the mission tree changes, one may need to consider the best way to manage these changes. For a forensic analysis and to also consider past, or future, events in a "what-if" analysis, the mission tree needs to be stored at various points in time. While each calculation is calculated for a given time t, we need to be able to quickly assess the estimate of the mission for various times as events occur. As such, the proper storage of the tree is critical. There are two ways in which we can store a Mission Tree:

Absolute—This is the most straightforward approach, but also the most data intensive. In this approach, the entire mission tree is logged any time a change occurs. The time at which the change is effective for is logged in the database. When a calculation

is needed for a given time, the entire mission tree is defined. This storage approach may make more sense in a mission tree that does not change frequently.
Relative—In this approach, the mission tree is stored in its entirety every so often. Every time a mission tree changes, instead of storing the entire mission tree, only the action is logged. When a calculation is needed for a given time, the closest complete mission tree is first queried, followed by actions that were logged for modifying that base tree into the tree applicable for that time.

This approach requires much less storage space, but also is slightly more inefficient as a list of changes must be processed to determine the actual mission tree. This approach is best suited to applications where the mission tree is frequently changing.

6 Mission Tree Example

The mission tree was originally developed as a domain-agnostic approach to assessing mission impact to one or multiple application domains. This definition enables the impact assessment of missions that span multiple domains. For example, ground assets can be assessed alongside cyber assets to create a combined mission impact estimate. However, each domain has a specific way to "map" the mission tree. In this section, we will discuss how the idea of a mission tree is applied to cyber networks, assets, and missions.

6.1 Input Data

As has been mentioned earlier, there has been a cornucopia of work focusing on the identification of various types of cyber attacks, which can be leveraged as inputs to the Mission Tree. The approaches applicable to provide input data to the Mission Tree must be able to:

1. Correlate multiple cyber events together to represent a single cyber attack.
2. Provide a generic categorization of each event and, ideally, whether the event was successful in its execution.
3. Identify one or more assets affected in some way by each event.

With these three capabilities, we can determine the damage to each asset on a Mission Tree, which will then enable us to calculate mission impacts. In this example we will use the Single State-Space Estimation technique with the following states:

1. Normal (0.0)—nothing abnormal has occurred on or to the asset, and the asset is functioning normally.
2. Attempted (0.25)—an abnormal event has occurred on or to the asset, but the event did not seem to have any negative effects on the asset.

3. Discovered (0.5)—the adversary has some knowledge of the asset or attributes thereof. However, the asset seems to be functioning normally.
4. Partially Compromised (0.75)—one or more roles on the asset are under the adversary's control. Therefore, while the asset may appear to be functioning properly, the asset cannot always be trusted.
5. Compromised (1.0)—the asset is assumed to be under complete control by the adversary, so we cannot trust the asset and will therefore have the greatest impact to the mission.

6.2 Mission Tree Example

In this section, we will create a dynamically changing mission tree for a simple computer network for a fictitious mission to be executed by a nation's military. Some of the missions can only be executed at night, while some missions can only be executed during the day. In this example, we will demonstrate how compromised assets can affect not only the current missions, but also missions requiring the asset in the future.

Our example assumes that the "night" missions must be executed between 18:00 and 05:00, while the "day" missions must be executed between 05:00 and 18:00. The time in between the missions is intended for rest and/or maintenance. In addition, the CommServer is shared between the missions, but during the night is only exclusively available to the night missions and exclusively available to the day missions during the day. In addition, during the day, one of the day missions requires the use of at least 2 of the available workstations. It has also been found that the sensors monitoring these workstations trigger many false positives for discovery attacks. So any value less than or equal to 0.5 is most likely a false positive and should not be taken into account. The night missions require the use of at least 2 of 3 available field units that communicate valuable information back to the soldiers. As such, these are important cyber assets that also need to be protected.

We can create a *static* mission tree as shown in Fig. 2. Note that immediately above the "At Least 2" aggregator for the workstations is a thresholding function. This is defined to filter out the false positives that are known to be caused by the sensors monitoring the network. As such, only values above 0.5 will ever propagate up to Day Mission 1. Also note that the CommServer is shared across all four missions. However, this is not entirely accurate given that anything that affects the CommServer during the day will only affect the day missions. We will be using a modified Dempster-Shafer calculation for each of the "And" nodes.

We can define two cyclic triggers to modify the mission tree to give an accurate assessment based on the intended availability of the CommServer.

At the respective times, each of the triggers perform their actions. Which allows for a more accurate assessment of mission impact. The two cyclic triggers fundamentally create two slightly different mission trees show in Figs. 3 and 4. It should be noted that any mission or aggregation node with no children will always be assumed to have

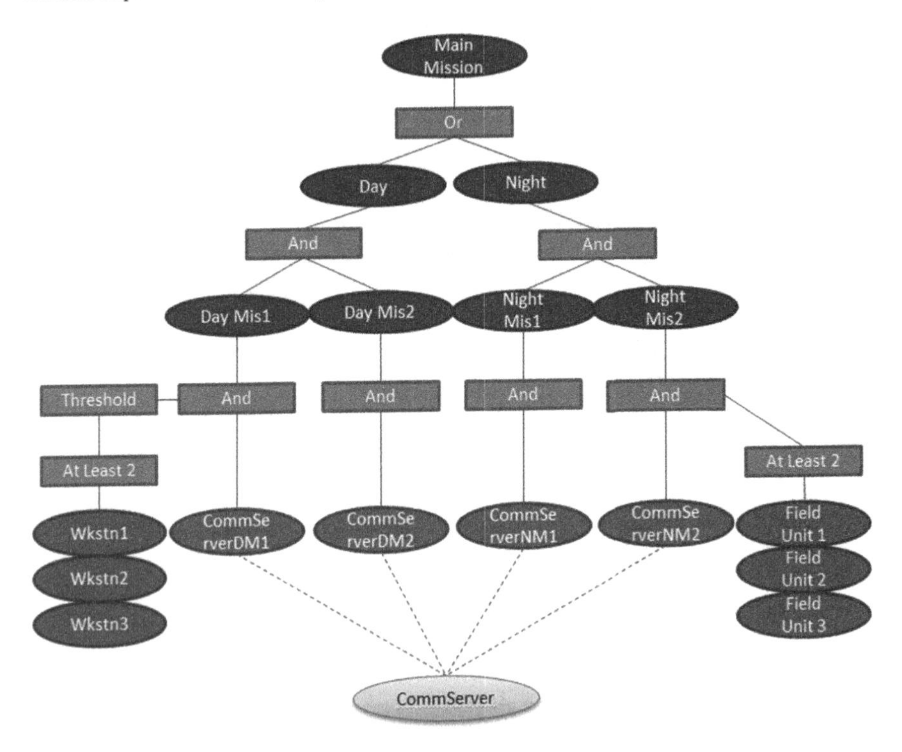

Fig. 2 Static mission tree for the basic day/night military operations

a zero impact. Also, the "And" aggregation nodes with singular children essentially just act as a pass through node. This minor detail is to ensure the data structure of the mission tree is maintained. It also allows for easier updates to the mission should any assets be added.

Figure 5 shows how the impact scores for each mission varied over time based on a set of cyber events defined in Table 5. Prior to the attack occurring, all missions had a 0.0 impact score. Note that due to the "at least 2" and thresholding aggregation nodes for the workstations, the impact to the workstations is not seen on Day Mission 1 until 13:00. It is important to note, however, that this does not imply that the events prior to 13:00 would have been ignored. It simply implies that it was not until 13:00 that the mission was negatively affected by the actions. In a realistic environment, the workstations would be monitored and the mitigation of workstation 1 should have immediately started when it was compromised. The different levels of the mission tree can be monitored by different people at the same time to try to prevent future events from causing higher level impacts.

By 15:00, the attacks to the workstations were resolved, so the impact decreases to 0. However, the impact spikes again when the CommServer is compromised. Note, however, that due to the dynamic nature of the missions, the night missions are not

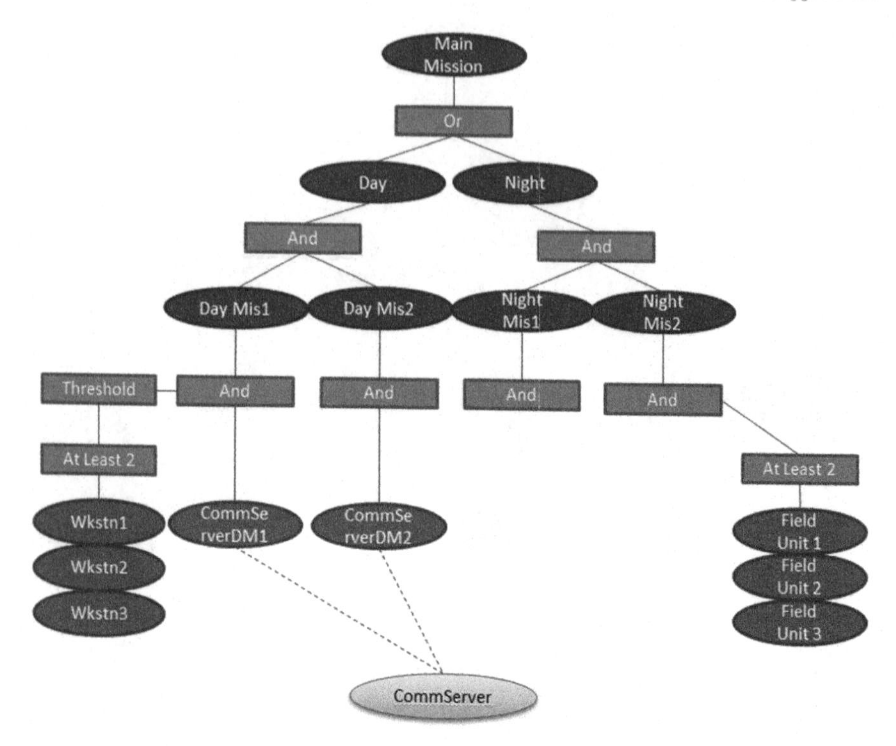

Fig. 3 Mission tree effective 8:00 M-F

yet affected. If the CommServer's compromise was mitigated prior to 18:00, there would have been no impact to the night missions.

6.3 Cyber Security Mission Tree Considerations

The previous section gives a simple example of a dynamically changing mission tree associated with a relatively small-scale computer network. The example demonstrates how a mission tree can be used to not only determine current impacts but also future impacts. Realistically, however, a typical enterprise network is comprised of hundreds, if not thousands of physical and virtual assets, so the practicality of the mission tree in such environments needs to be discussed.

Tools such as Snort® [13] or a more comprehensive tool such as HP's® Network Management Center [14] will be able to identify the states of assets which can then be fed to a state estimation process such as the simple state estimation process provided by FuSIA [11]. Each of these tools provides an initial data reduction to the point that the assessments indicate the damage to computers and their services.

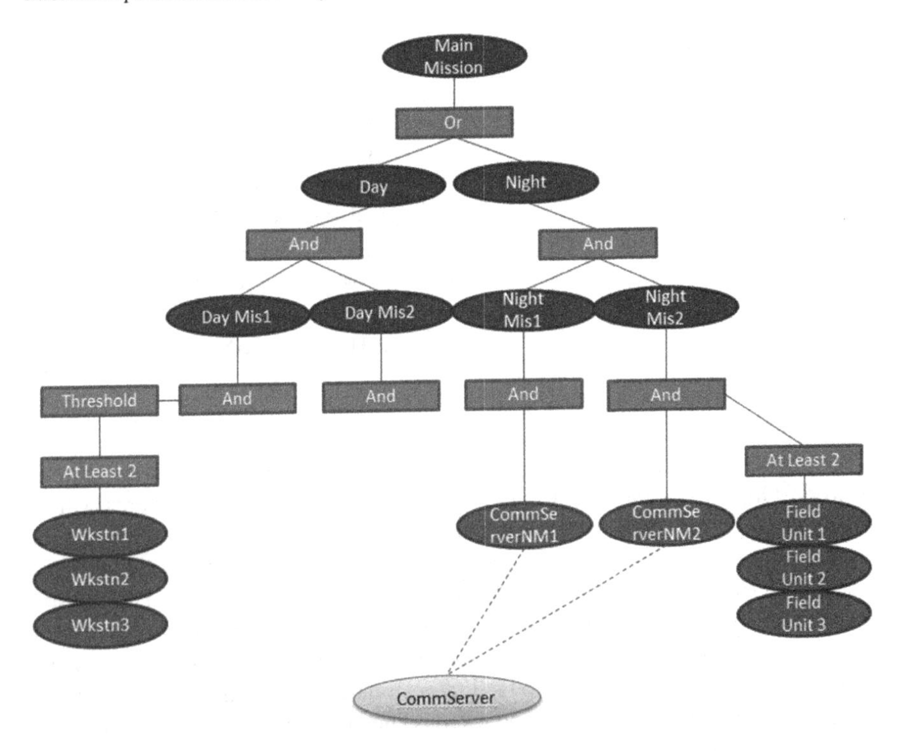

Fig. 4 Mission tree effective 18:00 M-F

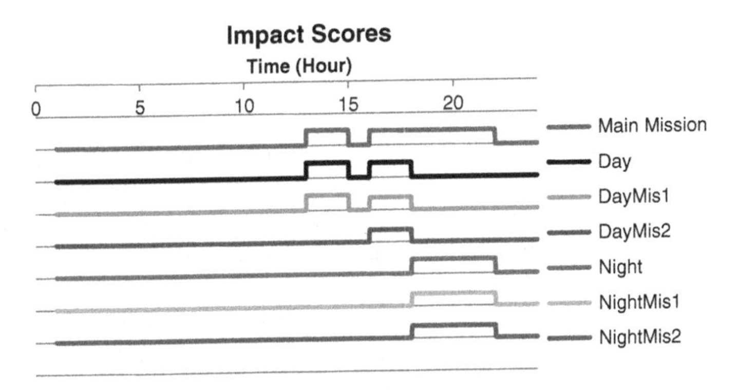

Fig. 5 Mission impacts esultant from Hacker attack over time

When modeling a mission tree for cyber warfare, an understanding of the computer network (e.g., the environment) is necessary for an adequate impact assessment. In addition, careful consideration must be taken into account to ensure that the granularity of the defined assets and roles is sufficient for accurate estimates, however, not so detailed that the provided assessments become unwieldy to manage.

Table 5 Triggers for mission tree

Trigger type	Time	Action list
Cyclic	8:00	Remove commServerNM1 Remove commServerNM2 Add commServerDM1 Add commServerDM2
Cyclic	18:00	Remove commServerDM1 Remove commServerDM2 Add commServerNM1 Add commServerNM2

Fundamentally, each analyst is looking to protect certain aspects of a computer network. Given the rapidly shifting focus to service-oriented architectures, whose resources can span one or more physical computers, we recommend modeling the services themselves as the assets—not the physical computers they reside on. Since many services can be provided redundantly, that can easily be modeled in the mission tree through the use of the "Or" aggregator where-ever that specific service is needed. Pre-processing of the events as described in the previous sub-section and knowledge of the computer network can help to facilitate estimates of impacts to specific services if only the physical computer being attacked is known. To further improve scalability, similar assets can be merged into a single asset to reduce complexity. For example, workstations (and their subsequent services) used by people of the same working group are often created from the same software image. Since each workstation would essentially be able to perform the same critical services, one could merge all of the individual workstation services into a single service on the mission tree.

If the "cyber missions" are ill-defined, the actual creation of the mission tree is not necessarily a trivial process. We recommend first building the top of the tree to mimic the structure of the organization, then creating specific tasks for each department and determining which services are necessary for the successful execution of these tasks.

It should also be noted that due to how generically the mission tree is defined, non-cyber assets can also be included on the mission tree. Although, of course, the assessment of those assets would require a different suite of situation assessment tools.

Using these considerations, a useful cyber mission tree can be developed that will allow for the quick estimate of mission impacts.

7 Challenges in Formal Evaluation of Mission Impact Assessment for Cyber Warfare

The previous sections presented a methodology and an example for estimating mission impact. There are, however, numerous challenges in practice to formally design an experiment and/or scenarios to evaluate the utility of this approach. This

Table 6 Cyber events

Time	Event	Damage
10:00	Discovery of Wkstn1	d(wkstn1)=0.5
11:00	Discovery of Wkstn2	d(wkstn2)=0.5
12:00	Compromise of Wkstn1	d(wkstn1)=1.0
13:00	Partial Compromise of Wkstn2	d(wkstn2)=0.75
15:00	Finished Mitigation of Wkstn1 & Wkstn2	d(wkstn1)=0.0 d(wkstn2)=0.0
16:00	Partial Compromise of CommServer	d(CommServerDM1)=0.75 d(CommServerDM2)=0.75
22:00	Finished Mitigation of CommServer	d(CommServerDM1)=0.75 d(CommServerDM2)=0.75

section details these challenges and suggests potential solutions so that one can work towards a better ability to not only evaluate the utility of mission impact assessment approaches, but also compare mission impact assessment approaches against each other.

7.1 Lack of Mission and Network Detail in Existing Cyber Warfare Scenarios

Perhaps the biggest challenge in evaluating mission impact assessment approaches is the lack of appropriate data. While there are some publicly available datasets to evaluate cyber warfare products, none of them currently provide any mission detail. This makes defining even somewhat accurate mission trees (and more specifically the aggregation nodes defining the relationships) impossible.Grimaila and Fortson [3] also argue that one of the limiting factors to perform damage assessment is due to the lack of documentation.

Even when it is available, the typical system information of an enterprise network is insufficient for accurate state estimation. For example, most data sets may provide a simple network diagram and a general list of services (not including version numbers). Most also do not include firewall or routing configurations. As such, state estimation may generate a large number of false positives, leading to poor perceived performance of the system—when in reality the poor performance is due to poor modeling.

Cyber warfare scenarios should include more detailed information about the computer networks, including (but not limited to) firewall rules, routing configurations, sensor placement, and as detailed of a list of services (including their versions) as possible. This information will significantly help the process of automated state estimation, thus leading to more accurate inputs to mission impact assessment. In addition, a detailed mission description should be included so that the user may better understand the goals for each part of the network. In an ideal scenario, the mission description would be formulated in a graphical manner, however, at a minimum a textual description would be useful to define an appropriate mission model.

7.2 Deviates Significantly from Current Cyber Defense Approaches

Cyber warfare analysts are generally concerned with the current state of the protected network. As soon as they identify a potential threat to any asset on the network, they seek to mitigate it. However, cyber attacks often have effects beyond just the computer network. These effects are often not known at the time of the attack and require extensive forensic analysis to determine.

Defining the necessary impact assessment models such as the mission tree or the state estimation rules may not be a trivial process given that it requires more information than most organizations currently document and diagram. While diagrams and other documents may be available for a network, many details are often just known by the IT personnel or discovered on-demand, since the models described this chapter are not necessary to maintain a reliable computer network.

Mission impact assessment could aid in the forensic analysis by providing quick estimates of impacts to other missions that would help to better define a starting point for a more detailed analysis, while ignoring the less likely entities to have been impacted. In addition, mission impact assessment could also provide a similar starting point for the mitigation strategy.

In order to alleviate this difficulty, automated discovery tools must be used to help develop the models necessary for state estimation and impact assessment. While there are cyber defense packages available that integrate discovery tools together to gather some of the required information, they are often cost-prohibitive for many smaller and medium-sized organizations. In addition, each organization would also need to document and diagram their mission structure(s) to allow for the development of a useful mission tree.

7.3 "Ground Truth" for Mission Impact Cannot Be Defined Numerically

Even if a scenario provided sufficient enough detail for accurate state estimation and the development of the mission tree, defining ground truth for the mission tree outputs is relative and arbitrary. How is the ground truth defined for the impact scores? Would distance metrics potentially skew results for approaches that for all intents and purposes had equal performance (i.e., if ground truth defines an impact of 0.8, and mission impact tools provide scores of 0.82, 0.76, and 0.99, are any of the scores really that much "better" than the others)? In addition, as evidenced in the simple scenario described above, the impact can vary significantly with time—even if no malicious or defense actions are taken in that period of time.

The nature of impact assessment often tends more towards educated opinion than facts, so defining ground truth for the purposes of collecting measures of performance may not be feasible due to the inability to quantitatively define the necessary values. As such, this problem is not well-suited towards any sort of formal "benchmarking", but there are other approaches focusing on the utility to the analyst that could be performed.

7.4 No Available Measures of Performance

Given that there is no ground truth that could be available for evaluating mission impact approaches, it is necessary to design an experiment for analysts identifying the threats and effects to the network. Such an experiment would require equally skilled analysts protecting the same network using different tools. Measures of effectiveness could then be collected to determine various metrics like attack response time and attack mitigation time.

8 Conclusion

In this chapter we have reviewed mission impact assessment for cyber warfare. Mission impact assessment can fundamentally be broken down into two processes: state estimation and impact assessment. The state estimation provides a damage score for each asset, and can be performed in multiple ways. The impact assessment provides quick estimates of mission health which can be calculated using a Mission Tree. Due to the dynamic nature of mission planning, the Mission Tree is also capable of dynamically changing over time to account for various mission changes. A simple military example was presented to discuss the potential utility for mission impact assessment. However, there still remain significant challenges in formalizing and deploying cyber warfare mission impact assessment processes due to the available technology and current network documentation processes.

References

1. Endsley, Mica R.: Toward a theory of situation awareness in dynamic systems. Hum. Factors J. **37**(1), 32–64 (1995)
2. Salerno, J.: Measuring situation assessment performance through the activities of interest score. In: Proceedings of the 11th International Conference on Information Fusion (2008)

3. Grimaila, M., Fortson, L.: Towards an information asset-based defensive cyber damage assessment process. In: Proceedings of the 2007 IEEE Symposium on Computational Intelligence in Security and Defense Applications, (2007)
4. Muccio, S., Kropa, B.: Cyber Mission Assurance", http://www.wpafb.af.mil/shared/media/document/AFD-110516-046.pdf
5. Ben-Bassat, M., Freedy, A.: Knowledge requirements and management in expert decision support systems for (Military) situation assessment. IEEE Trans. Syst. Man Cybern **12**(4): 479–490 (1982)
6. Musman, S., Temin, A., Tanner, M., Fox, D., Pridemore, B.: Evaluating the impact of cyber attacks on missions. Mitre Corp. http://www.mitre.org/work/tech_papers/2010/09_4577/09_4577.pdf
7. D'Amico, A., Buchanan, L., Goodall, J.: Mission impact of cyber events: scenarios and ontology to express the relationships between cyber assets, missions, and users. In: Proceedings of 5th International Conference on Information Warfare and Security, Wright-Patterson Air Force Base, OH (2010)
8. Jakobsen, G.: Mission cyber security situation assessment using impact dependency graphs. In: Proceedings of the 14th International Conference on Information Fusion, (2011)
9. Yang, S.J., Stotz, A., Holsopple, J., Sudit, M., Kuhl, M.: High level information fusion for tracking and projection of multistage cyber attacks. Elsevier Int J Infor Fusion, Spec Issue High-level Inf Fusion Situation Awareness **10**(1), 107–121 (2009)
10. Shafer, G.: A Mathematical Theory of Evidence. Princeton University Press, Princeton (1976)
11. Holsopple, J., Yang, S.J.: FuSIA: future situation and impact awareness. In: Proceedings of the 11th ISIF/IEEE International Conference on Information Fusion, Cologne, Germany, (2008)
12. Yager, R.R.: Generalized OWA aggregation operators. Fuzzy Optim. Decis. Making **2**, 93–107 (2004)
13. Snort®, http://www.snort.org
14. HP® Network Management Center, http://www.hpenterprisesecurity.com/

Uncertainty Modeling: The Computational Economists' View on Cyberwarfare

Suchitra Abel

Abstract The current research scenario shows considerable work on the fundamental considerations for Cybersecurity. The physical world will fuse with the digital world in the future through enhanced technologies. However, there still exists the problem of radical uncertainty, particularly in the form of information theft. In this project we provide an analysis of the critical factors affecting the security of internet-based businesses; we also present a casual model-based security system that affects and helps the central characteristics of contemporary internet-based businesses.

1 Introduction

The internet, including the businesses that operate via Internet, can be perceived as a contingent commodity market. However, the economy has only imperfect solutions for situations where information purchase or theft for the sake of making explosives, for example, is done via the Internet. In this chapter, I will present a re-formulation of the traditional Economists' viewpoint that can be applied to modern Cyberwarfare. This is a systematic expression since the traditional Economists could not have thought of the current situation of Cyberwarfare. I will show how intelligent systems can tackle this problem and deal with the different parameters effectively.

We have to be prepared to increase the internet-based business community's awareness of our efforts in the form of programs designed to prevent crimes. Our work here will make it possible for companies to have emergency escalation procedures, mass notifications, and supporting systems.

S. Abel (✉)
Department of Computer Engineering, Santa Clara University, Santa Clara, CA 95053, USA
e-mail: sabel@cse.scu.edu

R.R. Yager et al. (eds.), *Intelligent Methods for Cyber Warfare*,
Studies in Computational Intelligence 563, DOI 10.1007/978-3-319-08624-8_12

2 Background of Research and Brief Literature Survey

Businesses are often driven by their need to maximize their utility, thus influencing their policies and decisions according to that need. Researchers have developed economic and mathematical models that explore numerous aspects of businesses. In the context of this concern, I hereby present an uncertainty model that will be effective in advancing a method that assists such businesses.

There are researchers who facilitated the development of the foundations of our current research on modeling for CyberWarfare. Cartwright [1, 2] and Fine [3] have produced some of the classics. In more recent times, Pearl [4–6] has been researching about causal models and structural models that utilize probabilistic logic. These researchers have provided the background and the inspiration behind the current work.

There are also practical problems of Cyber-threat that arise with companies like Adobe and Microsoft. Adobe has recently released security updates for Adobe Flash Player to address multiple vulnerabilities. Adobe has also released security updates for Adobe Reader and earlier versions for Windows and Macintosh, in order to address multiple exposures. These susceptibilities could cause a crash and potentially allow an attacker to take control of an affected system [7].

Microsoft has released updates to address vulnerabilities in Microsoft Windows, Microsoft Office, Internet Explorer, and Microsoft Server Software.These weaknesses could allow increasing code execution, elevation of privilege, denial of service, or information exposé [8].

These problems can inspire one to do further research on finding a solution of threat identification and consequent engagement. The following section is concerned with these practical aspects.

3 The Cyberwarfare Scenario

There is a rational need for uncertainty models specially targeted towards cloud computing and mobile cloud computing. In general, Cloud computing enables convenient, on-demand network access to a shared pool of configurable computing resources (such as networks, servers, storage, applications, and services) that can be rapidly provisioned and released with minimal management effort or service provider interaction. There are businesses that rent Cloud services from Amazon [9], Google [10]. The businesses that rent Cloud services often have Research Divisions working on their security problems, inviting articles from outside the company too [11, 12].

Mobile Cloud security is another scenario that is becoming important. Smart phones, tablets, and cloud computing are converging in the new, briskly growing field of mobile cloud computing. In less than four years, there will be 1 trillion cloud-ready devices. One should learn about the devices (smart phones, tablets, Wi-Fi sensors, etc.), the trends (more flexible application development, changing work arrangements, etc.), the issues (device resource poverty, latency/bandwidth, security, etc.), and the enabling technologies that come along with a mobile cloud environment [13].

Companies like Nokia and Microsoft are interested in Cybersecurity issues. The author provided an invited talk to the Nokia Research Lab, on this topic [14].

There is also emphasis, in the current research world, on finding a solution to Cyberwarfare, in the domain of internet-based businesses *in general*. The focus of the current chapter is not on overall mobile cloud security or even overall cloud security. Even without taking direction towards the line of cloud security or mobile security, there are many general issues concerning the computational Economists' view on Cyberwarfare, geared towards internet-based business in general, that are worthy of discussion. Alarming and often intimidating Cyber-attacks on internet-based business have reached an all-time high. Cybercrime is costing corporations this year, much more than last year. The statistics are in accordance with definitions used by the Department of Homeland Security, which confirms that there is a significant emergency or a dangerous situation involving an immediate peril [15].

The overall discussion of security measures using Bayesian modeling is certainly worth researching into. The next section shows that though the traditional Economists handled uncertainty in the context of businesses, they could not anticipate the complexities of modern businesses, for example, internet-based businesses that are open to the public all over the world, and are vulnerable to Cyber-attacks.

4 Traditional Economists: How they Would have Handled this Problem

This section expounds the traditional Economist's method of handling uncertainty in the context of businesses, and shows the shortcomings of such a method. In this model, the Internet site owners do bear the risks of misusing their proprietary information; they need to use subjective probabilities in determining their structures. It is decentralized decision making. There are administrative rules, legal rules (for example, no insider trading), etc. Of course, price of the commodity plays a vital role in who is acquiring the products. The buyers and the sellers do not have to know each other. The concept of free market does not mean the absence of rules, but how the rules ensure their freedom, in the highly competitive economy.

The traditional way of approaching this problem is to pursue standard mathematical methods, such as formulation of utility functions. The likely arguments of a typical Internet business's utility function, u, are its overall assets, a, the regular purchases for peaceful purposes, p, and the individual actions, *Ind*, of the company in trying to be persistent with such purchases. The business's utility function has the memorable Von Neumann-Morgenstern properties [16]. It empowers them to formulate preferences on all the arguments of their utility function [17, 18].

The scheme to prevent unexpected disruption-causing actions and to carry on typical purchases is the payoff function called g(S). This scheme can be classified according to their fundamental characteristics in the following spheres: first, one has to consider the region of the presence or absence of individual choice. There is individual choice if the individual actions, *Ind*, that compose a challenging argument in the Utility function. This might represent a level of investment. Next, there is the region of sequencing of moves between individual business's actions and customer actions. Lastly, one must not overlook the information, monitored by the scheme to prevent the disturbance. This information state, which may be a vector, is a function of the act of people who might interrupt the normal activities, and the actions of the individual business.

The scheme functions by establishing the payoff—financial gain of the individual business if the customers lead to legitimate business that one wants.

This also depends on the monitoring of the information state, such as: are the incoming customers authentic, or do they have the possibility to be disorderly?

There should be a break-even fiscal anticipation, for example, that the interference does not really halt the business. The scheme is to maximize the individual business's expected utility subject to the constraint.

We consider first a simple situation in which there is no room for individual choice.

Suppose that there is no individual choice to intervene. The scheme already devised by the business is the one that works—and it monitors the customers' activities. In this case, the sole determinant of the individual business' utility is the uncertainty regarding the customers' actions, p. These are obtained from standardized data retrieval about such actions. The distribution of p is given by the disruptive actions' density function f (p). The notion of p can be treated as continuous or as discrete.

The scheme to deal with customer actions p monitors the possibilities of p. In this case, the information state $S = \{p\}$. The scheme gives the individual business a payoff, g (p). This payoff, added to the initial assets of the particular business, called a_0, gives the total current a, argument of its utility function. In a purely numerical work, the individual business' expected utility under this scheme will be

$$\int \mathrm{u}\,(a_0 + \mathrm{g(p)}, \mathrm{p})\, \mathrm{f(p)}\, \mathrm{dp} \tag{1}$$

It is to be interpreted as the integration of the utility u with the two arguments, payoff added to the initial assets and the customer actions, and together with the disruptive actions' density function, f (p); this provides the expected utility.

The "dp" term comes from the following: the function f(p) is continuous for $a \leq p \leq b$. The interval from a to b can be divided into n equal subdivisions, each of width Δp, so that $\Delta p = (b - a)/n$. The "dp" in the integral comes from the factor Δp.

The break-even constraint for this scheme is

$$\int \mathrm{g(p)}\, \mathrm{f(u)}\, \mathrm{d(u)} = 0 \tag{2}$$

This is the constraint, which should be obeyed, in order to maximize the individual business's expected utility. The "d (u)" term, with respect to the utility u, plays a role that is similar to the dp term in Eq. 1.

The scheme‘s objective is to maximize (1) with respect to (2). The scheme can employ the calculus of variations (calculating the maxima or minima of functional, which are often stationary). The business can employ the calculus of variations to derive the marginal efficiency condition for the optimal payoff function [19].

5 An Intelligent System to Address Critical Cases of Radical Uncertainty

5.1 Description of an Intelligent System

The model described in the previous section will not succeed in the case of radical uncertainty, since either there is not enough information available to use it as a parameter in a utility function, or its value is close to impossible to decipher. One can do immediate data analysis to give it some initial weight, but it really has no place in a calculus of variations. Instead of such calculations, we provide an AI based causal network, a solution that is well-suited to realizing the objective.

Bayesian causal networks represent independence (and dependence) relationships between variables. Thus, the links represent conditional relationships in the probabilistic sense.

My proposed system does not depend on the representative agent abstraction. There is no single type of consumer, nor is there a single type of economist who is analyzing the economy. Classically, models are used to generate quantitative statements. But the aggregate variables of a system can number up to hundreds, and the "representative consumer" or "representative economist" should be replaced by each economist/user of the system being represented as an individual.

For radical uncertainty, only immediately available knowledge can be used, and showing causal connections is critical. The cornerstone of our system is a causal model; such models are a system of processes that can account for the generation of the observed data. The ordering presented in the model respects the direction of time and causation. The judgments required in the construction of the model are meaningful, accessible and reliable. For example, we can assert that taking actions against the threat is independent of normal users accessing the site; we can translate this assertion into one involving causal relationships, once we know that the influence of normal business practices is mediated by the threat of the potential explosives-makers accessing the site. Dependencies that are not supported by causal links are spurious.

Conditional independence relationships are byproducts of stored causal relationships. So, representing these relationships directly would be a reliable way of expressing what we know about radical explosives-makers or material-purchasers.

5.2 Advantages of Bayesian Networks

An important point about building Bayesian networks on causal relationships is the ability to represent and respond to external or spontaneous changes, for example, sudden explosives-making purchase threats. Any local configuration of the mechanisms in the environment can be translated with only minor modification, into an isomorphic reconfiguration of the network topology. The use of causal relationships allows us to define the characteristics for the network topology.

As an example, suppose that in the process of doing normal business operations, suddenly the business schemes suspect an explosives maker's purchase threat. In this case, new nodes concerning suspected threat appear, with time stamp (before that, within a certain time period, normal purchases were completed and recorded). The previous nodes were connected to links; but now, when the abnormal nodes appear, we delete from the network all links incident to the node and its causal connections.

To represent the policy of *not* selling to this threat, we add necessary links and revise

P (buyers-nodes | requirement-nodes for purchase from this company).

Such changes would require much greater remodeling efforts if the network were not constructed in the causal direction but just having an associational order. This remodeling elasticity is the component that enables the agent to manage novel situations instantaneously.

It is quite conceivable to change certain node relationships without changing others. There is a modular configuration that permits one to deal with the effect of external interventions. The causal models are more informative than plain probability models. A joint distribution tells us how probable events are and how probabilities would change with subsequent observations. Causal models also tell us how these probabilities would change as a result of external interventions. Such changes cannot be deduced from a joint distribution, even if fully quantified.

Ideally, in the process of modeling, we need modularity. This is the ability of being made up of separate modules that can be rearranged, replaced, combined, or interchanged easily. The connection between modularity and involvements that are interventions is specified here. Instead of stating a new probability function for each of the many possible interventions, we indicate merely the immediate change implied by the intervention. We come to know the identity of the mechanism altered by the intervention, and the nature of the intervention.

A Bayesian network, in general, is a transporter of conditional independence relationships along the order of construction. The following product showing the distribution is:

$$P(x_1 \ldots x_n) = \pi\, P(x_i/pa_i) \tag{3}$$

pa_i are the select group of predecessors of x_i. The x's stand for the company components.

We can adjust this product's relevant factors and use the modified product to compute a new probability function.

If we have a distribution P defined on n discrete variables, ordered as $x_1, x_2, x_3 \ldots$ x_n, then, utilizing the chain rule of probability calculus, we can decompose P as the product of n conditional distributions.

Suppose that the group of x's is independent of all other predecessors once we know the value of a select group of predecessors called pa_j. Then one can write:

$$P(x_1 \ldots x_{j-1}) = P(x_j / pa_j) \tag{4}$$

This will considerably simplify the input information required. We need only the possible realizations of the set pa_j. This is a minimal set of predecessors of x_j that is sufficient for determining the probability of x_j.

5.3 *Causal Network Models*

We will examine how the sequencing of moves and the information state, described in the previous section, interact in the determination of optimal schemes. First, let us consider a general case displaying how a business works with the information state and exerts its choice based on the sequences of moves. This is a case in which a certain information state is used to increase the possibility of business without disruption (desired result) by the sequencing of moves, but may also have direct effect on the business, both beneficial and adverse.

Suppose that we wish to assess the total effects of the information state on the desired result, when the following factors are present: (a) controlled experiments are not feasible, since the individual businesses insist on deciding for themselves which scheme to use (b) the business's choice of schemes depends on the previously gathered sequence of moves, a quantity though not totally known (obtained by data mining and other forms of data analysis), but known to be correlated with the current sequence of moves.

Let Seq-Moves-Before-Choice and Seq-Moves-after-Choice be the following: the first is the quantity (sequence of moves) before the individual business exerted its choice. The second is the quantity after the individual business exerted its choice. One can assign *probabilities* of the total effect of the information state on the desired result, based on the causal model. The subsequent diagram (labeled as Fig. 1) demonstrates this process.

In order to build a complete picture, we have to note that a business needs at least the following information: (1) Initial assets or products data (including numbers and prices). Let us call this x_1. (2) Demand appraisal that it needs to do; this is called x_2. Consequently, the business has to actually perform the act of sale to customers, called x_3. As a result of sales, the business will have profits, called x_4.

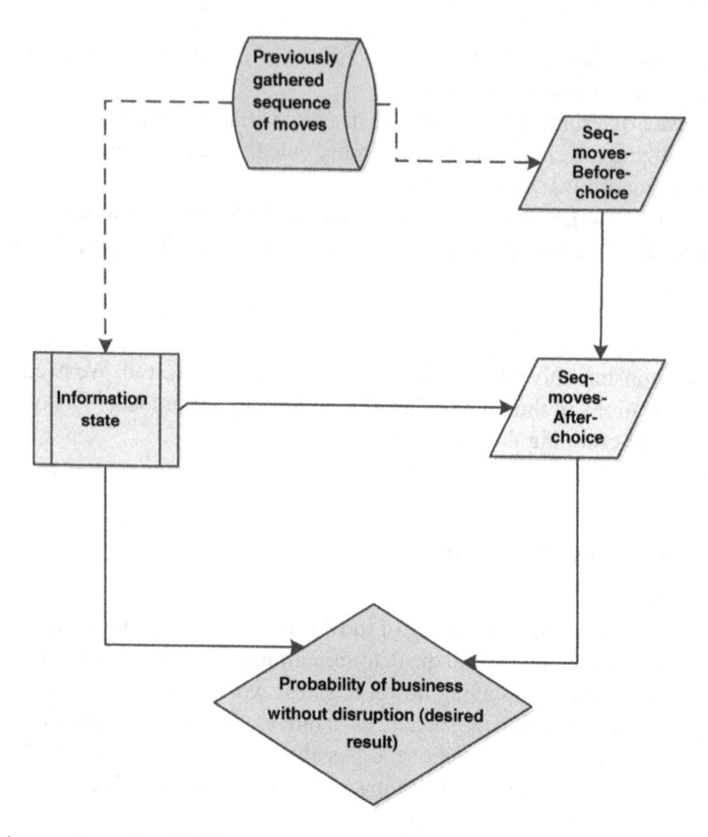

Fig. 1 A general causal model diagram showing the effect of the information state and the sequence of moves on the business

Therefore, next, let us draw a simple causal model, by constructing a directed acyclic graph (presented as Fig. 2). Suppose we know that two variables are dependent, data and demand appraisal (x_1 and x_2). In the case of suspected intervention, the arrows between x_2, x_3 and x_4 are removed, and the joint distribution also changes, leading to actions against the threat. y_1 through y_n are possible causal connections, with probability, of possible threats under radical uncertainty. (This is presented as Fig. 3).

As implied by our prior discussion, the principal concern in this chapter is to examine how the sequencing of moves and the information state interact in the determination of optimal business schemes.

In general, there often exist a set of schemes, implemented by a business, ensuring that the business is carried on, that is, that there are proper customers. This also includes the set of schemes to prevent the failure (built in by the business); the schemes ensure that the mechanisms are properly achieved, for example, by credit card monitoring, noting the buyer's involvement in the social media, etc.

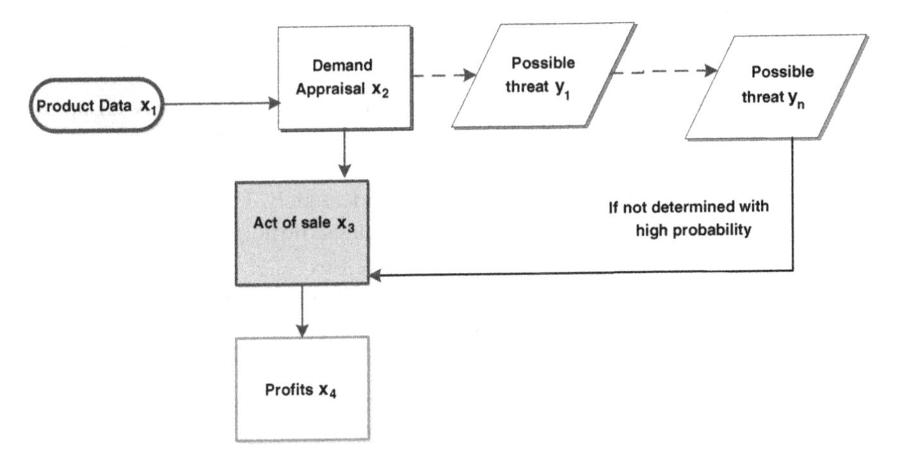

Fig. 2 Causal network model for uncertainty where the act of sale is completed

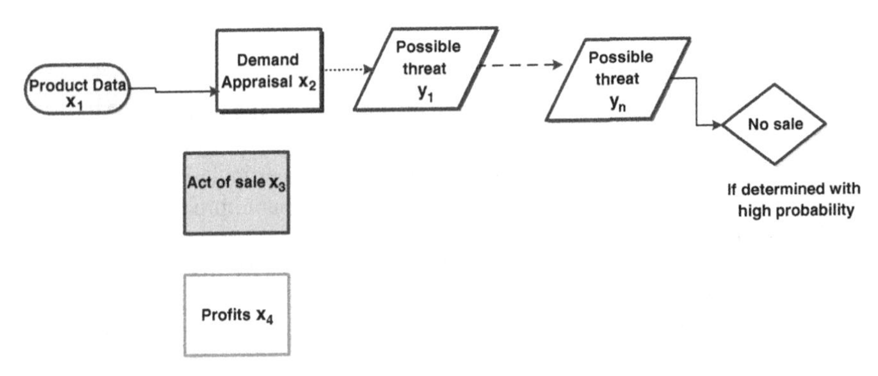

Fig. 3 Causal network model for uncertainty where threat is great, and therefore, no sale is generated

Next, there have to be, and indeed, there are, authentic internet based businesses. They might be, for example, businesses that supply materials for chemistry purposes.

What are the types of customers that the internet-based business has? There are non-disruptive customers who are using the businesses for peaceful purposes (the system might have some uncertainty about them). However, there are also distraction-causing customers or thieves—those that crash the system (they may be explosives makers). These cases cause radical uncertainty. They can be people or groups who are suspected of using these business websites to obtain material for warfare. Thus, there may be assumed unfavorable consequences. The operation of such markets provides the focus for our discussion. There is independence in the nature of these warfare schemed purchases, with respect to the internet businesses. We also assume that all such internet businesses have identical prospects, resources and utility functions. That is, they are not preferred businesses or have some pre-determined reputations.

We build a causal network model, of real world operations, in which the individual user (say, the Internet company owner) can formulate their own parameters of risk minimization and see how the values propagate to the ideal state. They all do not want the same solution. For some, a partial set of imperfect information might be enough.

Following are two diagrams of the causal network models. The first one represents the case where the act of sale is executed, since the threats do not have high probability. As a result, there are profits.

The second one represents the case of a causal network model for uncertainty where threat is great, as determined with high probability, and thus, no sale has come into effect. As a consequence, there are no profits from this particular action of "no sale".

6 Future Directions

In this section, the future directions of the current research are explored.

A new topic of research is the relevance of Bayesian modeling to Big Data.

Bayesian non-parametrics is an area in machine learning in which models grow in size and complexity as data accrue. As such, they are particularly relevant to the world of "Big Data", where it may be difficult or even counterproductive to fix the number of parameters a priori [20].

There is also a company [21] that is dealing with Big Data by producing a function called "BigData". The concept of Big Data is defined loosely as a data set that is too large for computer memory (RAM). A common strategy to deal with big data is to break it into smaller, manageable pieces, perform a function on those pieces, and combine the results back together. For this approach, the BigData function enables updating a model via Laplace Approximation.

The above mentioned work has been cited in several articles, such as [22].

Though Big Data is not the direct topic of this current project, it will ultimately be relevant to the current project, and therefore, I have mentioned it here. Big data-driven security system will be able to find the hidden patterns, the unexpected correlation, and the unexpected connections between data points tested under real-world conditions. Analyzing vast and complex data sets at high speed will allow us to spot the fake signal of an attack. This is because at some point, no matter how clever the attacker, they are liable to do something anomalous.

In a future direction of the work, in the new world of big data that provides cover for cyber attackers, we will concentrate on providing answers for devising a next-generation security system that can cope with emerging threats, The access controls will be smart in the new big data-driven security world. They will be able to inform or be informed by other controls [23].

My contribution in this regard will be substantial. Though the current work does not address any "self-learning" aspect, in the future, some aspects of "mutual learning" system have to be included. I think that the term "mutual learning" between the different controls is significant in this respect, rather than the traditional self-learning, which did not have the same direction as the prevention of destructive attacks executed through the Internet. It will be interesting to see how the payoff function changes as a result, or whether the payoff function is replaced by some other mathematical concept.

7 Conclusion

We need to create a system that is inspiring, persuading and enlightening. For that purpose, we need to program and test the proposed system, using credible manifestos. That will involve supporting real-time simulation that allows consumers to explore the influence of a causal network model towards CyberWarfare.

As the expected immediate results of the system, we will ascertain what is required in the current state of CyberWarfare. According to the Homeland Security report, spanning from 2011 to 2013, [24] cybercrime is costing corporations more than the previous year; the increase in costs is largely due to hackers using stealthier techniques. There are insidious kinds of attacks like malicious code, denial of service, stolen devices, Web-based attacks and malicious insiders. According to this report, the strategy has to change from watching the outside wall to trying to figure out what is happening inside the network. The current research is geared towards this goal of strategy change.

References

1. Cartwright, N.: Probabilities and experiments. J. Econom. **67**, 47–59 (1995)
2. Cartwright, N.: Causality: independence and determinism. In: Gammerman, A. (ed.) Causal Models and Intelligent Data Management, pp. 51–63. Springer, Berlin (1999)
3. Fine, K.: Reasoning with Arbitrary Objects. Blackwell, New York (1985)
4. Pearl, J.: Probabilistic Reasoning in Intelligent Systems. Morgan Kaufman, San Mateo (1988)
5. Pearl, J.: Belief networks revisited. Artif. Intell. **59**, 49–56 (1993)
6. Pearl, J.: From Bayesian networks to causal networks. In: Gammerman, A. (ed.) Bayesian Networks and Probabilistic Reasoning, pp. 1–31. Alfred Walter, London (1994)
7. United States Computer Emergency Readiness Team, Security updates available for Adobe flash player, Adobe reader, and Acrobat, http://www.us-cert.gov/ncas/current-activity?page=1. Accessed 10 Sept 2013
8. United States Computer Emergency Readiness Team, Microsoft releases September 2013 security bulletin, http://www.us-cert.gov/ncas/current-activity, Accessed 10 Sept 2013
9. What is Cloud Computing? http://aws.amazon.com/what-is-cloud-computing/
10. Google Elbows Into the Cloud, http://www.nytimes.com/2013/03/13/technology/google-takes-on-amazon-and-microsoft-for-cloud-computing-services.html
11. Technical Report, HP Bristol, UK, Cloud stewardship Economics (2012)
12. Abel, S.: Application of Bayesian causal model for threat identification in the context of cloud usage by cloud Steward businesses, unpublished manuscript (2013)
13. Cox, P.A.: Mobile cloud computing, http://www.ibm.com/developerworks/cloud/library/cl-mobilecloudcomputing/,. Accessed 11 March 2011

14. Abel, S.: Cybersecurity using Bayesian causal modeling in the context of mobile computing, including image usage, in mobile devices—invited talk and powerpoint presentation at Nokia Research Lab, Sunnyvale, CA, 13 Aug 2013
15. Executive Order 13636: Improving critical infrastructure cybersecurity department of homeland security, integrated task force incentives study, analytic report, http://www.dhs.gov/sites/default/files/publications/dhs-eo13636-analytic-report-cybersecurity-incentives-study.pdf. Accessed 12 June 2013
16. von Neumann, J., Morgenstern, O.: Theory of Games and Economic Behavior. Princeton University Press, Princeton (1944)
17. Mas-Colell, A., Whinston, M., Green, J.: Microeconomic Theory. Oxford University Press, Oxford (1995). ISBN 0-19-507340-1
18. Gilboa, I.: Theory of Decision Under Uncertainty. Cambridge University Press, Cambridge (2009)
19. Smith, D.R.: Variational Methods in Optimization. Dover, New York (1998)
20. Paisley, J.: Bayesian nonparametrics and big data, talk at the Columbia University School of Engineering, 22 Feb 2013
21. Statisticat, LLC, Big data and Bayesian inference, http://www.bayesian-inference.com/softwarearticlesbigdata. Accessed 22 Feb 2014
22. Maurya, M., Vishwakarma, U.K., Lohia, P.: A study of statistical inference tools for uncertainty reasoning in target tracking. Int. J. Comput. Networking Wirel. Mobile Commun. **3**(3), 1–10 (2013)
23. Intelligence-Driven Security: A new model using big data. Speaker: Mr. Art Coviello, Executive Vice President, EMC, Executive Chairman, RSA, In: The 3rd Annual International Cyber Security Conference, The Yuval Ne'eman Science, Technology & Security Workshop, Tel Aviv University, Israel
24. Homeland security report, http://www.oig.dhs.gov/assets/Mgmt/2013/OIG_13-42_Feb13.pdf

Printed in the United States
By Bookmasters